Dr. George Hart Wig

Subterranean World

Dr. George Hart Wig

Subterranean World

ISBN/EAN: 9783742845153

Manufactured in Europe, USA, Canada, Australia, Japa

Cover: Foto ©Klaus-Uwe Gerhardt /pixelio.de

Manufactured and distributed by brebook publishing software
(www.brebook.com)

Dr. George Hart Wig

Subterranean World

THE

SUBTERRANEAN WORLD.

PRINTED BY

SPOTTISWOODE AND CO., NEW-STREET SQUARE

LONDON

CARBONIFEROUS FOREST: CARBONIFEROUS PERIOD.

THE

SUBTERRANEAN WORLD.

BY

DR. GEORGE HARTWIG,

AUTHOR OF

'THE SEA AND ITS LIVING WONDERS,' 'THE TROPICAL WORLD,' 'THE POLAR WORLD,'
AND 'THE AERIAL WORLD.'

WITH THREE MAPS AND NUMEROUS ENGRAVINGS ON WOOD.

NEW EDITION

LONDON:

LONGMANS, GREEN, AND CO.

1892.

PREFACE

TO

THE SECOND EDITION.

———•◦•———

IN this new edition, besides correcting several errors, I have endeavoured to give an account of the production of the chief mining States in the world, up to the latest dates I could collect. Unfortunately it is extremely difficult to obtain good information on all points; no official report, for instance, on mines and mining East of the Rocky Mountains having been published in the United States for several years past. I must, therefore, beg the reader's indulgence, should he find some data omitted which it might interest him to know.

Dr. Hartwig.

Salon Villa, Ludwigsburg :
May 10, 1875.

PREFACE

THE FIRST EDITION.

NATURE displays her wonders not only in the starry heavens or in the boundless variety of animal and vegetable life on the surface of our earth. The dark regions underground likewise contain much that is remarkable or beautiful, and are the seats of gigantic operations, which are sometimes beneficent and sometimes disastrous to mankind.

Here lie concealed the mysterious laboratories of fire, which reveal to us their existence in' earthquakes and volcanic explosions. Here, too, in successive strata, repose the remains of extinct animals and plants. Here may be seen many a wonderful cavern, with its fantastic stalactites, its rushing waters, and its noble halls. Here have been deposited the rich stores of mineral wealth—the metals, the coals, the salt, the sulphur—without whose aid man would never have been more than a savage.

The aim of the present work has been to describe the wonders of this hidden world in their various relations to man, and to point out the methods he employs to make its treasures subservient to his wants.

The author trusts that he may have succeeded in giving a sketch of the phenomena resulting from the action of subterranean forces, which, with his account of the wonders of the sea, of the tropics, and of the frozen regions, may impart to the reader a fair idea of the history and present condition of the wonderful world in which we live.

Dr. Hartwig.

Salon Villa, Ludwigsburg:
   April 10, 1872.

CONTENTS.

CHAPTER I.

GEOLOGICAL REVOLUTIONS.

The Eternal Strife between Water and Fire—Strata of Aqueous Origin—Tabular View of their Chronological Succession—Enormous Time required for their Formation—Igneous Action—Metamorphic Rocks—Upheaval and Depression—Fossils—Uninterrupted Succession of Organic Life . . . . Page 1

CHAPTER II.

FOSSILS.

General Remarks — Eozoon Canadense—Trilobites — Brachiopods — Pterichthys Milleri — Oldest Reptiles —Wonderful Preservation of Colour in Petrified Shells — Primæval Corals and Sponges — Sea Lilies — Orthoceratites and Ammonites — Belemnites — Ichthyosarus and Plesiosaurus — Peterodactyli — Iguanodon — Tertiary Quadrupeds — Dinotherium — Colossochelys Atlas — Megatherium —Mylodon — Glyptodon —Mammoth — Mastodon — Sivatherium Giganteum—Fossil Ripple-marks, Rain-drops, and Footprints—Harmony has reigned from the beginning 8

CHAPTER III.

SUBTERRANEAN HEAT.

Zone of Invariable Temperature—Increasing Temperature of the Earth at a greater Depth—Proofs found in Mines and Artesian Wells, in Hot Springs and Volcanic Eruptions—The whole Earth probably at one time a fluid mass 31

CHAPTER IV.

SUBTERRANEAN UPHEAVALS AND DEPRESSIONS.

Oscillations of the Earth's Surface taking place in the present day—First ascertained in Sweden—Examples of Contemporaneous Upheaval and Depression in France and England—Probable Causes of the Phenomenon . . . 34

CHAPTER V.

SUBTERRANEAN WATERS AND ARTESIAN WELLS.

Subterranean Distribution of the Waters—Admirable Provisions of Nature—
Hydrostatic Laws regulating the Flow of Springs—Thermal Springs—Intermittent Springs—The Geysir—Bunsen's Theory—Artesian Wells—Le Puits
de Grenelle—Deep Borings—Various Uses of Artesian Wells—Artesian Wells
in Venice and in the Desert of Sahara.   .   .   .   .   .   Page 39

CHAPTER VI.

VOLCANOES.

Volcanic Mountains—Extinct and active Craters—Their Size—Dangerous Crater-
explorations—Dr. Judd in the Kilauea Pit—Extinct Craters—Their Beauty—
The Crater of Mount Vultur in Apulia—Volcanoes still constantly forming—
Jorullo and Isalco—Submarine Volcanoes—Sabrina and Graham's Island—
Santorin—Number of Volcanoes—Their Distribution—Volcanoes in a constant
state of eruption—Stromboli—Fumaroles—The Lava-lakes of Kilauea—Volcanic
Paroxysms—Column of Smoke and Ashes—Detonations—Explosion of Cones—
Disastrous Effect of Showers of Ashes and Lapilli—Mud Streams—Fish
disgorged from Volcanic Caverns—Eruptions of Lava—Parasitic Cones—
Phenomena attending the Flow of a Lava Stream—Baron Papalardo—Meeting
o Lava and Water—Scoriæ—Lava and Ice—Vast Dimensions of several Lava
Streams—Scenes of Desolation—Volcanoes considered as Safety-valves—Probable
Causes of Volcanoes 53

CHAPTER VII.

DESTRUCTION OF HERCULANEUM AND POMPEII.

State of Vesuvius before the eruption in the year A.D. 79—Spartacus—Premonitory
Earthquakes—Letter of Pliny the Younger to Tacitus, relating the death of his
uncle, Pliny the Elder—Benevolence of the Emperor Titus—Herculaneum and
Pompsii buried under a muddy alluvium—Herculaneum first discovered in
1713   81

CHAPTER VIII.

GAS SPRINGS AND MUD VOLCANOES.

Carbonic-acid Springs—Grotto del Cane—The Valley of Death in Java—Exagge-
rated Descriptions—Carburetted Hydrogen Springs—The Holy Fires of Baku—
Description of the Temple—Mud Volcanoes—The Macaluba in Sicily—Crimean
Mud Volcanoes—Volcanic Origin of Mud Volcanoes   88

CHAPTER IX.

EARTHQUAKES.

Extent of Misery inflicted by great Earthquakes—Earthquake Regions—Earthquakes in England—Great Number of Earthquakes—Vertical and Undulatory Shocks—Warnings of Earthquakes—Sounds attending Earthquakes—Remarkable Displacement of Objects—Extent and Force of Seismic Wave Motion—Effects of Earthquakes on the Sea—Enormous Waves on Coasts—Oscillations of the Ocean—Fissures, Landslips, and shattering Falls of Rock caused by Earthquakes—Causes of Earthquakes—Probable Depth of Focus—Opinions of Sir Charles Lyell and Mr. Poulett Scrope—Impressions produced on Man and Animals by Earthquakes . . . . Page 97

CHAPTER X.

THE GREAT EARTHQUAKE OF LISBON.

A dreadful All Saints' Day—The Victims of a Minute—Report of an Eye-witness—Conflagration—Banditti—Pombal brings Chaos into Order—Intrigues of the Jesuits—Damages caused by the Earthquake in other places; at Cadiz; in Barbary—Widespread Alarm—Remarks of Goethe on the Earthquake . 114

CHAPTER XI.

LANDSLIPS.

Igneous and Aqueous Causes of Landslips—Fall of the Diablerets in 1714 and 1749—Escape of a Peasant from his living Tomb—Vitaliano Donati on the Fall of a Mountain near Salenches—The Destruction of Goldau in 1806—Wonderful Preservation of a Child—Burial of Velleja and Tauretunum, of Plürs and Scilano—Landslip near Axmouth in Dorsetshire—Falling-in of Cavern-roofs—Dollinas and Jamas in Carniola and Dalmatia—Bursting of Bogs—Crateriform Hollows in the Eifel 121

CHAPTER XII.

ON CAVES IN GENERAL.

Their various Forms—Natural Tunnels—The Ventanillas of Gualgayoc—Eimeo—Torgatten—Hole in the Mürtschenstock—The Trebich Cave—Grotto of Antiparos—Vast Dimensions of the Cave of Adlesberg and of the Mammoth Cave—Discovery of Bauman's Cave—Limestone Caves—Causes of their Excavation—Stalactites and Stalagmites—Their Origin—Variety of Forms—Marine Caves—Shetland—Fingal's Cave—The Azure Cave—Cave under Bonifacio—Grotto di Nettuno, near Syracuse—The Bufador of Papa Luna—Volcanic Caves—The Fossa della Palomba—Caves of San Miguel—The Surtshellir 133

CHAPTER XIII.

CAVE RIVERS.

The Fountain of Vaucluse—The Fontaine-sans-fond—The Katabothra in Morea—Subterranean Rivers in Carniola—Subterranean Navigation of the Poik in the Cave of Planina—'The Stalactital Paradise'—The Piuka Jama . Page 149

CHAPTER XIV.

SUBTERRANEAN LIFE.

Subterranean Vegetation—Fungi—Enormous Fungus in a Tunnel near Doncaster—Artificial Mushroom-beds near Paris—Subterranean Animals—The Guacharo—Wholesale Slaughter—Insects in the Cave of Adelsberg—The Leptodirus and the Blothrus—The Stalita tænaria—The Olm or Proteus—The Lake of Cirknitz—The Archduke Ferdinand and Charon—The Blind Rat and the Blind Fish of the Mammoth Cave 156

CHAPTER XV.

CAVES AS PLACES OF REFUGE.

The Cave of Adullam—Mahomet in the Cave of Thaur—The Cave of Longara—The Cave of Egg—The Caves of Rathlin—The Cave of Yeermalik—The Caves of Grenada—Aben Aboo, the Morisco King—The Caves of Gortyna and Melidoni—Atrocities of French Warfare in Algeria—The Caves of the Dahra—The Cave of Shelas—St. Arnaud 169

CHAPTER XVI.

HERMIT CAVES—ROCK TEMPLES—ROCK CHURCHES.

St. Paul of Thebes—St. Anthony—His Visit to Alexandria, and Death—Numerous Cave Hermits in the East—St. Benedict in the Cave of Subiaco—St. Cuthbert—St. Beatus—Rock Temples of Kanara—The Wonders of Ellora—Ipsamboul—Rock Churches of Lalibala in Abyssinia—The Cave of Trophonios—The Grotto of St. Rosolia near Palermo—The Chapel of Agios Niketas in Greece—The Chapel of Oberstein on the Nahe—The repentant Fratricide . . . 178

CHAPTER XVII.

ICE-CAVES AND WIND-HOLES.

Ice-caves of St. Georges and St. Livres—Beautiful Ice-stalagmites in the Cave of La Baume—The Schafloch—Ice Cataract in the Upper Glacière of St. Livres—Ice Cavern of Eisenerz—The Cave of Yeermalik—Volcanic Ice-caves—Æolian Caverns of Terni—Causes of the low Temperature of Ice-caves . . 192

CHAPTER XVIII.

ROCK-TOMBS AND CATACOMBS.

Biban-el-Moluk, the Royal Tombs of Thebes—The Roman Catacombs—Their Extent—Their Mode of Excavation—Touching Sepulchral Inscriptions—Antony Bosio, the Columbus of the Catacombs—The Cavaliere di Rossi—The Catacombs of Naples and Syracuse—The Catacombs of Paris . . . . Page 202

CHAPTER XIX.

CAVES CONTAINING REMAINS OF EXTINCT ANIMALS.

The Cave Hyena and the Cave Bear—The Cavern of Kirkdale—The Moa Caves in New Zealand—Various Species of Moas—Their enormous size . . 217

CHAPTER XX.

SUBTERRANEAN RELICS OF PREHISTORIC MAN.

The Peat Mosses of Denmark—Shell-mounds—Swiss Lacustrine Dwellings—Ancient Mounds in the Valley of the Mississipi—The Caves in the Valley of the Meuse—Dr Schmerling—Human Skulls in the Cave of Engis—Explorations of Sir Charles Lyell in the Cave of Engihoul—Caverns of Brixham—Caves of Gower—The Sepulchral Grotto of Aurignac—Flint Implements discovered in the Valley of the Somme—Gray's Inn Lane an ancient Hunting-ground for Mammoths 221

CHAPTER XXI.

TROGLODYTES OR CAVE-DWELLERS. CANNIBAL CAVES.

Cave Dwellings in the Val d'Ispica—The Sicanians—Cannibal Caves in South Africa—The Rock City of the Themud—Legendary Tale of its Destruction . 232

CHAPTER XXII.

TUNNELS.

Subterranean London—The Mont Cenis Tunnel—Its Length—Ingenious Boring Apparatus—The Grotto of the Pausilippo—The Tomb of Virgil . . 237

CHAPTER XXIII.

ON MINES IN GENERAL.

Perils of the Miner's Life—Number of Casualties in British and Foreign Coal Mines—Life in a Mine—Occurrence of Ores—Extent and Depth of Metallic Veins—Mines frequently discovered by Chance—The Divining Rod—Experi-

mental Borings—Stirring Emotions during their Progress—Sinking of Shafts—
Precautions against Influx of Water—Expense—Shaft Accidents—Various
Methods of working Mineral Substances—Working in Direct and Reverse Steps—
Working by Transverse Attacks—Open Quarry Workings—Pillars and Stall Sys-
tem—Long Wall System—Dangerous Extraction of Pillars—Mining Implements
—Blasting—Heroes in Humble Life—Firing in the Mine of Rammelsberg—
Transport of Minerals underground—Modern Improvements—Various Modes of
Descent—Corfs—Wonderful Preservation of a Girl at Fahlun—The Loop—
Safety Cage—Man Machines—Timbering and Walling of Galleries—Drainage
by Adit Levels—Remarkable Adit—The Great Cornish Adit—The Georg
Stollen in the Hartz—The Ernst August Stollen—Steam-pumps—Drowning of
Mines—Irruption of the Sea into Workington Colliery—Hubert Goffin—
Irruption of the River Garnock into a Mine—Ventilation of Mines—Upcast
Shafts—Fire-damp—Dreadful Explosions—The Safety-lamp—The Choke-damp
—Conflagrations of Mines—The Burning Hill in Staffordshire . . Page 244

CHAPTER XXIV.

GOLD.

The Golden Fleece—Golden Statutes in ancient Temples—A Free-thinking
Soldier—Treasures of ancient Monarchs—First Gold Coins—Ophir—Spanish
Gold Mines—Bohemian Gold Mines—Discovery of America—Siberian Gold
Mines—California—Marshall—Rush to the Placers—Discovery of Gold in
Australia—The Chinaman's Hole—New Eldorados—Hydraulic Mining in Cali-
fornia—Quartz-crushing 255

CHAPTER XXV.

SILVER.

Its ancient Discovery—Its Uses among the luxurious Romans—The Mines of
Laurium—Silver Mines of Bohemia, Saxony, and Hungary—Colossal Nuggets—
Silver Ores—Silver Production of Europe—Mexican Silver Mines—The Veta
Madre of Guanaxuato—The Conde de la Valenciana—Zacatecas and Catorce—
Adventures of a Steam-engine—La Bolsa de Dios Padre—The Conde de la
Regla—Ill-fated English Companies—Indian Carriers—The Dressing of Silver
Ores—Amalgamating process—Enormous Production of Mexican Mines—Potosi
—Cerro de Pasco—Gualgayoc—The Mine of Salcedo—Hostility of the Indians
—The Monk's Rosary—Chilian Mines—The Comstock Lode . . . 297

CHAPTER XXVI.

COPPER.

Its Valuable Qualities—English Copper Mines—Their comparatively recent
Importance—Dreary Aspect of the Cornwall Copper Country—Botallack—Sub-
marine Copper Mines—A Blind Miner—Swansea—Smelting Process—The Mines
of Fahlun—their Ancient Records—Alten Fjord—Drontheim—The Mines of
Röraas—The Mines of Mansfeldt—Lake Superior—Mysterious Discoveries—
Burra Burra—Remarkable Instances of Good Fortune in Copper-mining . 315

CONTENTS.

CHAPTER XXVII.

TIN.

Tin known from the most remote antiquity—Phœnician Traders—The Cassiterides—Diodorus Siculus—His Account of the Cornish Tin Trade—The Age of Bronze—Valuable Qualities of Tin—Tin Countries—Cornish Tin Lodes—Tin Streams—Wheal Vor—A Subterranean Blacksmith—Huel Wherry, a Tin Mine under the Sea—Carclaze Tin Mine—Dressing of Tin Ores—Smelting—The Cornish Miner . . . . Page 332

CHAPTER XXVIII.

IRON.

Iron the most valuable of Metals—Its wide Diffusion over the Earth—Meteoric Iron—Iron very anciently known—Extension of its Uses in Modern Times—British Iron Production—Causes of its Rise—Hot Blast—Puddling—Coal Smelting—The Cleveland District—Rapid Rise of Middlesborough—British Iron Ores—Production of Foreign Countries—The Magnetic Mountain in Russia—The Eisenerz Mountain in Styria—Dannemora—Elba—The United States—The Pilot Knob—The Cerro del Mercado 345

CHAPTER XXIX.

LEAD.

Its Properties and Extensive Uses—Alston Moor—Belgian Lead Mines—Galena in America—Extraction of Silver from Lead Ores—Pattison's Process—A great part of our Wealth is due to the Laboratory 364

CHAPTER XXX.

MERCURY.

Not considered as a true Metal by the Ancients—Its Properties and Uses—Almaden—Formerly worked by Convicts—Diseases of the Miners—Idria—Its Discovery—Conflagration of the Mine—Its Produce—Huancavelica—New Almaden 370

CHAPTER XXXI.

THE NEW METALS.

Zinc—The Ores, but not the Metal, known to the Ancients—Rapid increase of its Production—Chief Zinc-producing Countries—Platinum—Antimony—Bismuth—Cobalt and Nickel—Wolfram—Arsenic—Chrome—Manganese—Cadmium—Titanium—Molybdenum—Aluminium—Aluminium Bronze—Magnesium—Sodium—Palladium—Rhodium—Thallium 380

CHAPTER XXXII.

COAL.

The Age of Coal—Plants of the Carboniferous Age—Hugh Miller's Description of a Coal-forest—Vast Time required for the Formation of the Coal-fields—Derangements and Dislocations—Faults—Their Disadvantages and Advantages—Bituminous Coals—Anthracites—Our Black Diamonds—Advantageous Position of our Coal Mines—The South Welsh Coal-field—Great Central and Manchester Coal-fields—The Whitehaven Basin and the Dudley Area—Newcastle and Durham Coal-fields—Costly Winnings—A Ball in a Coal-pit—Submarine Coal Mines —Newcastle—View from Tynemouth Priory—Hewers—Cutting Machines—Putters—Onsetters—Shifters—Trapper-boys—George Stephenson—Rise of Coal Production—Probable Duration of our Supply—Prussian Coal Mines—Belgian Coal Mines—Coal Mines in various other Countries—Maunch Chunk . Page 390

CHAPTER XXXIII.

BITUMINOUS SUBSTANCES.

Formation of Petroleum—Enormous Production of the Pennsylvanian Wells—Asphalt used by the Ancients—Asphalte Pavements—The Pitch Lake or Trinadad—Jet—Its Manufacture in Whitby 426

CHAPTER XXXIV.

SALT.

Geological Position of Rock Salt—Mines of Northwich—Their immense Excavations—Droitwich and Stoke—Wielicza—Berchtesgaden and Reichenhall—Admirable Machinery—Stassfurt—Processes employed in the Manufacture of Salt—Origin of Rock-salt Deposits 431

CHAPTER XXXV.

SULPHUR.

Sulphur Mines of Sicily—Conflagration of a Sulphur Mine—The Solfataras of Krisuvick—Iwogasima in Japan—Solfatara of Puzzuoli—Crater of Teneriffe —Alaghez—Büdöshegy in Transylvania—Sulphur from the Throat of Popocatepetl—Sulphurous Springs—Pyrites Mines of San Domingo in Portugal—The Baron of Pommorão. 441

CHAPTER XXXVI.

AMBER.

Various Modes of its Collection on the Prussian Coast—What is Amber?—The extinct Amber Tree—Insects of the Miocene Period enclosed in Amber—Formidable Spiders—Ancient and Modern Trade in Amber . . . 449

CHAPTER XXXVII.

MISCELLANEOUS MINERAL SUBSTANCES USED IN THE INDUSTRIAL ARTS.

Alum—Alum Mines of Tolfa—Borax—The Suffioni in the Florentine Lagoons—China-clay: how formed—Its Manufacture in Cornwall—Plumbago—Emery —Tripolite . . . . Page 458

CHAPTER XXXVIII.

CELEBRATED QUARRIES.

Carrara—The Pentelikon—The Parian Quarries—Rosso antico and Verde antico— The Porphyry of Elfdal—The Gypsum of Montmartre—The Alabaster of Volterra—The Slate Quarries of Wales—'Princesses' and 'Duchesses,' Ladies' and 'Fat Ladies'—St. Peter's Mount near Maestricht—Egyptian Quarries—Haggar Silsilis—The Latomiæ of Syracuse—A Triumph of Poetry . . . 464

CHAPTER XXXIX.

PRECIOUS STONES.

Diamonds—Diamond-cutting—Rose Diamonds—Brilliants—The Diamond District in Brazil—Diamond Larras—The great Russian Diamond—The Regent—The Koh-i-Noor—Its History—The Star of the South—Diamonds used for Industrial Purposes—The Cape Diamond-fields—The Oriental Ruby and Sapphire—The Spinel—The Chrysoberyl—The Emerald—The Beryl—The Zircon—The Topaz —The Oriental Turquoise—The Garnet—Lapis Lazuli—The Noble Opal— Inferior Precious Stones—The Agate-cutters of Oberstein—Rock Crystal—The Rock-crystal Grotto of the Galenstock 477

LIST OF ILLUSTRATIONS.

MAPS.

Of the World, showing the Distribution of Volcanoes and the
    Districts visited by Earthquakes . . . . *to face page* 60
Of Great Britain, showing the Coal-fields and Chief Mining
    Districts . . . . " 400
Of America, showing the Coal-fields and Mineral Districts . " 410

WOODCUTS.

		PAGE
Carboniferous Forest . . . *engraved by* G. Pearson, *to face title*		
Tabular Geological Profile of Strata, with correspond-		
ing Fossils . . . . . *engraved by* G. Pearson		3
Aqueous Strata disturbed by Igneous Formations . " "		4
Ammonites Henleyi (Middle Lias) *from* Haughton's 'Manual of Geology'		9
Tribolite . . . . *from* Kemp's 'Phasis of Matter'		11
Magnified Eye of Tribolite . . " " "		11
Pterygotus acuminatus (Eurypterid) *from* Haughton's 'Manual of Geology'		12
Spirifer princeps (Brachiopod) . " " "		13
Pterichthys Milleri, restored (Old Red Sandstone of		
Scotland) . . . . . " "		14
Ventriculites, Fossil Sponge (Chalk) . . . " "		16
Siphonia costata, Fossil Sponge (Green Sand, War-		
minster) . . . . . . " "		16
Encrinus liliiformis (Muschelkalk, Germany) . " "		17
Pentacrinus briareus . . . . . " "		17
Marsupites ornatus (Chalk) . . . . " "		18
Turrilites tuberculatus . . . . " "		19
Restored Belemnite . . . . . " "		19
Ichthyosaurus communis . . . . " "		20

PAGE

Plesiosaurus dolichodeirus British Museum—(found in the
Lias of Street, near Glastonbury) *from* Haughton's 'Manual of Geology' — 21
Glyptodon clavipes . . . . " " " 25
Diagram illustrating action of Syphon . *engraved by* G. Pearson 44
Section of an Intermittent Spring . . . " " 45
Geysirs of Iceland . . . . . " " 46
Porous Strata, Artesian Well sunk in the London
Basin . . . . . . " " 49
Middle and Valley Lake Craters, Mount
Gambier, South Australia *from* Wood's 'Australia' *to face page* 53
Extinct Crater of Haleakala . . *from* Webb's 'Celestial Objects' 57
Eruption of Vesuvius, Bay of Naples . *engraved by* G. Pearson, *to face page* 81
Mud Volcanoes of Trinidad . . . *engraved by* G. Pearson 94
Great Earthquake at Lisbon . . *engraved by* G. Pearson, *to face page* 114
Axmouth Landslip . . . . *engraved by* G. Pearson 128
Stalactital Cavern at Aggetelek: the Cave
of Borodla . . *engraved by* G. Pearson, *to face page* 133
Entrance to the Cave of Adelsberg . . *engraved by* G. Pearson 137
Stalactital Cavern in Australia . . *from* Wood's 'Australia' 141
Cave under Bonifacio . . . . *from* Forester's 'Corsica' 145
Leptodirus Höchenwartii . . . *engraved by* G. Pearson 163
The *Proteus anguinus* . . . " " 166
Blind Fish (*Amblyopsis spelæus*) . . " " 168
Indian Rock-cut Temple . . *engraved by* G. Pearson, *to face page* 178
Rock Temples of Ajunta (general view) . *engraved by* G. Pearson 182
Lower Glacière of St. Livres . . *from* Browne's 'Ice Caves' 193
Ice Streams in the Upper Glacière of St. Livres " " 196
Entrance to the Glacière of St. Georges . " " 201
Gallery with Tombs *from* Northcote and Brownlow's 'Roma Sotterranea' 208
Cave in Dream Lead Mine, near Wirksworth, Derby-
shire . . . . . *engraved by* G. Pearson 216
Boring Machine in the Tunnel, Mont Cenis {*taken from the* 'Illustrated London News,' *by permission*} 238
Boring Machine in the Second Working
Gallery, Mont Cenis Tunnel . . " " 230
Process of Boring . . . . *engraved by* G. Pearson 251
Section of a Lead Mine in Cardiganshire {*from* Ure's 'Dictionary of Arts, Manufactures, and Mines'} 252
Part of a Colliery laid out in four panels . " " 255
General View of Mining Operations . . *engraved by* G. Pearson 257
Tools used by Miners in Cornwall . {*from* Ure's 'Dictionary of Arts, Manufactures, and Mines'} 258
Conveyance of Minerals underground . *engraved by* G. Pearson 262
Miners descending Shaft in Owen's Safety Cage " " 265
Timbering of a Mine . . . {*from* Ure's 'Dictionary of Arts, Manufactures, and Mines'} 268
Transverse Sections of Walled Drain Galleries " " 269
Drainage of a Mine by Adit Levels . " " 269
Safety Lamp . . . . " " 270
Gold-washing in Australia . . . " " 280
Stamping Mill *from* Ure's 'Dictionary of Arts, Manufactures, and Mines' 292
Grinding Mill " " " " 306
307

PAGE

The Botallack Mine, Cornwall . . . *engrav.d by* G. Pearson 317

St. Michael's Mount, Cornwall . . . „ „ 303

Blast Furnace *from* Ure's 'Dictionary of Arts, Manufactures, and Mines' 352

Pecopteris adiantoides . . *from* Haughton's 'Manual of Geology' 391

Sphenopteris affinis . . „ „ 391

Lepidodendron elegans . . „ „ 392

Asterophyllites comosa . . „ . „ 392

Sigilaria oculata . . . „ „ 392

Calamites nodosus . . „ „ 393

Coalbeds rendered available by elevation, *from* 'Our Coal and Our Coal Pits' 397

Section of Coal-field south of Malmesbury { *from* 'Ure's Dictionary of Arts, Manufactures, and Mines' } 398

Coal-basin of Clackmannanshire . . „ „ 403

Dudley Coal-field . *.from* Howitt's 'Visits to Remarkable Places' 407

Shipping Coal . . „ „ „ 412

Coal-hewers at Work . . *engraved by* G. Pearson 415

Pitch Lakes of Trinidad . . *engraved by* G. Pearson, *to face page* 429

Insects and Vegetable Substances enclosed in Amber, *engraved by* G. Pearson 452

Penrhyn Slate Quarry, North Wales . *engraved by* G. Pearson, *to face page* 469

DIRECTIONS to BINDER for placing the full-page illustrations :—

1. Carboniferous Forest . . . . *to face title*

2. Middle and Valley Lake Craters, Mount Gambier,
   South Australia . . . . *to face page* 53

3. Eruption of Vesuvius, Bay of Naples . . „ 81

4. Great Earthquake at Lisbon . . . „ 114

5. Stalactital Cavern at Aggtelek : the Cave of Borodla „ 133

6. Indian Rock-cut Temple : Porch of the Chaitya Cave
   Temple, Ajunta . . . . „ 178

7. Pitch Lake of Trinidad . . . . „ 429

8. Penrhyn Slate Quarry, North Wales . . „ 469

CHAPTER I.

GEOLOGICAL REVOLUTIONS.

The Eternal Strife between Water and Fire—Strata of Aqueous Origin—Tabular View of their Chronological Succession—Enormous Time required for their Formation—Igneous Action—Metamorphic Rocks—Upheaval and Depression—Fossils—Uninterrupted Succession of Organic Life.

GEOLOGY teaches us that, from times of the remoteness of which the human mind can form no conception, the surface of the earth has been the scene of perpetual change, resulting from the action and counter-action of two mighty agents—water and subterranean heat.

Ever since the first separation between the dry land and the sea took place, the breakers of a turbulent ocean, the tides and currents, the torrents and rivers, the expansive power of ice, which is able to split the hardest rock, and the grinding force of the glacier, have been constantly wearing away the coasts and the mountains, and transporting the spoils of continents and islands from a higher to a lower level.

During our short historical period of three or four thousand years, the waters, in spite of their restless activity and the considerable local changes effected by their means, have indeed produced no marked alteration in the great outlines of the sea and land; but when we consider that their influence has extended over countless ages, we can no longer wonder at the enormous thickness of the stratified rocks of aqueous origin which, superposed one above the other in successive layers, constitute by far the greater part of the earth-rind.

Our knowledge of these sedimentary formations is indeed as yet but incomplete, for large portions of the surface of the globe have never yet been scientifically explored; but a

B

careful examination and comparison of the various strata composing the rocky foundations of numerous countries, have already enabled the geologist to classify them into the following chronological systems or groups, arranged in an ascending series, or beginning with the oldest.

1. Laurentian, named from its discovery northward of the River St. Lawrence in Canada.

2. Cambrian ⎤ These three groups owe their name to their
3. Silurian ⎬ occurrence in Wales and Devonshire,
4. Devonian ⎦ where they were first scientifically explored.

5. Carboniferous. In this group the most important coalfields are found.

6. Permian, from the Russian province of Permia.

7. Triassic.

8. Lias.

9. Oolite.

10. Cretaceous.

11. Tertiary; subdivided into Eocene, Miocene, and Pliocene.

12. Recent marine and lacustrine strata.

Each of these systems consists again of numerous sections and alternate layers, sometimes of marine, sometimes of freshwater formation, the mere naming of which would fill several pages.

When we reflect that the Laurentian system alone has a thickness of 30,000 feet; that many of the numerous subdivisions of the Triassic or Oolitic group are 600, 800, or even several thousand feet thick, and that each of these enormous sedimentary formations owes its existence to the disintegration of pre-existing mountain masses—we can form at least a faint notion of the enormous time which the whole system required for its completion.

Had the levelling power of water never met with an antagonistic force, there can be no doubt that the last remains of the dry land, supposing it could ever have risen above the ocean, must long since have been swept into the sea. But while water has been constantly tending to reduce the irregularities of the earth's surface to one dull level, the expansive force of subterranean heat has been no less unceasingly active in various species of stone which geologists include under the name of metamorphic rocks.

Besides the more paroxysmal and violent revolutions resulting from the action of subterranean fire, we find that the earth-rind has at all times been subject to slow oscillatory movements of upheaval and subsidence, frequently alternating on the same spot with long periods of rest. The greater part of the actual dry land has been deep sea, and then again land and ocean many times in succession; and doubtless the actual sea bottom would exhibit similar alternations were we able to explore it. The same materials have repeatedly been exposed to all these changes—now raised or poured out by subterranean fires, and then again swept away by the waters; now changed from solid rock into sand and mud, and then again converted, by pressure or heat, into solid rock. Thus the history of the earth-rind opens to us a vista into time no less grand and magnificent than the vista into space afforded by the contemplation of the starry heavens.

The oldest and the newest stratified rocks are composed of the same mineral substances; for clay, sandstone, and limestone occur in the Silurian and in the Carboniferous formation; in the Cretaceous and Triassic systems; in the Tertiary and in the Alluvial deposits, which have immediately preceded the present epoch.

Where then, it may be asked, does the geologist find a chronological guide to lead him through the vast series of strata which, in the course of countless ages, have been deposited in the water? How is he able to distinguish the boundaries of the various periods of creation? Where are the precise indications which enable him to decipher the enigmas which the endless feuds of fire and water have written in the annals of our globe?

The fossil remains of animals and plants wonderfully furnish the guidance which he needs. The corals and shells, the ferns and conifera, the teeth and bones found in the various strata of the earth-rind, are the landmarks which point out to him his way through the labyrinth of the primitive ages of our globe, as the compass directs the mariner over the pathless sea. Every leading fossil has its fixed chronological character, and thus the age of the formation

in which it occurs may be ascertained, and its place determined in the geological scale. It would, however, be erroneous to suppose that each successive formation has been
the seat of a totally distinct creation, and that the organic
remains found in one particular stratum are separated by an
impassable barrier from those which characterise the preceding or following sedimentary deposits.

As on the surface of the earth or along the shore of the
sea, each land or each coast has not only its peculiar plants or
animals, but also harbours many of the organic forms of the
neighbouring countries or conterminous shores; as the tropical
organisations gradually pass into those of the temperate zones,
and these again merge into those of the polar regions, so also
the stream of life has from the first flowed uninterruptedly,
in gradually changing forms, through every following age.
New genera and species have arisen, and others have disappeared, some after a comparatively short duration, others
after having outlasted several formations; but every extinct
form has but made way for others, and thus each period has
not only witnessed the decay of many previously flourishing
genera and species, but has also marked a new creation.

No doubt the numerous local disturbances above mentioned
have frequently broken the chain of created beings; but a
gradual progress, a continuous development from lower to
more highly organised species, genera, orders, and classes,
has from the beginning been the general and constant law
of organic life. Universal destructions of existing forms,
revolutions covering the whole surface of the earth with
ruin, have most assuredly never occurred in the annals of
our globe.

Nor must it be supposed that the whole scale of sedimentary formations is to be found superimposed in one spot;
for as in our times new strata are chiefly growing at the
mouths of rivers, or where submarine currents deposit at the
bottom of the ocean the fine mud or sand which is conveyed
into the sea by the disintegration of distant mountain chains,
so also from the beginning each stratum could only have
been deposited in similar localities; and while it was slowly
increasing, and not seldom acquiring colossal dimensions in
some parts of the globe, others remained comparatively but

little altered, until new oscillatory movements produced a change in their former position, and opening new paths to the rolling waters, here set bounds to the progress of one formation, and there favoured the deposition of another.

A complete study of all the various transformations by fire or water which the surface of our earth has undergone would require an elaborate treatise on geology, and lies far beyond the scope or the pretensions of a popular volume which is chiefly devoted to the description of caves and mines. But I should be neglecting some of the most interesting features of the subterranean world, were I to omit all mention of the fossils imbedded in its various strata; of its internal heat; of the upheavals and subsidences which have played so conspicuous a part in the history of the earth-rind, and are still proceeding at the present day; of the water percolating or flowing beneath the earth's crust, and finally of the volcanoes and earthquakes, which prove to us that the ancient subterranean fires, far from being extinct, are still as powerful as ever in remodelling its surface.

CHAPTER II.

FOSSILS.

General Remarks — Eozoon Canadense — Trilobites — Brachiopods — Pterichthys Milleri — Oldest Reptiles — Wonderful Preservation of Colour in Petrified Shells — Primæval Corals and Sponges — Sea-lilies — Orthocerntites and Ammonites — Belemnites — Ichthyosaurus and Plesiosaurus — Pterodactyli — Iguanodon — Tertiary Quadrupeds — Dinotherium — Colossochelys Atlas — Megatherium — Mylodon — Glyptodon — Mammoth — Mastodon — Sivatherium giganteum — Fossil Ripple-marks, Rain-drops, and Footprints — Harmony has reigned from the beginning.

THE fossil remains of plants and animals, which have successively flourished and passed away since the first dawn of organic life, occupy a prominent place among the wonders of the subterranean world. A medal that has survived the ruin of empires is no doubt a valuable relic, but it seems to have been struck but yesterday when compared with a shell or a leaf that has been buried millions of years ago in the drift of the primeval ocean, and now serves the geologist as a waymark through the past epochs of the earth's history.

If we examine the condition in which the fossils have been preserved in the strata successively deposited on the surface of our globe, we find that in general only parts of the original plant or animal have escaped destruction, and in these fragments also the primitive substance has often been replaced by other materials, so that only their form or their impression has triumphed over time. While soft and delicate textures have either been utterly swept away, or could only be preserved under the rarest circumstances (as, for instance, the insects and flowers inclosed in amber), a greater degree of hardness or solidity naturally gave a better chance of escaping destruction. Thus among plants the most frequent fossil remains are furnished by stems, roots, branches, fruit stones, leaves; and, among animals, by corals, shells, calcareous crusts, teeth, scales, and bones. But the few memorials that have thus survived the lapse of ages enable us to form some idea of the multitudes that have entirely perished; and the petrified shell of the Ammonite, or the

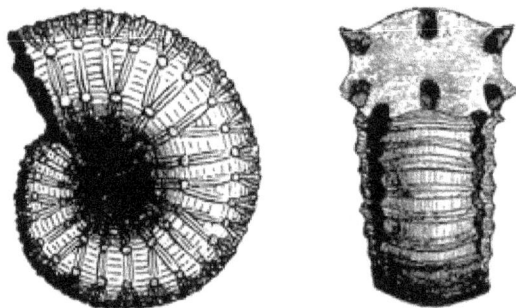

AMMONITES HENLEYI (MIDDLE LIAS).

jointed arms of the Encrinite, are proofs of the existence of the world of tiny beings which served them for their nourishment and have been utterly swept away. If we consider that the number of all the known species of fossil plants hardly amount to 3,000, while the Flora of the present day, as far as it has been examined by systematical botanists, numbers at least 250,000 species; that the host of living insects is probably still more numerous, although not much more than 1,500 extinct species of this class are known to us; and that, finally, the remains of all the extinct crustaceous fishes, reptiles, and warm-blooded animals are far outnumbered by the species actually living—we may form some idea of the vast multitudes that have left no trace behind, and whose total loss will for ever confine within narrow limits our knowledge of the past phases of organic creation. This loss appears still greater when we consider the enormous extent of time during which the fossils known to us have successively existed, and that a part only of the comparatively small number of the orders, genera, and species to which they belong existed at one and the same epoch. But as, owing to the hard texture and mode of life which are so eminently favourable for the preservation of shells, we have been enabled to collect about 11,000 fossil species, a number not much

inferior to that of the molluscs of the present day, we may
justly conclude that the more perishable forms of life, of
which, consequently, fewer vestiges have been preserved,
were comparatively as numerous, and that ever since the
first dawn of organic life our earth has borne an immense
variety of plants-and animals.

Though comparatively but few species have been preserved,
yet sometimes the accumulation of fossil remains is truly
astonishing. In the carboniferous strata we not seldom find
more than one hundred beds of coal interstratified with
sandstones, shales, and limestones, and extending for miles
and miles in every direction. How luxuriant must have
been the growth of the forests that could produce masses
such as these, and what countless multitudes of herbivorous
insects must have fed upon their foliage or afforded food to
carnivorous hordes scarcely less numerous than themselves !
The remains of corals, encrinites, and shells often form the
greater part of whole mountain ranges, and, what is still
more remarkable, mighty strata of limestone or flint are not
seldom almost entirely composed of the aggregated remains
of microscopical animals.

After these remarks on fossils in general, I will now
briefly point out some of the most striking of the species so
preserved to us as they successively appeared upon the stage
of life.

In the Lower Laurentian Rocks, the most ancient strata
known, only one fossil has hitherto been found. The *Eozoon
canadense*, as it has been called, belonged to the Rhizopods,
which occupy about the lowest grade in the scale of animal
existence. Its massive skeletons, composed of innumerable
cells, would seem to have extended themselves over submarine
rocks, their base upwards of twelve inches in width and their
thickness from four to six inches. Such is the antiquity of
the Eozoon that the distance of time which separated it
from the Trilobites of the Cambrian formation may be equal
to the vast period which elapsed between these and the Tertiary ages. In other words, it is beyond our imagination to
conceive.

In the next following Cambrian formation we find, besides
some zoophytes and shells, a number of Trilobites, which,

however, appear to have been most abundant in the Silurian seas, where they probably swarmed as abundantly as the crabs and shrimps in the waters of the present age. Few fossils are more curious than these strange crustaceans,

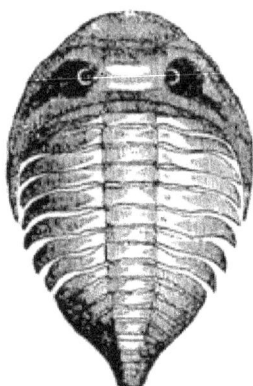

TRILOBITE.   MAGNIFIED EYE OF TRILOBITE.

which so widely differ from their modern relatives. The jointed carapace is divided into three lobes, the middle prominent one forming the axis of the body, while the lateral ones were free appendages, under which the soft membranaceous swimming feet were concealed. Large eyes, resembling those of a dragon-fly, projected from the odd crescent-shaped head, and, being composed of many hundred spherical facets, commanded a wide view of the horizon. Provided with such complicated organs of vision, the helpless animal could betimes perceive the approaching enemy, or more easily espy its prey, consisting, most likely, of the smaller marine annelides or molluscs. From the structure of these remarkable eyes we may conclude that the waters of the old Cambrian or Silurian Ocean were as limpid as those of the present seas, and that the natural relations of light to the eye and of the eye to light cannot have greatly changed since that period. Many, if not all, of the Trilobites were capable of rolling themselves up into a ball, like wood-lice; and accordingly it is found that in many of them the contour of the head and tail is so constructed that they fit accurately when rolled up. Most probably the Trilobites either swam in an inverted position, the belly upwards, or crawled slowly along at the bottom of the shallow coast waters, where they lived gregariously in vast numbers.

Contemporaneous with the Trilobites were the Eurypterids, which vary from one foot to five or six feet in length. One of the most striking characteristics of this remarkable order of crustaceans is the formidable pair of pincers with which they were armed. As their whole structure shows them to

PTERYGOTUS ACUMINATUS (EURYPTERID).   SPIRIFER PRINCEPS (BRACHIOPOD).

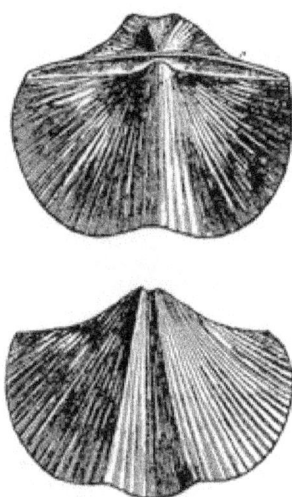

have been active swimmers, they must have made considerable havoc among the smaller fry of the Devonian and Silurian seas.

Then also abounded in hundreds of species the Brachiopods, a class of molluscs now but feebly represented by a scanty remnant. The greater part of the interior of the shell, consisting of two unequal valves, is occupied with branching arms, furnished with cilia, which cause a constant current to flow towards the mouth of the mollusc, and thus provide for its nourishment. The arms, as in the family of the Spiriferidæ, are sometimes supported by calcareous skeletons, arranged like loops or spirals.

Some Brachiopods are attached to stones, like oysters; in others the larger valve is perforated, and a sinewy kind of foot, passing through the aperture, serves as a holdfast to the animal.

Most of these helpless creatures did not survive the Carboniferous period, but the Terebratulæ, which still have their representatives in the modern seas, existed even then, so that their genealogical tree may justly boast of a very high antiquity.

The fishes, of which the oldest known specimen has been found in the Upper Silurian group (Lower Ludlow), become more frequent in the next following Devonian epoch, where they appear in a variety of wonderful forms, widely different from those of the present day. While in nearly all the existing fishes the scales are flexible, and generally either of a more or less circular form (cycloid), as in the salmon, herring, roach, &c., or provided with comb-like teeth, projecting from the posterior margin (ctenoïd), as in the sole or perch, the fishes of the Devonian, Permian, and Carboniferous periods were decked with hard bony scales, either covered with a brilliant enamel, as in our sturgeons (ganoid), and arranged in regular rows, the posterior edges of each slightly overlapping the anterior ones of the next, or irregular in their shape, and separately imbedded in the skin (placoid), as in the sharks and rays of the present day. With rare exceptions their skeleton was cartilaginous; but the less perfect ossification of their bones was amply compensated by the solid texture of their enamelled coat of mail, which afforded them a better protection against enemies and injuries from without than is possessed by any bony-skeletoned fish of our days. They were, in fact, comparatively as well prepared for a hostile encounter as an ancient knight in armour, or as one of our modern iron-plated war ships. One of the most remarkable of these mail-clad Ganoids was the *Pterichthys Milleri* of the Old Red Sandstone of Scotland. In most of our fishes the pectoral fins are but weakly developed ; here they constitute real arms, moved by strong muscles, and resembling the paddle of the turtle.

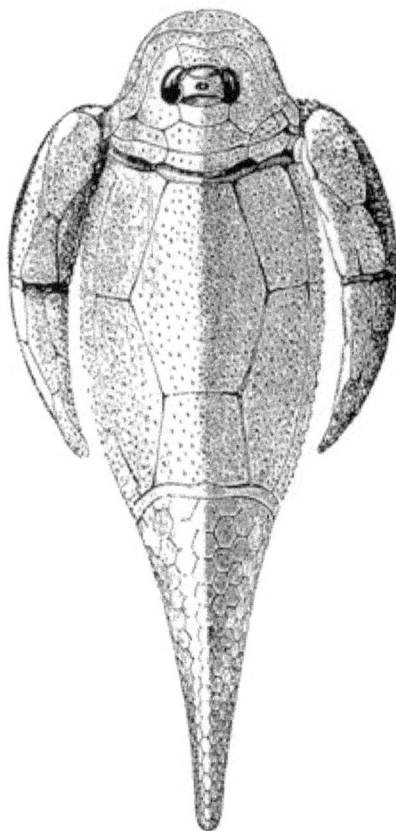

PTERICHTHYS MILLERI—RESTORED. (OLD RED SANDSTONE OF SCOTLAND.)

Besides the enormous masses of vegetable matter which distinguish the Carboniferous period, the stone beds of that formation likewise contain a vast number of animal remains. From the reptiles and fishes down to the corals and sponges, many new families, genera, and species crowd upon the scene, while many of the previously flourishing races have either entirely disappeared, or are evidently declining. Thus the Trilobites, formerly so numerous, are reduced to a few species in the Carboniferous period, and vanish towards its close.

In 1847 the oldest known reptiles were found in the coal field of Saarbrück, in the centre of spheroidal concretions of

clay iron-stone, which not only faithfully preserved the
skulls, teeth, and the greater portions of the skeletons of these
ancient lizards, but even a large part of their skin, consisting of long, narrow, wedge-shaped, tile-like, and horny
scales, arranged in rows.   What a lesson for human pride!
The pyramid of the Pharaoh Cheops, reared by the labour of
thousands of slaves, has been unable to preserve his remains
from spoliation even for the short space of a few thousand
years, and here a vile reptile has been safely imbedded in a
sarcophagus of iron ore during the vast period of many
geological formations.

Still more recently (1854) other wonders have been brought
to light in the clay iron-stone of Saarbrück.   The wing of a
grasshopper, with all its nerves as distinctly marked as if
the creature had been hopping about but yesterday, some
white ants or termites (now confined to the warmer regions
of the globe), a beetle, and several cockroaches, give us some
idea of the insects that lived at the time when our coal-beds
were forming.   Another highly interesting circumstance, relating to the fossils of that distant period, is that in several of
them the patterns of their colouring have been preserved.
Thus *Terebratula hastata* often retains the marks of the
original coloured stripes which ornamented the living shell.
In *Aviculopecten sublobatus* dark stripes alternate with a light
ground, and wavy blotches are displayed in *Pleurotomaria
carinata*.   From these facts Professor Forbes inferred that
the depth of the seas in which the Mountain Limestone was
formed did not exceed fifty fathoms, as in the existing seas
the Testacea, which have shells and well-defined patterns,
rarely inhabit a greater depth.

The Magnesian Limestone or Permian group is remarkable
chiefly for the vast number of fishes that have been found in
some of its members, such as the marl slate of Durham and
the Kupferschiefer, or copper slate of Thuringia.   From the
curved form of their impressions, as if they had been spas-
modically contracted, the fossil fish of the latter locality are
supposed to have perished by a sudden death before they sank
down into the mud in which they were entombed.   Probably
the copper which impregnates the stratum in which they
occur is connected with this phenomenon.   Mighty volcanic

eruptions corrupted the water with poisonous metallic salts, and destroyed in a short time whole legions of its finny inhabitants.

From the earliest ages the corals play a conspicious part in fossil history; and as in our days we find them encircling islands and fringing continents with hugh ramparts of lime-

VENTRICULITES—FOSSIL SPONGE (CHALK).

stone, so many an ancient reef, now far inland, and raised several thousand feet above the level of the sea, bears witness to the vast terrestrial changes that have taken place since it was first piled up by the growth of countless zoophytes.

SIPHONIA COSTATA—FOSSIL SPONGE (GREEN SAND, WARMINSTER).

With regard to the dimensions of the fossil corals, we do not find that any of them exceeded in size their modern relatives; but their construction was widely different.

The fossil sponges of the primitive seas are likewise very unlike those of the present day.

Thus in all the ancient strata we find abundant spongidæ with a stony skeleton, while all the modern sponges possess a horny frame. The Petrospongidæ, or stone sponges, which have long since disappeared, are frequently shapeless masses ; but a large number are cup-shaped, with a central tubular cavity, lined, as well as the outer surface, with pores more or less regularly arranged.

The Crinoids, or Sea-lilies, now almost entirely extinct, were

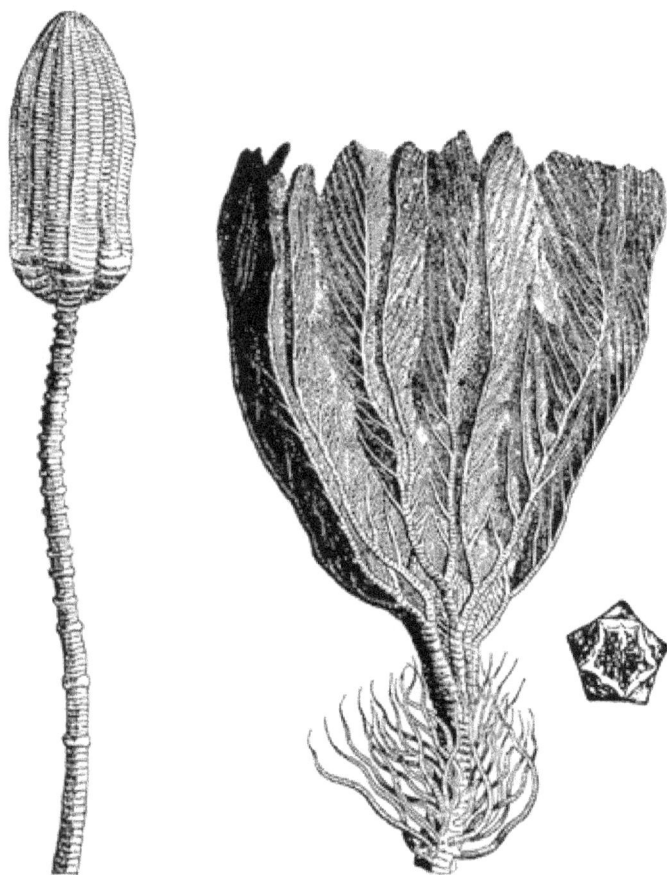

ENCRINUS LILIIFORMIS.
(Muschelkalk, Germany.)

PENTACRINUS BRIARREUS.

extremely common in the primeval seas. Unlike our modern sea-stars, to which they are allies, they did not move about

c

freely from place to place, but were affixed, like flowers, to a slender flexible stalk, composed of numerous calcareous joints connected together by a fleshy coat. The Carboniferous Mountain Limestone is loaded with their remains, and the *Encrinus liliiformis* is one of the leading fossils of the Muschelkalk of the Triassic group. The *Pentacrinus briareus* is of more modern date, and occurs in tangled masses, forming thin beds of considerable extent in the Lower Lias. This beautiful Crinoid, with its innumerable tentacular arms, appears to have been frequently attached to the drift wood of the Liassic sea, like the floating barnacles of the present day. In the still more recent Chalk group is found a remarkable form of star-fish, the *Marsupites ornatus*, which resembles in all respects the Crinoids, except that it is not and never was provided with a stem. It seems to have been rolled lazily to and fro, by the influence of the waves, at the bottom of the sea, and to have been anchored in its place by the action of gravity alone.

Of all the changes that have taken place in organic life, none perhaps are more remarkable than the transformations which the Cephalopod molluscs have undergone during the various geological eras. In the more ancient Palæozoic seas flourished the Orthoceratites, or straight-chambered shells, resembling a nautilus uncoiled. In the Carboniferous

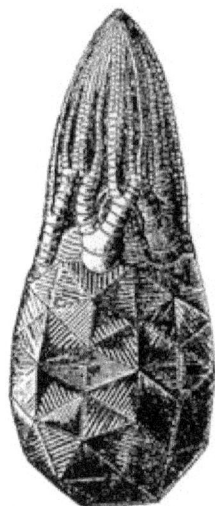

MARSUPITES ORNATUS. CHALK.

ages the Goniatites acquired their highest development. These shells were spirally wound, having the lobes of the chambers free from lateral denticulations or crenatures, so as to form continuous and uninterrupted outlines.

Both Orthoceratites and Goniatites disappear in the Triassic times, and are replaced by hosts of Ammonites, which successively flourished in more than 600 species, and are characterised by an external siphon and chambers of complicated, often foliated, pattern. This foliated structure gives a remarkable character to the intersection of the chamber partitions with the shell, and must have added greatly to the strength of the shell, which was always delicate and often very beautiful. The Ammonites, which made their first appearance towards the end of the Triassic period, abounded in the Oolitic and Cretaceous periods, and were replaced by new forms before the Tertiary beds were deposited. Among these we find the *Ancyloceras gigas*, which may be regarded as an Ammonite partially unrolled, and the *Turrilites tuberculatus*, which has the form and peculiar symmetry of a univalve shell.

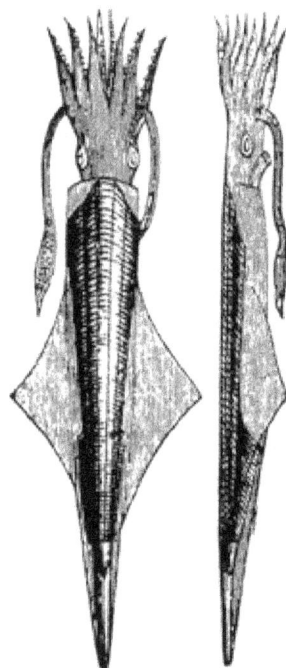

TURRILITES TUBERCULATUS.   RESTORED BELEMNITE.

In several of the older rocks, especially the Lias and Oolite, Belemnites are frequently met with. These singular dart- or arrow-shaped fossils were supposed by the ancients to be the thunderbolts of Jove, but are now known to be the petrified internal bones of a race of voracious cuttle-fishes, whose importance in the Oolitic or Cretaceous Seas may be judged of by the frequency of their remains and the 120 species that have been hitherto discovered.

Belemnites two feet long have been found, so that, to judge by analogies, the animals to which they belonged as cuttle-

bones must have measured eighteen or twenty feet from end
to end. Provided with prehensile hooks on their long arms,
and with a formidable parrot-like bill, these huge creatures
must have proved most dangerous antagonists, even to the
well-protected fishes that lived in the same seas. But of all
the denizens of the Mesozoic Ocean none were more powerful
than the large marine or enaliosaurian reptiles, which, flourishing throughout the whole of the Triassic period, were lords
of all they surveyed down to the end of the Cretaceous
epoch. First among these monsters appears the gigantic
Ichthyosaurus, which has been found no less than forty feet

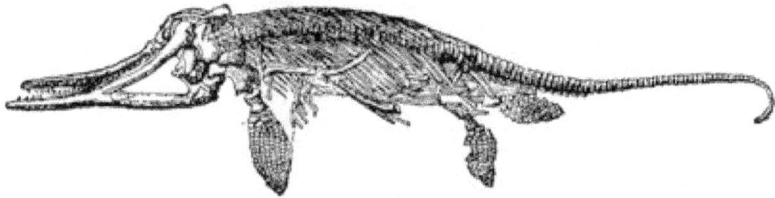

ICHTHYOSAURUS COMMUNIS.

long—a creature half fish, half lizard, and combining, in
strange juxtaposition, the snout of the porpoise, the teeth of
the crocodile, and the paddles of the whale. But the most
remarkable of its features is the eye, surpassing a man's head
in size, and wonderfully adapted for vision both far and
near.

In the quarries of Caen in Normandy, at Lyme Regis in
Dorsetshire, and particularly at Kloster Banz in Franconia,
where the largest known specimen has been discovered, entire
skeletons of the formidable Ichthyosaurus have been exhumed from the Liassic shale—memorials of the ages long
since past, when lands now far removed from the ocean still
lay at the bottom of the sea, and formed the domain of
gigantic lizards. The enormous jaw-bones of the Ichthyosauri, which in the full-grown animal could be opened seven
feet wide, were armed along their whole length with powerful
conical teeth, showing them to have been carnivorous, and
the half-digested remains of fishes and reptiles found within
their skeletons indicate the precise nature of their food. The
size of the swallowed object proves also that the cavity of

the stomach must have corresponded with the wide opening
of the jaws. Thus powerfully equipped for offensive warfare;
excellent swimmers from their compressed cuneiform trunk,
their long broad paddles, and their stout vertical tail-fin;
provided, moreover, with
eyes capable of piercing
the dim light of the
ocean depths, they must
have been formidable
indeed to the contem-
poraneous fishes.

The Ichthyosaurus
was admirably formed
for cleaving the waves
of an agitated sea; but
the Plesiosaurus was
equally well organised
for pursuing its prey in
shallow creeks and bays
defended from heavy
breakers. Its long
swan-like neck no doubt
enabled it to drag many
a victim from its hid-
ing-place. While these
huge lizards were the
terror of the seas, the
Pterodactyles, a race of
winged lizards, armed
with long jaws and
sharp teeth, hovered in
the air. With the ex-
ception of the greatly
elongated fifth finger,

PLESIOSAURUS DOLICHODEIRUS.
(British Museum—Found in the Lias of Street, near Glastonbury.)

to which, as well as to the whole length of the arm and
body, the membranous wing or organ of flight was attached,
the fingers of this strange animal were provided with sharp
claws, so that it was probably enabled, like the bat, to sus-
pend itself from precipitous rock-walls.

It is a remarkable fact, that, whereas the Pterodactyles of

the older Lias beds did not exceed ten or twelve inches in length, the later forms, found fossil in the Greensand and Wealden beds of the Lower Cretaceous formation, must have been at least 16½ feet long. That these reptiles were not the only vertebrated animals capable of hovering in the air at the time when the huge Ichthyosaurus was lord of the seas, is proved by a bird about the size of a rook, which was discovered in 1862, in the lithographic slate of Solenhofen in Bavaria, a stone-bed belonging to the period of the Upper Oolite. The skeleton of this valuable specimen, now in the British Museum, is almost entire, with the exception of the head, and retains even its feathers. Still older fossil mammalia have been found near Stuttgard, in the uppermost bed of the Triassic deposits, and in the Lower Oolite of Oxfordshire. These interesting remains, which carry back the existence of the mammals to a very remote period, belong to small marsupial, or opposum-like, animals. The jaws, which are the principal parts preserved, are exceedingly minute, and remarkable for the number and distribution of their teeth, which prove them to have been either insectivorous or rodent.

The remains of the Ichthyosauri and Plesiosauri occur chiefly in the Liassic group, but the more recent Cretaceous (Wealden) formation is distinguished by the presence of still more enormous land saurians. On their massive legs and unwieldy feet these monsters stood much higher than any reptile of our days, and resembled in bulk and stature the elephants of the present world.

The carnivorous Megalosaurus (for its sharply serrated teeth indicate its mode of life) appears to have preceded the gigantic Iguanodon, whose dentition denotes a vegetable food. Like the giant sloths of South America—the Megatherium and the Mylodon—the Iguanodon was provided with a long prehensile tongue and fleshy lips to seize the leaves and branches on which it fed. Professor Owen estimates its probable length at between fifty and sixty feet, and to judge by the proportions of its extremities, and particularly of its huge feet, it must have exceeded the bulk of the elephant eightfold.

During the following Upper Cretaceous epoch flourished

the Mosasaurus, a marine saurian, first discovered in the quarries of St. Peter's Mount, near Maestricht,* and supposed to have been twenty-four feet in length. But the supremacy of the reptiles was now drawing to its close, and in the Tertiary period we at length see the Mammalia assume a prominent place on the scene of life. The oldest of these tertiary quadrupeds differ so widely from those of the present day as to form distinct genera. The Palæotheriums, for instance, of which there are seventeen species, varying in dimension from the size of a rhinoceros to that of a hog, combine in their skeleton many of the characters of the tapir, the rhinoceros, and the horse, while the Anoplotheriums, whose size varied from that of a hare to that of a dwarf ass, resembled in some respects the rhinoceros and the horse, and in others the hippopotamus, the hog, and the camel.

In the Miocene epoch many of these more ancient quadrupeds no longer appear upon the scene, while others still flourish in its upper period along with still existing genera, and with forms long since extinct, such as the Dinotherium. This huge animal is particularly remarkable for its two large and heavy tusks, placed at the extremity of the lower jaw, and curved downwards like those in the upper jaw of the walrus. It was formerly supposed to be an herbivorous cretacean, and to have used its anterior limbs principally in the act of digging for roots. The remains on which these speculations were founded were the huge jaws and shoulder-blade discovered at Epplesheim in Hesse Darmstadt; but an immense pelvis of the animal, measuring six feet in breadth and four and a quarter feet in height, discovered by Father Sanno Solaro, in the department of the Haute Garonne, proves that this supposed aquatic pachyderm was a gigantic marsupial, and that the dependent trunks of the unwieldy animal, instead of serving the purpose of anchoring it to the banks of rivers, answered the more homely, but equally important office, of lifting the young into the maternal pouch. 'The remarkable history of the successive discovery of its bones,' says Professor Haughton, ' and the change of views consequent thereupon, should teach geologists modesty in the expression of

* Chapter XXXVIII.

their opinions.' During this period also flourished in India, along with many other strange forms of life, the Colossochelys Atlas, a tortoise of the most gigantic proportions, measuring, probably, nearly twenty feet on the curve of the carapace, and dwarfing into insignificance the great Indian tortoise of the present day.

The nearer we approach our own times, the greater becomes the proportion of still existing genera and species; and it is remarkable that as early as the Pliocene epoch we find a geographical distribution of mammalian life analogous to that which now characterises the various regions of the earth.

Thus the fossil monkeys of South America have the nostrils wide apart like all the existing simiæ of the new world, and fossil monkeys with approximated nostrils, the characteristic mark of all the old world quadrumana, are exclusively found in Asia and in Europe, where now a small species of monkey is confined to the Rock of Gibraltar, but where, in the Upper Miocene times, large long-armed apes, equalling man in stature, lived in the oak forests of France. Thus also South America, where alone sloths and armadilloes exist at the present day, is the only part of the world where, in the younger tertiary rocks, the remains of analogous mammals—the Megatherium, the Mylodon, and the Glyptodon—have been found.

The Mylodon was a colossal sloth, eleven feet long and with a corresponding girth. When we consider the huge size of the pelvis and the massiveness of the limbs, we must needs conclude that Professor Owen could not possibly have given the unwieldy animal a more appropriate surname than that of *robustus*.

The Megatherium was of still larger size. Its length was as much as eighteen feet, the breadth of its pelvis was six feet, and the tail, where it was attached to the body, must have measured six feet in circumference. The thigh bone was nearly three times as great as that of the largest known elephant, the bones of the instep and those of the foot being also of corresponding size. The general proportions both of the Megatherium and Mylodon resembled those of the elephant, the body being relatively as large, the legs shorter and thicker, and the neck very little longer. The Megatherium may have had a short proboscis, but the Mylodon exhibits no mark of such contrivance.

It is evident, from the bulk and construction of these huge animals, that they did not, like the sloths of the present day, crawl along the under side of the boughs till they had reached a commodious feeding place, but that, firmly seated on the strong tripod of their two hind legs and powerful tail, they uprooted trees or wrenched off branches with their fore limbs, which were well adapted for grasping the trunk or larger branches of a tree. The long and powerful claws were also, no doubt, useful in the preliminary process of scratching away the soil from the roots of the trees to be prostrated. This task accomplished, the long and curved fore claws would next be applied to the opposite sides of the loosened trunk. 'The tree being thus partly undermined and firmly grappled with, the muscles of the trunk, the pelvis, and hind limbs, animated by the nervous influence of the unusually large spinal cord, would combine their forces with those of the anterior members in the efforts at prostration. If now we picture to ourselves the massive frame of the Megatherium, convulsed with the mighty wrestling, every vibrating fibre reacting upon its bony attachment with a force which the sharp and strong crests and apophyses loudly bespeak, we may suppose that that tree must have been strong indeed which, rocked to and fro, to right and

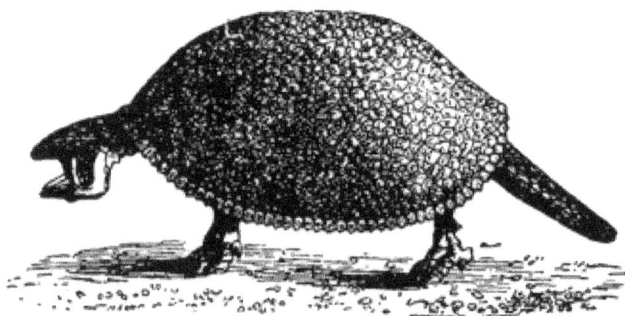

GLYPTODON CLAVIPES.

left, in such an embrace, could long withstand the efforts of its ponderous assailant.'

The Glyptodon, a colossal armadillo of the size of an

ox, was covered with a thick heavy tessellated bony armour, which, when detached from the body, resembled the section of a large cask. This harness measured on its curve from head to tail at least six feet, and four feet from side to side, so that a Laplander might have squatted comfortably under its roof.

In the superficial deposits of diluvial drift, in Germany and England, in Italy and Spain, in Northern Asia as well as in North America, between the latitudes of 40° and 75°, the bones of the large extinct Pachyderms have been found, and become more and more abundant as we approach the ice-bound regions within the Arctic Circle. The Siberian tundras, and the islands in the Polar Sea beyond, are, above all, so rich in the fossil remains of the Mammoth, or primitive elephant, that its tusks form a not unimportant branch of commerce. From the presence of so large an animal in treeless wilds, where now only small rodents or their persecutors, the Arctic fox and snow owl, find the means of subsistence, it has been inferred that Siberia must in those times have enjoyed a tropical climate ; but many weighty arguments have been arrayed against this opinion. The musk-ox, it is well known, prefers the stinted herbage of the Arctic regions, while the allied buffalo can only thrive in a warm country, and different species of bears are found in all zones ; so also the primitive elephant was formed for a temperate or cold climate. Instead of being naked, like his living Asiatic and African relations, the Mammoth was covered with a warm clothing, well fitted to brave a low temperature, a fact sufficiently proved by the carcass of one of these animals which was found, in the year 1803, imbedded in a mass of ice on the bank of the Lena in latitude 70°. Its skin was covered first with black bristles, thicker than horse-hair, from twelve to sixteen inches in length, secondly with hair of a reddish-brown colour, about four inches long, and thirdly with wool of the same colour as the hair, about an inch in length.

The discoveries of Middendorff on the banks of the Taymur likewise show that in those times the climate of Siberia was by no means tropical, for in latitude 75° 15′ he found the trunk of a larch imbedded with the bones of a Mammoth in an alluvial stratum fifteen feet above the level of the sea. Fragments of pine leaves have likewise been extracted from cavities in the molar teeth of a fossil rhinoceros, discovered on the banks of the Wiljui, in latitude 64°. The numerous land and freshwater shells accompanying the Mammoth in the highest latitudes are also, almost without exception, identical with those now existing in Siberia.

The Mastodon, though not uncommon among the fossils of the old world, is more abundantly found in North America. The molar teeth of this huge animal, whose grinding surfaces had their crowns studded with conical eminences, more or less resembling the teats of a cow, differed greatly from the flat-crowned grinders of the Mammoth; but both had twenty ribs like the living elephant, and must have been similar in size and general appearance. The body of the Mastodon would seem to have been longer, its limbs thicker and shorter, and, perhaps, its form, on the whole, rather approaching that of the hippopotamus, which it probably resembled also in some of its habits. Its mouth was broader than that of the elephant, and although it was certainly provided with a long trunk, it must have lived on soft succulent food, and it seems to have rarely left the marshes and muddy ponds, in which it would find ample food.

The most complete, and probably the largest, specimen of the Mastodon ever found was exhumed in 1845, in the town of Newbury, New York, the length of the skeleton being twenty-five feet, and its height twelve feet. From another specimen, found in the same year, in Warren County, New Jersey, the clay in the interior within the ribs, just where the contents of the stomach might naturally have been looked for, furnished some bushels of vegetable substance. A microscopic examination proved this matter to consist of pieces of small twigs of a coniferous tree of the cypress family, probably the young shoots of the white cedar (*Thuja occidentalis*) which is still a native of North America.

This interesting discovery likewise proves that the climate of North America was then, like that of Siberia, not very different from that of the present day.

The most remarkable of the fossil Ruminants are found among the deer tribe. The largest of these is the *Sivatherium giganteum,* discovered in the Tertiary beds of the sub-Himalayan hills. It was a deer with four horns, and, to judge by the size of its bones, must have exceeded the elephant in its dimensions. Near this huge 'antlered monarch of the waste' the extinct *Cervus megaceros,* found in the bogs and shallow marls of Ireland, appears as a mere dwarf, in spite of its large branching palmate horns, often weighing eighty pounds, and a corresponding stature far exceeding that of our modern deer.

The colossal size of many of the extinct plants and animals might seem to favour the belief that organic life has degenerated from its former powers; but a survey of existing creation soon proves the vital principle to be as strong and flourishing as ever.

No fossil tree has yet been found to equal the towering height of the huge Sequoias and Wellingtonias of California; and though the Horsetails and Clubmosses of the Carboniferous ages may well be called colossal when compared with their diminutive representatives of the present day, yet their height by no means exceeded that of the tall bamboos of India. No fossil bivalve is as large as the Tridacna of the tropical seas; and though our nautilus is a mere pigmy when compared with many of the Ammonites, our naked cuttlefishes are probably as bulky as those of any of the former geological formations. The living crustaceans and fishes are not inferior to their predecessors in size, and though the giant saurians of the past were much larger than our crocodiles, yet they do not completely dwarf them by comparison. The extinct Dinornis * far surpassed the ostrich in size, but the Mammoth and the Mastodon find their equal in our elephant; and though the sloths of the present day are mere pigmies when compared with the Megatherium, yet no extinct mammal attains the size of the Greenland whale.

The perfect preservation of so many fossil remains of animals and plants, which enables us to trace the progress of organic life on earth from one vast epoch to another, is surely wonderful enough; but we must consider it as a still greater wonder that phenomena usually so evanescent as

* Chapter XIX.

foot-prints, ripple-marks, and rain-prints, should in some cases have been permanently engraved in stone, and appear as distinct after millions of years as if their traces had been left but yesterday. All these marks were at first printed on soft argillaceous mud, on the sea-shore, or on the borders of lakes and rivers, which retained them as they became dry. Sand or clay having then been drifted into the mould by the wind, or deposited in its cavity by the next tide, a permanent cast was made, indented in the lower stratum and standing out in relief on the upper one.

Thus rain-drops on greenish slates of the Coal period, with several worm tracks, such as usually accompany rain-marks on the recent mud of modern beaches, have been discovered near Sydney, in Cape Breton. As the drops resemble in their average size those which now fall from the clouds, we may presume that the atmosphere of the Carboniferous period corresponded in density with that now investing the globe, and that different currents of air varied then as now in temperature, so as, by their mixture, to give rise to the condensation of aqueous vapour.

In like manner it has been possible to detect the foot-prints of reptiles, even in shales as old as the Cambrian formation, and to follow their trail as they walked or crawled along.

In the Upper New Red Sandstone (Lower Trias), near Hildburghausen, in Saxony, a strange unknown animal, supposed to belong to the frog order, has left foot-prints bearing a striking resemblance to the impressions made by a human hand; and in the still older red sandstone of Connecticut, a gigantic bird has marked a foot four times larger than that of the ostrich. It existed long before the Ichthyosaurus was seen on earth, and yet by a singular chance its traces, printed on a foundation proverbially unstable, have outlived the wreck of so many ages.

However brief and defective the foregoing review of the fossil world may have been, it has still sufficed to point out the existence on our planet of so many habitable surfaces, each distinct in time, and peopled with its peculiar races of aquatic and terrestrial beings, all admirably fitted for the new states of the globe as they arose, or they would not have increased and multiplied and endured for indefinite periods.

'The proofs now accumulated,' says Sir Charles Lyell, ' of the close analogy between extinct and recent species are such as to leave no doubt on the mind that the same harmony of parts and beauty of contrivance which we admire in the living creation has equally characterised the organic world at remote periods. Thus, as we increase our knowledge of the inexhaustible variety displayed in living nature, our admiration is multiplied by the reflection that it is only the last of a great series of pre-existing creations, of which we cannot estimate the number or limit in times past.'

CHAPTER III.

SUBTERRANEAN HEAT.

Zone of invariable Temperature—Increasing Temperature of the Earth at a greater Depth - Proofs found in Mines and Artesian Wells, in Hot Springs and Volcanic Eruptions—The whole Earth probably at one time a fluid mass.

BORN neither to soar into the air, nor to inhabit the deep waters, nor to pass his life in subterranean darkness, man is unable to depart to any considerable distance from the earth's surface. If he ascends in a balloon, he soon reaches the limits where the rarefied atmosphere renders breathing impossible ; a few thousand feet limit his efforts to pierce the earth's crust ; and should he be cast out into the sea, he is soon drowned. But beyond the limits to which his body is confined, his mind soars into space, and plunging into the unknown interior of our globe, seeks to unravel the mystery of its formation. In the following pages I purpose briefly to point out the circumstances which guide him in his speculations, and enable him to roam, at least in spirit, through the profound abysses of the subterranean world.

As we all know, the temperature of the atmosphere soon communicates its changes to the surface of the earth ; and our meadows, which when warmed by the rays of the sun are green and covered with flowers, harden in winter into a lifeless plain. But the influence of the sun's heat upon the soil is merely superficial, so that in the temperate zones the annual fluctuations of the thermometer are no longer perceptible at a depth of from 60 to 80 feet.

Thus, in the cellars of the Parisian observatory, a thermometer, placed many years ago 86 feet below the surface, invariably indicates + 11°7 Celsius ; the summer above may be ever so intensely hot, or the winter ever so cold, the column of mercury never deviates a hair's breadth from the height it has once attained. Below these limits the warmth of the earth gradually increases—a fact placed beyond all doubt by the innumerable observations that have been made in mines, and during the boring of Artesian wells. For wherever sinkings have been made, a rising of the thermometer has always been found to take place as the auger penetrates to a greater depth below the surface. Thus, to cite but a few examples, the temperature of the Artesian well of Grenelle in Paris, which, at a depth of 917 French feet, amounted to + 22°2 C., increased at the depth of 1,555 feet to + 26°43, and the water, which now gushes forth from the depth of 1,684 feet, constantly maintains the same lukewarm temperature of + 27°70.

During the boring of the well of Neusalzwerk, in Westphalia, the temperature rose at the various depths of 580, 1,285, and 1,935 feet from +19°7 C. to +27°5 and +31°4, until, finally, when the depth of 2,144 feet was attained, the saline spring issued forth with a constant temperature of +33°6. As from the experience acquired in mines and Artesian wells, the temperature is found to increase by one degree for about every successive 80 or 100 feet, the internal warmth of the earth, supposing it to increase in the same proportion towards the centre, would, at the depth of 10,000 feet, be equal to that of boiling water, and at that of 80 or 100 miles sufficiently great to melt the hardest rock.

Whether this steady increase really takes place is of course only matter of conjecture; but the history of hot springs and volcanic eruptions shows us that everywhere a very high degree of heat exists at considerable depths below the surface.

Most springs in the temperate zone, without being warm in a remarkable degree, still possess a higher temperature than the average warmth of the air in the locality where they gush forth, while in the tropical zone they are frequently cooler—a proof that in both cases they issue from a depth independent of the fluctuating atmospherical influences of the surface. While these cool or cold springs, spread in immense numbers over the earth, attest the existence everywhere of a subterranean source of heat, the warm and hot springs remind us of its intensity at more considerable depths. These thermal sources are confined to no climate, for in the cold land of the Tschuktschi, where the soil must be perpetually frozen to a depth of several hundred feet, boiling water is found to gush forth, as well as in the tropical Feejee Islands.

The hot springs, though of frequent occurrence in all parts of the world, are not the only or principal vents of subterranean heat. Far greater quantities of caloric are constantly pouring forth from the numerous volcanoes and solfataras, which are likewise distributed all over the surface of the globe. The violent convulsions which attend every outflow of lava are proofs that these torrents of liquid stone must have been forced upwards from a far greater depth than the water of the hot springs. The temperature necessary for their production likewise points to this fact, for to melt stones a heat of at least 2,000° C. is required. But volcanoes, like hot springs, are found in every zone; beyond the Arctic Circle, as well as in the most southern land attained by Sir James Ross in his memorable voyage. They line the coasts of the Pacific, as well as those of the Sea of Kamtschatka. They desolate Iceland as they devoured Pompeii and Herculaneum; and everywhere they pour forth the same masses of fluid stone; so that the geologist is not able to distinguish the lavas of the Andes chain from those of Etna or Vesuvius. But phenomena so much alike in character, common to all parts of the globe, can hardly be dependent upon mere local circumstances, and speak loudly in favour of the theory which supposes our earth to have been at one time a ball of liquid fire. Wandering through space during a course of unnumbered ages, this huge mass of molten stones and metals gradually cooled, and at length got covered with a solid crust, below which the ancient furnaces are still burning, and striving to burst their fetters. Well may we say with Horace—

Incedimus per ignes
Suppositos cineri doloso.'

CHAPTER IV.

SUBTERRANEAN UPHEAVALS AND DEPRESSIONS.

Oscillations of the Earth's Surface taking place in the present day—First ascertained in Sweden—Examples of Contemporaneous Upheaval and Depression in France and England—Probable Causes of the Phenomenon.

WHILE the sea and the atmospheric ocean are subject to perpetual fluctuations, and the poet justly compares the uncertain tenure of human prosperity with the restless wave or the inconstant wind, the solid earth is generally regarded as the emblem of stability. But an examination of the various strata of aqueous origin which constitute by far the greater part of the actual dry land soon shows the fallacy of this opinion.

The fossils of marine origin which occur in so many of our oldest rocks, now situated far above the level of the ocean, must necessarily have been raised from the deep. On the towering Andes, fifteen thousand feet above the tide-marks of the Pacific, the geologist finds sea-shells imbedded in the rock, and high above the snow-line the chamois-hunter of the Alps wonders at the sight of spirally-wound Ammonites that once enjoyed life at the bottom of the Liassic Sea. In strata of a more modern date, we find, on the banks of the river Senegal, far inland, large deposits of the *Arca senilis*; a mollusc still living on the neighbouring coast. On the borders of Loch Lomond, twenty feet above the level of the sea, shells of the edible cockle and sea-urchin repose in a layer of brown clay, and the banks of the Forth and of the Clyde, thirty feet higher than the storm tides, inclose remains of common shells of the present period, such as the oyster, the mussel, and the limpet. Along the shores of the Mediterranean, at Monte Video and at Valparaiso, in the isles of the Pacific and at the Cape, in California and Haïti, we meet with similar instances of elevation, which, though geologically recent, may yet be of a sufficiently ancient date to have preceded the appearance of man on earth. But proofs are not wanting that the upheaving power which has wrought so many changes in the past is still actively employed in remodelling the surface of the earth.

This important geological fact was first ascertained on the coast of Sweden, where the peculiar configuration of the shore makes it easy to appreciate slight changes in the relative level of land and water. For the continent is fringed with countless rocky islands called the 'skär,' within which boats and small vessels sail in smooth water even when the sea without is strongly agitated. But the navigation is very intricate, and the pilot must possess a perfect knowledge of the breadth and depth of every narrow channel, and the position of innumerable sunkenr ocks. On such a coast even a slight change of level could not fail to become known to the mariner, and to attract the attention of the learned, as soon as the book of nature began to be more accurately studied.

Early in the last century the Swedish naturalist Celsius collected numerous observations, all pointing to the fact of a slow elevation of the land. Rocks both on the shore of the Baltic and the German Ocean, known to have been once sunken reefs, were in his time above water; small islands in the Gulf of Bothnia had been joined to the continent, and old fishing grounds deserted, as being too shallow, or entirely dried up. These changes of level, which he estimated at about three feet in a century, Celsius attributed to a sinking of the waters of the Baltic, owing possibly to the channel, by which it discharges its surplus waters into the Atlantic, having been gradually widened and deepened by the waves and currents. But the lowering of level would in that case have been uniform and universal over that inland sea, and the waters could not have sunk at Torneo while they retained their former level at Copenhagen, Wismar, Stralsund, and other towns which are now as close to the water's edge as at the time of their foundation. Playfair (1802) and Leopold von Buch (1807) first attributed the change of level to the slow and insensible rising of the land, and the subsequent investigations of Sir Charles Lyell in 1834 have placed the fact beyond a doubt.

The attention of geologists having once been directed to the partial upheaval of the Scandinavian peninsula, similar facts were soon pointed out in other countries. At Bourg-neuf, near La Rochelle, the remains of a ship wrecked on an oyster bank in the year 1752, now lie in a cultivated field, fifteen feet above the level of the sea; and within a period of twenty-five years the parish has gained at least 1,500 acres, a very acceptable gift of the subterranean plutonic power. Port Bahaud, where formerly the Dutchmen used to take in cargoes of salt, is now 9,000 feet from the sea, and the Island of Olonne is at present surrounded only by swamps and meadows. These and similar phenomena, such as the constant rise of the chalk cliffs at Marennes, cannot possibly be explained by recent driftings, but evidently proceed from a slow upheaval of the coasts and the adjacent sea-bed.

On the opposite shores of the Atlantic, we find New-foundland undergoing a similar process of elevation ; for cliffs over which, thirty or forty years ago, schooners used to sail with perfect safety, are now quite close to the surface; and in the Pacific the depth of the channel leading to the port or Honolulu is gradually decreasing from the same cause.

While many coasts thus show signs of progressive elevation, others afford no less striking proofs of subsidence, frequently in close proximity to regions of upheaval.

Thus on the south-west coast of England, in Cornwall, Devon, and Somerset, submarine forests, consisting of the species still flourishing in the neighbourhood, are of such frequent occurrence that, according to Sir Henry de la Beche, 'it is difficult not to find traces of them at the mouths of all the numerous valleys which open upon the sea.' Sometimes they are covered with mud or sand, and generally the roots are found in the situation where they originally grew, while the trunks have been horizontally levelled. At Bann Bridge, specimens of ancient Roman pottery have been discovered twelve feet below the level of the sea, and the remains of an old Roman road, now submerged six feet deep, prove that the subsidence of the land has been going on since the times of Julius Cæsar and Agricola.

On the east coast the phenomenon is still more striking, particularly in the Wash, that shallow bay between Lincolnshire and Norfolk, on whose opposite shores a submarine forest extends, the trunks and stubbles of which become apparent at ebb-tide. On the coasts of Normandy and Brittany we likewise find traces of depression, pointing to some future time when perhaps many a bluff headland, now boldly fronting the ocean, may have disappeared beneath the waves.

Huts of the Esquimaux and of the early Danish colonists on the coast of Greenland, now submerged at high tide, could not possibly have been originally constructed in so inconvenient a situation; and at Puynipet, in the South Sea, habitations sunk beneath the water likewise prove a gradual subsidence of the land.

On many coasts and islands modern scientific explorers have hewn marks in the rock, to enable future generations to judge of the changes which are slowly and surely altering the configuration of the land and tracing new boundaries to the ocean. Had our forefathers left us similar memorials, we should know much more about the oscillatory movements of the earth-rind than we know now; but, unfortunately, experimental natural philosophy is but of recent date, and the marks chiselled out upon the Swedish rocks in the years 1731 and 1752 are the earliest records by which the chronological progress of elevation or subsidence can be distinctly ascertained.

This phenomenon, which has played so important a part in the physical annals of our globe, having once been accurately determined, enables the geologist to explain many facts for which, before it became known, it was impossible to account.

We now need not wonder at seeing sea-shells imbedded in the highest mountains or buried hundreds of fathoms under the ground, at alternating layers of marine and sweet-water deposits being frequently storied one above the other, or at originally horizontal strata being now found at every possible angle of inclination.

The imperceptible slowness with which many of these vast changes are actually taking place warrants the inference that violent volcanic revolutions have no doubt been far less instrumental in moulding the earth-rind to its present form than the slow oscillatory movements of elevation and depression which from time immemorial have been constantly altering its surface.

The causes of these oscillatory movements are still very imperfectly known, though a probable hypothesis attributes them to the expansion by increased temperature of extensive deep-seated masses of matter. As the elevation of some tracts seem to coincide with the proportionate depression of others at a greater or less distance, these alternating upheavals and subsidences may possibly be the result rather of the lateral shifting of the flow of heat from one mass of subterranean matter to a neighbouring mass than of its positive increase on the whole. 'Such a lateral division of the outward flow of heat,' says Mr. Poulett Scrope, 'we may presume to be caused by the deposition over certain areas of thick newly-formed beds of any matter imperfectly conducting heat, like sedimentary sands, gravels, clays, shales, or calcareous mud, by which the outward transmission of heat being checked, it must accumulate beneath, while a portion of it will pass off laterally to augment the temperature of mineral matter in neighbouring areas; just as the water of a spring, if its usual issue is blocked up, will accumulate in the fissures or pores of the rock containing it, until it finds a vent on either side and at a higher level. Owing to this increase, the resistance opposed by the overlying rocks in that quarter may be sooner or later overcome, and their elevation brought about, through the dilatation of the mineral matter beneath.'

CHAPTER V.

SUBTERRANEAN WATERS AND ARTESIAN WELLS.

Subterranean Distribution of the Waters—Admirable Provisions of Nature—Hydrostatic Laws regulating the Flow of Springs—Thermal Springs—Intermittent Springs—The Geysir—Bunsen's Theory—Artesian Wells—Le Puits de Grenelle—Deep Borings—Various Uses of Artesian Wells—Artesian Wells in Venice and in the Desert of Sahara.

IN every zone the evaporating power of the sun raises from the surface of the ocean vapours, which hover in the air until, condensed by cold, they descend in rain upon the earth. Here part of them are soon restored to the sea by the swollen rivers; another part is once more volatalised; but by far the larger quantity finds its slow way into the bowels of the earth, where it serves for the perennial supply of wells and springs.

The distribution of these subterranean waters, and the simple laws which regulate their circulation, afford us one of the most interesting glimpses into the physical economy of our globe. We know that the greater part of the earth's surface is composed of stratified rocks, or alternate beds of impermeable clay and porous limestone or sand, which were originally deposited in horizontal layers, but have since been more or less displaced and set on edge by upheaving forces. Wherever permeable beds of limestone or sand crop out on the surface of the land, the residuary portions of rain-water which are not disposed of by floods or by evaporation must necessarily penetrate into the pores and fissures, and descend lower and lower, until they finally reach an impermeable stratum which forbids their further progress to a greater depth.

The granite, gneiss, porphyry, lava, and other unstratified and crystalline rocks of igneous origin, which cover about a

third part of the habitable globe, are likewise intersected by
innumerable fissures and interstices, which in a similar
manner, collect'and transmit rain-water.

Thus the plutonic or volcanic forces which have gradually
moulded the dry land into its present form have also
provided it with the necessary filters, drains, reservoirs, and
conduits, for the constant replenishment of springs, brooks,
and rivers. As every porous layer is more or less saturated
with moisture, the stratified rocks are frequently traversed
at various depths by distinct sheets of water, or rather, in
most cases, by permanently drenched or waterlogged sheets
of chalk or sand. Thus, in a boring undertaken in search of
coal at St. Nicolas d'Aliermont, near Dieppe, no less than
seven very abundant aquiferous layers or beds of stone were
met with from about 75 to 1,000 feet below the surface.
In an Artesian boring at Paris, five distinct sheets of water,
each of them capable of ascension, were ascertained; and
similar perforations executed in the United States, and other
countries, have in the same manner traversed successive
stages of aqueous deposits.

Thus there can be no doubt that vast quantities of water
are everywhere accumulated in the porous strata of which a
great part of the superficial earth-rind is composed, the
rapidity with which they circulate varying of course with the
amount of hydrostatic pressure to which they are subjected,
and the more or less porous and permeable nature of the
beds through which they percolate. Were the ground we
stand on composed of transparent crystal, and the subterranean water-courses tinged with some vivid colour, we
should then see the upper earth-crust traversed in every
direction by aqueous veins, and frequently as saturated with
water as the internal parts of our body are with blood. But
Nature not only perennially feeds our springs and brooks
from the inexhaustible fountains of the deep; it is also one
of her infinitely wise provisions that the same water which,
if placed in casks or open tanks, becomes putrid, continues
fresh so long as it remains in the cavities and interstices of
the terrestrial strata. While filtering through the earth, it
is generally cleansed of all the organic substances whose
decay would inevitably taint its purity, and comes forth

salubrious and refreshing, a source of health and enjoyment to the whole animal creation.

The extreme limits to which the waters descend into the earth of course escape our direct observation, as the lowest point to which the subterranean regions have been probed is less than 2,000 or 2,500 feet below the level·of the sea; but as we know from the formation of many basins that the strata of which they are composed attain in many cases a thickness of from 20,000 to 30,000 feet, there can hardly be a doubt that they are permeated by water to an equal depth.

As steam plays so great a part in volcanic phenomena, the seat or effective cause of which must needs be sought for at an immense distance below the surface of the earth, we have another proof of the vast depth to which the subterranean migrations of water are able to attain.

After this brief glimpse into the reservoirs of the deep, we have to ascertain the power which raises their liquid contents and forces them to reappear upon the surface of the earth. If we pour water into a tube, bent in the form of the letter ʊ, it will rise to an equal level in both branches. We will now suppose that the left branch of the tube opens at the top into a vast reservoir, which is able to keep it constantly filled, and that the right branch is cut off near the bottom, so that only a small vertical piece remains. The pressure of the water column in the left branch will in this case force the liquid to gush out of the orifice of the shortened right branch to the level which it occupied while the branch was still entire.

These two hydrostatic laws, or rather these two modifications of the same law, have been frequently put to practical uses; as, for instance, in the communicating tubes which distribute the waters of an elevated source or reservoir to the various districts of a town, or in the subterranean conduits which serve to create fountains, such as those of Versailles or the Crystal Palace.

When the Romans intended to lead water from one hill to another, they constructed, at a vast expense, magnificent aqueducts across the intermediate valley; but the Turks, whom we look upon as ignorant barbarians, obtain the same result in a much more economical manner, and in this respect far

surpass the ancients, who, had they been better acquainted
with the first principles of hydrostatics, would indeed have
left us fewer specimens of their architectural skill, but would
at the same time have saved themselves a great deal of unnecessary expense.

Down the slope of the hill from which the water is to be
conducted, the Turks lay a tube of brick or metal, which,
crossing the valley, moulds itself to its different inflections,
and ultimately ascends the declivity of the hill on the
opposite side, where, in virtue of the law above cited, the
water rises as high as on that from which it descended. If
we suppose the descending branch of the tube to be prolonged only as far as the level of the valley, with a superficial
orifice, then the liquid will of course gush forth in a vertical
column, and form a *jet d'eau*, or fountain, its height being
determined by the elevation of the sheet of water by which
it is fed, and the consequent degree of pressure which acts
upon it. This is the principle on which all artificial fountains are constructed. The conduit, for instance, which feeds
the grand fountain of the Tuileries receives its water from
a reservoir situated on the heights of Chaillot.

Whatever the form of the tube may be in which the liquid
is contained, the simple hydrostatic law which regulates its
level remains unmodified. Let the tube be circular, elliptic,
or square, with a single orifice or with many—let it be open
or choked with pebbles or permeable sand—in every case the
water will invariably rise to the same height, provided the
tube be perfectly water-tight; or else gush forth wherever
it finds an opening below the highest level.

This hydrostatic principle so perfectly illustrates the origin
of springs, that it is almost superfluous to enter into any
further details on the subject.

When we consider that porous or absorbent strata, alternating with impermeable strata, frequently crop out on
the back or on the slope of hills or mountains, and then,
having reached their base, extend horizontally beneath the
plain, there can be no doubt that they are placed in the
same hydrostatic conditions as ordinary water-conducting
tubes, and that wherever any fissure or opening occurs in
the superincumbent impervious strata at any point below

the highest level of the water, springs must necessarily be formed.

As the same strata often extend over many hundreds of miles, we cannot wonder that sources frequently issue from the centre of immense plains, for the hydrostatic pressure which causes them to gush forth may have its seat at a very considerable distance.

As the waters by which the springs are fed have often vast subterranean journeys to perform, their temperature is naturally independent of that of the seasons or of the changes of the atmosphere. Thus, cold springs occur in a tropical climate, when their subterranean channels descend from high mountains, and boiling sources gush forth in the Arctic regions when forced upwards from a considerable depth.

While the waters filter through the earth, they also naturally dissolve a variety of substances, and hence all springs are more or less impregnated with extraneous particles. But many of them, particularly such as are of a higher temperature, contain either a large quantity or so peculiar a combination of mineral substances as to acquire medicinal virtues of the highest order.

The geological phenomena which favour the production of thermal springs are extremely interesting, and point to a deep-seated origin. By far the greater number of these fountains arise near the scene of some great subterranean disturbance, either connected with volcanic action, or with the elevation of a chain of mountains, or ascend through clefts and fissures caused by disruption. Thus the thermal springs of Matlock and Bath accompany great natural crevices in the mountain limestone, and the hot springs of Wiesbaden and Ems, of Carlsbad and Toeplitz, all lie contiguous to remarkable dislocations, or to great lines of elevation, or to the neighbourhood of a volcanic focus.

One of the most remarkable phenomena of thermal springs is the constant invariableness of their temperature and their mineral impregnations. During the last fifty or sixty years, ever since accurate thermometrical observations and chemical analyses have been made, the most celebrated mineral sources of Germany have been found to contain the same proportion of mineral substances. This is truly astonishing when we consider that the latter are merely dissolved by the waters while passing through the bowels of the earth, and that a considerable number of them are frequently found together in the same source.

Another remarkable fact is, that, even in countries exposed to violent and frequent earthquakes, so many subterranean watercourses have remained unaltered for 2,000 years at least. The sources of Greece still flow apparently as in the times of Hellenic antiquity. The spring of Erasinos, two leagues south of Argos, on the declivity af the Chaonian mountains, is mentioned by Herodotus. At Delphi the Cassotis (now Wells of Saint Nicholas) still flows under the ruins of the temple of Apollo, and the hot baths of Aidepsos still exist in which Sylla bathed during the Mithridatic war.

Many springs exhibit the singular phenomenon of an intermittence which is independent of the quantity of rain falling in the district, or of the flux and reflux of the tide in a neighbouring river. In many cases the simple and well-known hydrostatical law exemplified in the common siphon* affords a very ready and sufficient explanation of the phenomenon.

In the annexed diagram the vessel a communicates, by a tube c, with the siphon tube b, and it is manifest that

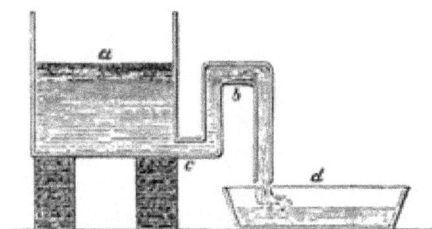

when the water in a rises above the level of the top of b, it

---

* A siphon, as is well known, is a bent tube, having one leg longer than the other. When this tube is filled with any liquid, and the shorter end is immersed in a vessel containing liquid of the same kind, the weight of the column in tho longer leg will cause the liquid to begin to run out, and it will continue running till the vessel is emptied. This arises from the pressure of air on the exposed surface of fluid, forcing it up through the tube to prevent vacuum, which would otherwise be formed at the highest point; and the extreme limit of length at which the siphon will act is therefore determined by the height of a column of tho fluid equal to the pressure of the atmosphere (fifteen pounds on the square inch). The limit in the case of water is something more than thirty feet.

will begin to flow over and escape, as at *d*. But as soon as this is the case the tube *b* begins to act as a siphon, and draws off all the water in *a*, so that if a constant supply is poured into *a*, but at a rate slower than the rate of the discharge at *d*, there will be an intermittent discharge, the interval depending on the relation of the rate of filling to that of emptying.

The case of a subterranean cavity in a limestone rock, slowly fed by drainage from the cracks and fissures of the rock above, and communicating at a distant point with the surface by a bent or siphon tube, is evidently strictly analogous.

SECTION OF AN INTERMITTENT SPRING.

Iceland, pre-eminently the land of volcanic wonders, possesses in the Great Geysir the most remarkable intermittent fountain in the world. 'At the foot of the Laugarfjall hill, in a green plain, through which several rivers meander like threads of silver, and where chains of dark-coloured mountains, overtopped here and there by distant snow-peaks, form a grand but melancholy picture, dense volumes of steam indicate from afar the site of a whole system of thermal springs congregated on a small piece of ground not exceeding twelve acres in extent. In any other spot the smallest of these boiling fountains would arrest the traveller's attention, but here his whole mind is absorbed by the Great Geysir. In the course of countless ages, this monarch of springs has formed out of the silica which it deposits a mound which rises to about thirty feet above the general surface of the

plain, and slopes on all sides, to a distance of a hundred
feet or thereabouts, from the border of a large circular basin
situated in its centre, and measuring about fifty-six feet in
the greatest diameter and fifty-two feet in the narrowest.
In the middle of this basin, forming as it were a gigantic

GEYSIRS OF ICELAND.

funnel, there is a pipe or tube, which at its opening in the
basin is eighteen or sixteen feet in diameter, but narrows
considerably at a little distance from the mouth, and then
appears to be not more than ten and twelve feet in diameter.
It has been probed to a depth of seventy feet, but it is more
than probable that hidden channels ramify further into the
bowels of the earth. The sides of the tube are smoothly
polished, and so hard that it is not possible to strike off a
piece of it with a hammer. Generally the whole basin is
found filled up to the brim with sea-green water as pure
as crystal, and of a temperature of from 180° to 190°.
Astonished at the placid tranquillity of the pool, the traveller
can hardly believe that he is really standing on the brink of
the far-famed Geysir; but suddenly a subterranean thunder
is heard, the ground trembles under his feet, the water in
the basin begins to simmer, and large bubbles of steam rise
from the tube and burst on reaching the surface, throwing
up small jets of spray to the height of several feet. Every

instant he expects to witness the grand spectacle which has chiefly induced him to visit this northern land ; but soon the basin becomes tranquil as before, and the dense vapours produced by the ebullition are wafted away by the breeze. These smaller eruptions are regularly repeated every eighty or ninety minutes, but frequently the traveller is obliged to wait a whole day or even longer before he sees the whole power of the Geysir. A detonation louder than usual precedes one of these grand eruptions ; the water in the basin is violently agitated ; the tube boils vehemently ; and suddenly a magnificent column of water, clothed in vapour of a dazzling whiteness, shoots up into the air with immense impetuosity, to the height of eighty or ninety feet, and, radiating at its apex, showers water and steam in every direction. A second eruption and a third rapidly follow, and after a few minutes the fairy spectacle has passed away like a fantastic vision. The basin is now completely dried up, and on looking down into the shaft, the traveller is astonished to see the water about six feet from the rim, and as tranquil as in an ordinary well. After about thirty or forty minutes it again begins to rise, and after a few hours reaches the brim of the basin. Soon the subterranean thunder, the shaking of the ground, the simmering above the tube begin again— a new gigantic explosion takes place, to be followed by a new period of rest—and thus this wonderful play of nature goes on, day after day, year after year, and century after century. The mound of the Geysir bears witness to its immense antiquity, as its water contains but a minute portion of silica.'[*]

The explanation of these wonderful phenomena has exercised the ingenuity of many natural philosophers ; but Professor Bunsen's theory seems the most plausible. Having first ascertained, by experiment, that the water at the mouth of the tube has a temperature, corresponding to the pressure of the atmosphere, of about 212° F., he found it much hotter at a certain-depth below ; a thermometer, suspended by a string in the pipe, rising to 266° F., or no less than 48° above the boiling point. By letting down stones, suspended by strings, to various depths, he next came to the conclusion

* ' The Polar World,' p. 54.

that the tube itself is the main seat or focus of the mechanical power which forces the huge water column upwards. For the stones which were sunk to greater distances from the surface were not cast up again when the next eruption of the Geysir took place, whereas those nearer the mouth of the tube were ejected to a considerable height by the ascending water-column. Other experiments also were made, tending to demonstrate the singular fact that there is often scarcely any motion below when a violent rush of steam and water is taking place above. It seems that when a lofty column of water possesses a temperature increasing with the depth, any slight ebullition, or disturbance of equilibrium, in the upper portion may first force up water into the basin, and then cause it to flow over the edge. A lower portion, thus suddenly relieved of part of its pressure, expands, and is converted into vapour more rapidly than the first, owing to its greater heat. This allows the next subjacent stratum, which is much hotter, to rise and flash into a gaseous form ; and this process goes on till the ebullition has descended from the middle to near the bottom of the funnel.*

In many geological basins the deep subterranean waters are frequently inclosed over a surface of many square miles between impermeable beds of clay or hard rock, which nowhere permit them to escape; but if a hole be bored deep enough to reach a permeable bed, it is evident that they will then gush forth more or less violently, according to the degree of hydrostatic pressure which acts upon them. This is the simple theory of the Artesian Wells, so called from the French province of Artois, where, as far back as the beginning of the twelfth century, springs of water were artificially obtained by perforating the soil to a certain depth in places where no indication of springs existed at the surface. The barbarous inhabitants of the Sahara seem, however, to have long preceded the Artesians in the art of sinking deep wells; for Olympiodorus, a writer who flourished at Alexandria about the middle of the sixth century, mentions pits sunk in the oasis to the depth of 200 or 300 yards, and pouring forth streams of water, used for irrigation.

* Liebig's 'Annalen,' translated in 'Reports and Memoirs of the Cavendish Society,' London, 1848, p. 351.

By the aid of geological science, and of greater mechanical skill, Artesian borings* are at present frequently undertaken in civilised countries, wherever the nature of the ground promises success, and the want of water is sufficiently great to warrant the attempt. Sometimes the water is reached at a moderate distance from the surface, but not seldom it has been found necessary to bore to a depth of 200 or 300 fathoms. Often efforts, even on this large scale, have proved vain, and the work has been abandoned in despair.

One of the most remarkable instances on record of a successful sinking for water is that of the Artesian well of Grenelle, one of the Parisian suburbs.

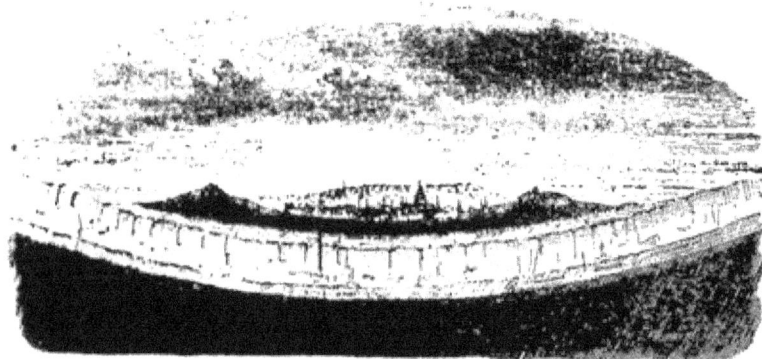

POROUS STRATA. ARTESIAN WELL SUNK IN THE LONDON BASIN.

The work was begun with an auger of about a foot in diameter, and the borings showed successively the alluvial soil and subsoil, and the tertiary sands, gravels, clays, lignite, &c., until the chalk was reached. The work was then carried on regularly through the hard upper chalk down to the lower chalk with green grains, the dimension of the auger being reduced at 500 feet to a nine-inch, and at 1,300 feet to a six-inch aperture. When the calculated depth of 1,500 feet had been reached, and as yet no result appeared, the Government began to be disheartened. Still, upon the urgent representations of the celebrated Arago, the sinking was continued, until at length, at the depth of 1,800 feet, the auger, after a

* See Chapter on Mines in general, for a short account of earth-boring operations.

violent shock which made the ground tremble, suddenly
turned without an effort. 'Either the auger is broken, or
we have gained our end,' exclaimed the director of the work ;
and a few moments after, a large column of water gushed
out of the orifice. It took more than seven years to accomplish this grand work (1833–41), which was retarded by
numberless difficulties and accidents. About half-a-million
gallons of perfectly limpid water of a temperature of 82°
Fahr. are daily supplied by the Puits de Grenelle, and amply
repay its cost (862,432 fr. 65 centimes = 14,500l.).

The high temperature of Artesian springs, when rising
from considerable depths, has been turned to various practical uses. Thus, near Canstadt, in Wurtemberg, several
mills are kept in work, during the severest cold of winter,
by means of the warm water of Artesian wells which has
been turned into the mill-ponds, and at Heilbronn several
proprietors save the expense of fuel by leading Artesian
water in pipes through their green-houses. In some
localities the pure and constantly temperate Artesian waters
are made use of for the cultivation of cress. The vigorous
growth of this salutary herb in the beds of rivulets, where
natural springs gush forth, gave the idea of this application, which is so profitable that the cress nurseries of Erfurt
yield a produce of 12,000l. a year. Fish ponds have also
been improved by such warm springs being passed through
them.

Among the localities benefited by the boring of Artesian
wells, Venice deserves to be particularly noticed. Formerly the City of the Doges had no other supply of water
but that which was conveyed by boats from the Brenta, or
obtained from the rain collected in cisterns. Hence the joy
of the inhabitants may be imagined, when, in 1846, an Artesian boring in the Piazza San Paolo began to disgorge its
water at the rate of forty gallons per minute, and when other
undertakings of the same kind proved equally successful.

Wherever a well gushes forth in the Sahara, it brings life
into the wilderness; the date-tree flourishes as far as its
fertilising waters extend, and the wandering Arab changes
into a sedentary cultivator of the soil. Thus the boring of
Artesian wells on the desert confines of South-Algeria has

been the means of wonderful improvement; and if the French have too often marked their dominion in Africa by a barbarous oppression of the Arabs, they, in this instance at least, appear in the more amiable light of public benefactors.

A boring apparatus was first landed at Philippeville in April 1856, and conveyed with immense difficulty to the Oasis Wad Rir at Tamerna. The work was begun in May, and on the 19th of June, a spring, to which the grateful inhabitants gave the name of the 'Well of Peace,' gushed forth. Soon after another source was tapped at Tamelhat, in the Oasis Temacen, and received the name of the 'Well of God's Blessing.'

The beneficent instrument of abundance was now conveyed to the Oasis Sidi Rasched, fifteen miles beyond Tuggurt. Here the auger had scarcely reached a depth of 120 feet when a perfect stream gushed forth, which, according to the praiseworthy Arab custom, received the name of the 'Well of Thanks.' The opening of this wonderful source gave rise to many touching scenes. The Arabs came in throngs to witness the joyful spectacle: each of them poured some of the water over his head, and the mothers bathed their children in the gushing flood. An old scheik, unable to conceal his emotion, fell down upon his knees, and shedding tears of joy, fervently thanked God for having allowed him to witness such a day.

The next triumph was the boring of four wells in the desert of Morran, where previously no spring had existed. In the full expectation of success, everything had been prepared to turn this new source of wealth to immediate use, and part of a nomadic tribe instantly settled on the spot, and planted 1,200 date-trees. A dreary solitude was changed, as if by magic, into a scene of busy life.

These few examples suffice to show the vast services which Artesian wells are destined at some future time to render to many of the arid regions of Africa. Both in the Sahara and in the basin-shaped deserts, which extend, under various names, from the Cape Colony to the neighbourhood of Lake Ngami, there are, beyond all doubt, numberless spots where water, the fertilising element, may be extracted from the bowels of the earth.

In the droughty plains of Australia also a vast sphere of
utility is reserved to the Artesian wells. Here, also, they will
subdue the desert, unite one coast to another by creating
stations in the wilderness, and, with every new source which
they call to life, promote both material progress and intellectual improvement.

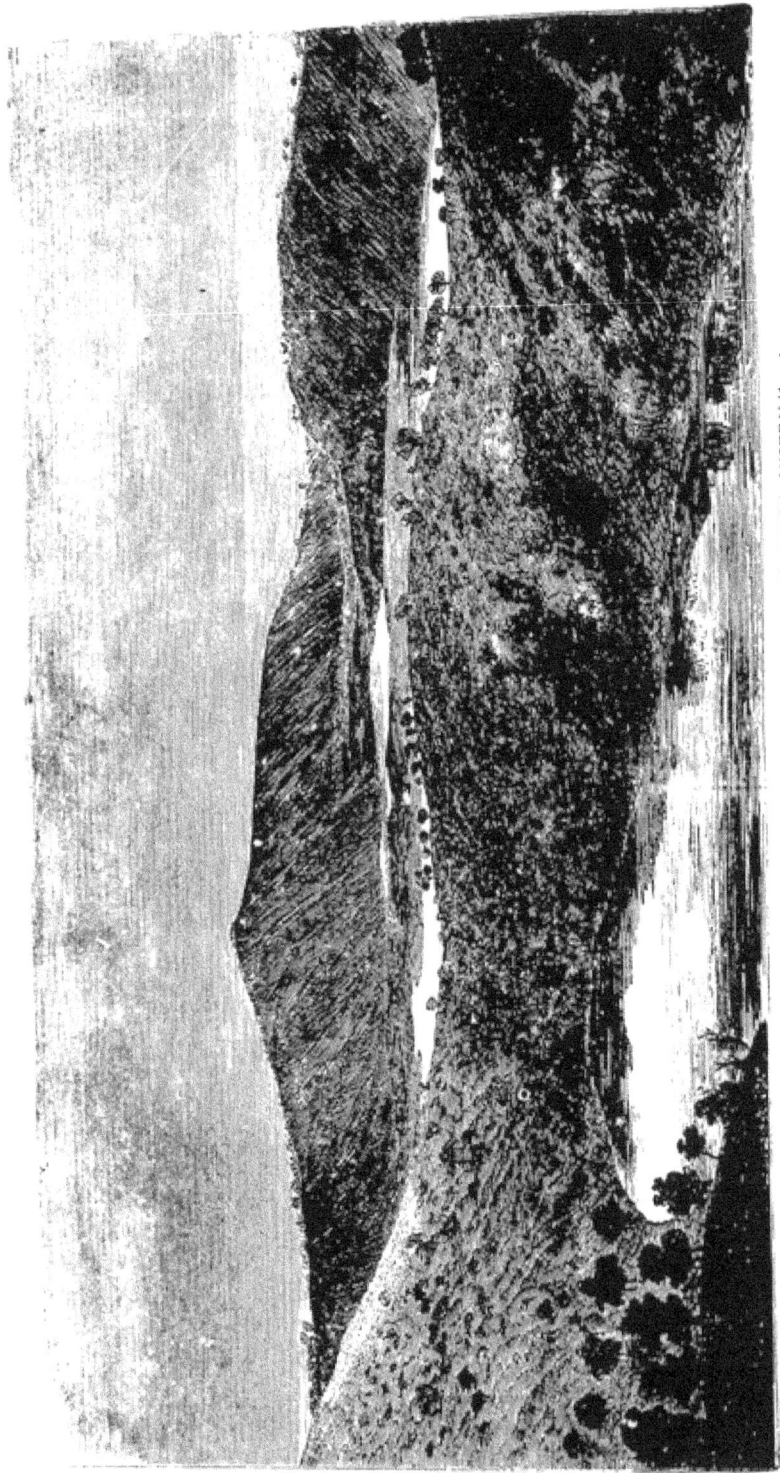

MIDDLE AND VALLEY LAKE CRATERS, MOUNT GAMBIER, SOUTH AUSTRALIA.

CHAPTER VI.

VOLCANOES.

Volcanic Mountains—Extinct and Active Craters—Their Size—Dangerous Crater-explorations—Dr. Judd in the Kilauea Pit—Extinct Craters—Their Beauty—The Crater of Mount Vultur in Apulia—Volcanoes still constantly forming—Jorullo and Isalco—Submarine Volcanoes—Sabrina and Graham's Island—Santorin—Number of Volcanoes—Their Distribution—Volcanoes in a constant state of Eruption—Stromboli—Fumaroles—The Lava Lakes in Kilauea—Volcanic Paroxysms—Column of Smoke and Ashes—Detonations—Explosion of Cones—Disastrous Effects of Showers of Ashes and Lapilli—Mud Streams—Fish disgorged from Volcanic Caverns—Eruption of Lava—Parasitic Cones—Phenomena attending the Flow of a Lava Stream—Baron Papalardo—Meeting of Lava and Water—Scoriæ—Lava and Ice—Vast dimensions of several Lava Streams—Scenes of Desolation—Volcanoes considered as Safety-valves—Probable Causes of Volcanoes.

VOLCANOES are vents which either have communicated, or still communicate, by one or several chimney-like canals or shafts, with a focus of subterranean fire, emitting, or having once emitted, heated matter in a solid, semi-liquid, or gaseous state. The first eruption of a volcano necessarily leaves a mound of scoriæ and lava, while numerous eruptions at length raise mountains, which are frequently of an amazing extent and height. These mountains, which are generally called volcanoes, though in reality they are but an effect of volcanic action situated far beneath their base, are called *extinct* when for many centuries they have exhibited no signs of combustion—*active*, when, either perpetually or from time to time, eruptions or exhalations of lava, scoriæ, or gases take place from their summits, or from vents in their sides. Their shape is generally that of a more or less truncated cone; but while some, like Cotopaxi or the Peak of Teneriffe, rise with abrupt declivities in the shape of a sugar-loaf, others, like Mauna Loa in the island of

Hawaii, gradually, and almost imperceptibly, ascend from a vast base embracing many miles in circuit.

Their heights also vary greatly. While some, like Madana in Santa Cruz, or Djebel Teir on the coast of the Red Sea, scarcely raise their summits a few hundred feet above the level of the ocean, others, like Chuquibamba (21,000 feet) or Aconcagua (22,434 feet), hold a conspicuous rank among the giant mountains of the earth.

The summit of a volcano generally terminates in a central cavity or crater, where the eruptive channel finds its vent. Craters are sometimes regularly funnel-shaped, descending with slanting sides to the eruptive mouth, but more commonly they are surrounded with high precipitous rock-walls, while their bottom forms a plain, which is frequently completely horizontal, and sometimes of a considerable extent. Its surface is rough and uneven, from the mounds of volcanic sand, or scoriæ, or of hardened lava with which it is covered, and generally exhibits a scene of dreadful desolation, rendered still more impressive by the steam and smoke, which, as long as the volcano continues in an active state, issues from its crevices.

Within this plain the eruptive orifice or mouth of the volcano is almost universally surrounded by an elevation, composed of ejected fragments of scoriæ thrown from the vent. Such cones are forming constantly at Vesuvius, one being no sooner destroyed by any great eruption, before another begins to take shape and is enlarged, till often it reaches a height of several hundred feet.

Thus the crater of an active volcano is the scene of perpetual change—of a continual construction and re-construction, and the sands of the sea do not afford a more striking image of inconstancy.

The various craters are of very different dimensions. While the chief crater of Stromboli has a diameter of only fifty feet, that of Gunong Tenger, in Java, measures four miles from end to end; and though the depth of a crater rarely exceeds 1,000 or 1,500 feet, the spectator, standing on the brink of the great crater of Popocatepetl, looks down into a gulf of 8,000 feet.

From the colossal dimensions of the larger craters, it may

well be imagined that their aspect exhibits some of the sub-
limest though most gloomy scenery in nature—the picture of
old Chaos with all its horrors.

The volcano Gunong Tjerimai, in Java, which rises to the
height of 9,000 feet, is covered with a dense vegetation up to
the crater's brink. On emerging from the thicket, the
wanderer suddenly stands on the verge of an immense excavation encircled with naked rocks. He is obliged to hold
himself by the branches of trees, or to stretch himself flat
upon the ground, so as to be able to look down into the
yawning gulf. The inaccessible bottom of the crater loses
itself in misty obscurity, and glimmers indistinctly through
the vapours slowly and incessantly ascending from its
mysterious depths. All is desolate and silent, save when
a solitary falcon, hovering over the vast chasm, awakes with
her discordant screech the echoes of the precipice. Through
a telescope may be seen, in various parts of the huge
crater walls, swarms of small swallows, which have there
built their nests, flying backwards and forwards. The wandering eye can detect no other signs of life, the attentive
ear distinguish no other sound.

Humboldt describes the view down the crater of the Rucu-
Pichincha—a volcano which towers above the town of Quito
to a height of 15,000 feet—as the grandest he ever beheld
during all his long wanderings. Guided by an Indian, he
ascended the mountain in 1802, and after scaling, with great
difficulty and no small danger, its steep and rocky sides, he
at length looked down upon the black and dismal abyss,
whence clouds of sulphurous vapour were rising as from the
gates of hell.

The descent into the crater of an active volcano is at all
times a difficult and hazardous enterprise, both from the
steepness of its encircling rock walls, and the suffocating
vapours rising from its bottom ; but it is rare indeed that a
traveller has either the temerity or the good fortune to penetrate as far as the very mouth of the eruptive channel, and to
gain a glimpse of its mysterious horrors. When M. Housel
visited Mount Etna in 1769, he ventured to scale the cone
of stones and ashes which had been thrown up in the centre
of the crater, where thirty years before there was only a

prodigious chasm or gulf. On ascending this mound, which emitted smoke from every pore, the adventurous traveller sunk about mid-leg at every step, and was in constant danger of being swallowed up. At last, when the summit was reached, the looseness of the soil obliged him to throw himself down flat upon the ground, that so he might be in less danger of sinking, while at the same time the sulphurous exhalations arising from the funnel-shaped cavity threatened suffocation, and so irritated his lungs as to produce a very troublesome and incessant cough. In this posture the traveller viewed the wide unfathomable gulf in the middle of the crater, but could discover nothing except a cloud of smoke, which issued from a number of small apertures scattered all around. From time to time dreadful sounds issued from the bowels of the volcano, as if the roar of artillery were re-bellowed throughout all the hollows of the mountain. They were no doubt occasioned by the explosions of pent-up gases striking against the sides of these immense caverns, and multiplied by their echoes in an extraordinary manner. After the first unavoidable impression of terror had been overcome, nothing could be more sublime than these awful sounds, which seemed like a warning of Etna not to pry too deeply into his secrets.

Dr. Judd, an American naturalist, who, in 1841, descended into the crater of Kilauea, on Mauna Loa, in Hawaii, well-nigh fell a victim to his curiosity. At that time the smallest of the two lava pools which boil at the bottom of that extraordinary pit appeared almost inactive, giving out only vapours, with an occasional jet of lava at its centre. Dr. Judd, considering the quiet favourable for dipping up some of the liquid with an iron ladle, descended for the purpose to a narrow ledge bordering the pool. While he was preparing to carry out his idea, his attention was excited by a sudden sinking of its surface; the next instant it began to rise, and then followed an explosion, throwing the lava higher than his head. He had scarcely escaped from his dangerous situation, the moment after, by the aid of a native, before the lava boiled up, covered the place where he stood, and, flowing out over the northern side, extended in a stream a mile wide to a distance of more than a mile and a half!

In extinct volcanoes, the picture of desolation originally shown by their craters has not seldom been changed into one of charming loveliness. Tall forest-trees cover the bottom of the Tofua crater in Upolu, one of the Samoan group; and in the same island, a circular lake of crystal purity, belted with a girdle of the richest green, has formed in the depth of the Lanuto crater.

The lakes of Averno near Naples, and of Bolsena, Bracciano, and Ronciglione, likewise fill the hollows of extinct craters, constituting scenes of surpassing beauty, rendered still more impressive by the remembrance of the stormy past which preceded their present epoch of tranquillity and peace. Mr. Mallet describes, with glowing colours, the singular beauty

EXTINCT CRATER OF HALEAKALA.

of the forest scenery around the two extinct craters of Mount Vultur in Apulia, which time has converted into two deep circular lakes.

' I descend amongst aged trunks and overarching boughs, and pass over masses of rounded lava-blocks and cemented lapilli. All is quietude; the soft breeze of a quiet winter's afternoon fans across the embosomed water, from the early wheat-fields and the furrowed acres of the opposite steep slopes, and brings the gentle ripple lapping amongst the roots of the old hazels at my feet.

' Off before me, and to my left, crowning the slope, are the grey ruins of some ancient church or castle, and far above me to the right, nestled against the lava crags, behind and above it, standing out white and clear, I see the strong buttressed mass of the monastery of St. Michael. How hard it it is to realise that this noble and lovely scene, full of every leafy beauty, was once the innermost bowl of a volcano ; that every stone around me, now glorious in colour with moss and lichen, sedum and geranium, was once a glowing mass, vomited from out that fiery and undiscovered abyss, which these placid waters now bury in their secret chambers.'

The line of demarcation between active and extinct volcanoes is not easily drawn, as eruptions have sometimes taken place after such long intervals of repose as to warrant the belief that the vents from which they issued had long since been completely obliterated. Thus, though nearly six centuries have passed since the last eruption of Epomeo in the island of Ischia, we are not entitled to suppose it extinct, since nearly seventeen centuries elapsed between this last explosion and the one which preceded it. Since the beginning of the fourteenth century Vesuvius also enjoyed a long rest of nearly three hundred years. During this time the crater got covered with grass and shrubs, oak and chestnut trees grew around it, and some warm pools of water alone reminded the visitor of the former condition of the mountain, when, suddenly, in December 1631, it resumed its ancient activity, and seven streams of lava at once burst forth from its subterranean furnaces.

While, in many volcanic districts, such as that of the Eifel on the left bank of the Rhine, and of Auvergne in Central France, the once active subterranean fires have long since been extinguished, and no eruption of lava has been recorded during the whole period of the historic ages, new volcanoes, situated at a considerable distance from all previously active vents, have arisen from the bowels of the earth, almost within the memory of living man. From the era of the discovery of the New World to the middle of the last century, the country between the mountains Toluca and Colima, in Mexico, had remained undisturbed, and the space, now the site of Jorullo, which is one hundred miles distant from each of the above-mentioned volcanoes, was occupied by fertile fields of sugar-cane and indigo, and watered by two brooks. In the month of June 1769, hollow sounds of an alarming nature were heard, and earthquakes succeeded each other for two months, until, at the end of September, flames issued from the ground, and fragments of burning rocks were thrown to prodigious heights. Six volcanic cones, composed of scoriæ and fragmentary lava, were formed on the line of a chasm, running in the direction of N.E. to S.W. The least-of the cones was 300 feet in height, and Jorullo, the central volcano was elevated 1,600 feet above the level of the plain. The ground where now, in Central America, Isalco towers in proud eminence, was formerly the seat of an estancia or cattle-estate. Towards the end of the year 1769 the inhabitants were frequently disturbed by subterranean rumblings and shocks, which constantly increased in violence until, on February 23, 1770, the earth opened, and pouring out quantities of lava, ashes, and cinders, gave birth to a new volcanic mountain.

Besides those volcanic vents which are situated on the dry land, there are others which, hidden beneath the surface of the sea, reveal their existence by subaqueous eruptions. Columns of fire and smoke are seen to rise from the discoloured and agitated waters, and sometimes new islands are gradually piled up by the masses of scoriæ and ashes ejected from the mouth of the submarine volcano. In this manner the island of Sabrina rose from the bottom of the sea, near St. Michael's in the Azores, in the year 1811; and still more recently, in 1831, Graham's Island was formed in the Mediterranean, between the coast of Sicily and that projecting part of the African coast where ancient Carthage stood. Slight earthquake shocks preceded its appearance, then a column of water like a water-spout, 60 feet high and 800 yards in circumference, rose from the sea, and soon afterwards dense volumes of steam, which ascended to the height of 1,800 feet. Then a small island, a few feet high with a crater in its centre, ejecting volcanic matter, and immense columns of vapour, emerged from the agitated waters, and in a fortnight swelled to the ample proportions of a height of 200 feet, and a circumference of three miles. But both Sabrina and Graham's Island, being built of loose scoriæ, were soon corroded by the waves, and their last traces have long since disappeared under the surface of the ocean.

Near Pondicherry, in India; near Iceland, in the Atlantic Ocean; half a degree to the south of the equator in the prolongation of a line drawn from St. Helena to Ascension; near Juan Fernandez, &c., similar phenomena have occurred within the last hundred years, but probably nowhere on a grander scale than in the Aleutian Archipelago, where, about thirty miles to the north of Unalaska, near the isle of Umnack, a new island, now several thousand feet high and two or three miles in circumference, was formed in 1796. The whole bottom of the sea between this new creation of the volcanic powers and Umnack has been raised by the eruptive throes which gave it birth; and where Cook freely sailed in 1778, numberless cliffs and reefs now obstruct the passage of the mariner.

The famous subaqueous volcano which, in the year 186 before the Christian era, began its series of historically recorded eruptions, by raising the islet of Hiera (the 'Sacred') in the centre of the Bay of Santorin, opened two new vents in 1866. Amid a tremendous roar of steam and the shooting up of prodigious masses of rock and ashes, two islets were formed, which ultimately rose to the height of 60 and 200 feet. The eruption continued for many months, to the delight and wonder of the numerous geologists who came from all sides to witness the instructive spectacle. Thus, in many parts of the ocean, we see the submarine volcanic fires laying the foundations of new islands and archipelagos, which, after repeated eruptions following each other in the course of ages, will probably, like Iceland, extend over a considerable space and become the seats of civilised man.

As a very considerable part of the globe has never yet been scientifically explored, it is of course impossible to determine the exact number of the extinct and active volcanoes which are scattered over its surface. Werner gives a list of 193 volcanoes, and Humboldt mentions 407, of which 225 are still in a state of activity. The newest computation of Dr. Fuchs, of Heidelberg,* increases the number to a total of 672, of which 270 are active. Future geographical discoveries will, no doubt, make further additions to the list, and show that at least through a thousand different vents

* 'Die vulcanischen Erscheinungen der Erde.' Leipzig, 1865.

the subterranean fires have, at various periods of the earth's history, piled up their cones of scoriæ and lava.

The volcanoes are very unequally distributed over the surface of the globe, for, while in some parts they are thickly clustered together in groups or rows, we find in other parts vast areas of land without the least sign of volcanic action.

An almost uninterrupted range of volcanoes extends in a sinuous line from the Gulf of Bengal, through the East Indian Archipelago, the Moluccas, the Philippines, Formosa, Japan, and the Kuriles, to Kamtschatka. This desolate peninsula is particularly remarkable for the energy of its subterranean fires, as Ermann mentions no less than twenty-one active volcanoes, ranged in two parallel lines throughout its whole length, and separated from each other by a central range of mountains, containing a large and unknown number of extinct craters.

In Java, where more than thirty volcanoes are more or less active, the furnaces of the subterranean world are still more concentrated and dreadful.

The immense mountain-chains which run parallel to the western coast of America are likewise crowned with numerous volcanic peaks. Chili alone has fourteen active volcanoes, Bolivia and Peru three, Quito eleven. In Central America we find twenty-one volcanoes, which are chiefly grouped near the Lake of Nicaragua, and to the west of the town of Guatemala.

The peninsula of Aljaska, and the chain of the Aleütes, possess no less than thirty-six volcanoes, scattered over a line about 700 miles long; and thus we find the eastern, western, and northern boundaries of the Pacific encircled with a girdle of volcanic vents, while the subterranean fires have left the western shores of the Atlantic comparatively undisturbed.

With the exception of Iceland, which is famous for the widely devastating eruptions of its burning mountains, the volcanic energies of Europe are at present limited to the submarine crater of Santorin, and to the small area of Etna, Vesuvius, and the Lipari Islands. But, situated in the centre of the ancient seats of civilisation, and for so many centuries the object of the naturalist's researches, of the traveller's curiosity, and of the poet's song, they surpass in renown all other volcanic regions in the world. Most other volcanoes vent their fury over lands either so wild or so remote that the history of their eruptions almost sounds like a legend from another planet; but thousands of us have visited Etna and Vesuvius, and the explosion of their rage menaces towns and countries which classical remembrances have almost invested with the interest of home.

Some volcanoes are in a continual state of eruption. Isalco, born, as we have seen, in 1770, has remained ever since so active as to deserve the name of the Faro (lighthouse) of San Salvador. Its explosions occur regularly, at intervals of from ten to twenty minutes, and throw up a dense smoke and clouds of ashes and stones. These, as they fall, add to the height and bulk of the cone, which is now about 2,500 feet high. For more than two thousand years the fires of Stromboli have never been extinct, nor has it ever failed to be a beacon to the mariner while sailing after nightfall through the Tyrrhenian Sea. Mr. Poulett Scrope, who visited Stromboli in 1820, and looked down from the edge of the crater into the mouth of the volcano, some 300 feet beneath him, found the phenomena precisely such as Spallanzani described them in 1788. 'Two rude openings show themselves among the black chaotic rocks of scoriform lava which form the floor of the crater. One, is to appearance, empty, but from it there proceeds, at intervals of a few minutes, a rush of vapour, with a roaring sound, like that of a smelting furnace when the door is opened, but infinitely louder. It lasts about a minute. Within the other aperture, which is perhaps twenty feet in diameter, and but a few yards distant, may be distinctly perceived a body of molten matter, having a vivid glow even by day, approaching to that of white heat, which rises and falls at intervals of from ten to fifteen minutes. Each time that it reaches in its rise the lip of the orifice, it opens at the centre, like a great bubble bursting, and discharges upwards an explosive volume of dense vapour with a shower of fragments of incandescent lava and ragged scoriæ, which rise to a height of several hundred feet above the lip of the crater.'

The volcanoes of Masaya, near the lake of the same name

in Nicaragua ; of Sion, in the Moluccas ; and of Tofua, in the Friendly Islands, are also, like Stromboli, in a state of permanent eruption. But far more commonly the volcanoes burst forth only from time to time in violent paroxysms, separated from each other by longer phases of moderate activity, during which their phenomena are confined to the exhalation of vapours and gases, sometimes also to the ejection of scoriæ or ashes ; to the oscillations of lava rising or subsiding in the shaft of the crater, to the gentle outflow of small streams of lava from its eruptive cone, and to slight commotions of its border. A continual or periodical exhalation of steam and gases from the shaft of the crater or from chasms and fissures in its bottom, is the commonest phenomenon shown by an active volcano while in a state of tranquillity. Aqueous vapours compose the chief part of these exhalations, and along with other volatile substances, such as sulphuretted hydrogen, sulphurous acid, muriatic acid, and carbonic acid, form the steam-jets or *fumaroles*, which escape with a hissing or roaring noise from all the crevices and chasms of the crater, and, uniting as they ascend in a single vapour-cloud, ultimately compose the lofty column of steam which forms so conspicuous a feature in the picturesque beauty of Etna or Vesuvius. High on the summit of Mauna Loa, where all vegetation has long since ceased, the warm steam of the fumaroles gives rise to a splendid growth of ferns in crevices sheltered from the wind ; and on the island of Pantellaria, the shepherds, by laying brushwood before the fumaroles, condense the steam, and thus procure a supply of water for their goats.

The gentle fluctuations of lava in a crater while in a state of moderate activity are nowhere exhibited on a grander scale than in the pit of Kilauea on Mauna Loa. The mountain rises so gradually as almost to resemble a plain, and the crater appears like a vast gulf excavated in its flanks. The traveller perceives his approach to it by a few small clouds of steam, rising from fissures not far from his path. While gazing for a second indication, he stands unexpectedly upon the brink of the pit. A vast amphitheatre seven miles and a half in circuit has opened to view. Beneath a gray rocky precipice of 650 feet, a narrow plain of hardened lava extends, like a vast gallery, around the whole interior. Within this gallery, below another similar precipice of 340 feet, lies the bottom, a wide plain of bare rock more than two miles in length. Here all is black monotonous desolation, excepting certain spots of a blood-red colour, which appear to be in constant yet gentle agitation.

When Professor Dana visited Kilauea (December 1840), he was surprised at the stillness of the scene. The incessant motion in the blood-red pools was like that of a cauldron in constant ebullition. The lava in each boiled with such activity as to cause a rapid play of jets over its surface. One pool, the largest of the three then in action, was afterwards ascertained by survey to measure 1,500 feet in one diameter and 1,000 in another; and this whole area was boiling, as seemed from above, with nearly the mobility of water. Still all went on quietly. Not a whisper was heard from the fires below. White vapours rose in fleecy wreaths from the pools and numerous fissures, and above the large lake they collected into a broad canopy of clouds, not unlike the snowy heaps or cumuli that lie near the horizon on a clear day, though their fanciful shapes changed more rapidly.

On descending afterwards to the black ledge or gallery at the verge of the lower pit, a half-smothered gurgling sound was all that could be heard from the pools of lava. Occasionally, there was a report like that of musketry, which died away, and left the same murmuring sound, the stifled mutterings of a boiling fluid.

Such was the scene by day—awful, melancholy, dismal—but at night it assumed a character of indescribable sublimity. The large cauldron, in place of its bloody glare, now glowed with intense brilliancy, and the surface sparkled with shifting points of dazzling light, occasioned by the jets in constant play. The broad canopy of clouds above the pit, which seemed to rest on a column of wreaths and curling heaps of lighted vapour, and the amphitheatre of rocks around the lower depths, were brightly illuminated from the boiling lavas, while a lurid red tinged the distant parts of the inclosing walls and threw their cavernous recesses into deeper shades of darkness. Over this scene of restless fires and glowing vapours, the heavens by contrast seemed unnaturally black, with only here and there a star, like a dim point of light.

A paroxysmal eruption is generally announced by the intensification of the phenomena above described. Slight earthquakes are felt in the neighbourhood of the volcano, and follow each other in more rapid succession and with greater violence as the catastrophe draws near. A deep noise like the rolling of thunder, or like the roar of distant artillery, is heard under the ground; the white steam from the crater ascends in denser clouds, which soon acquire a darker tinge; and now the bottom of the crater suddenly bursts with a terrific crash, and with the rapidity of lightning an immense column of black smoke shoots up into the air, and expanding at its upper end into a broad horizontal canopy, assumes a shape which has been compared with that of the Italian pine, the graceful tree of the South. As the column of smoke spreads over the sky, it obscures the light of the sun and changes day into night. Along with the smoke, showers of glowing lava are cast high up into the air, and, rising like rockets, either fall back into the crater or rattle down the declivity of the cone.

At night the scene assumes a character of matchless grandeur, when the column of smoke—or, more properly speaking, of scoriæ, vapour, and impalpable dust—is illuminated by the vivid light of the lava glowing in the crater beneath. It then appears as an immense pillar of fire, rising with steady majesty in the midst of the uproar of all the elements, and ever and anon traversed by flashes of still greater brilliancy from the masses of liquid lava hurled forth by the volcano.

The detonations which accompany an eruption are sometimes heard as single crashes, at others as a rolling thunder or as a continuous roaring. They are frequently audible at an astonishing distance, over areas of many thousand square miles, and with such violence that they may be supposed to proceed from the immediate neighbourhood. Thus, during the eruption of Cosiguina in Nicaragua, which took place in the year 1834, the detonations were heard as loud as a thunderstorm in the neighbourhood of Kingston in Jamaica,

and even at Santa Fè de Bogota, which is a thousand miles
distant from the volcano. With the increase of steam generated during an eruption, the quantity of ejected scoriæ likewise increases in an astonishing manner, so that the volcano's
mouth resembles a constantly discharging mine of the most
gigantic dimensions.

The stones and ashes projected during a volcanic eruption
vary considerably in size, from blocks twelve or fifteen feet
in diameter to the finest dust. Both their immense quantity,
and the force with which they are hurled into the air, show
the utter insignificance of the strength displayed by the
most formidable engines invented by man when compared
with elementary power. Huge blocks are shot forth, as from
the cannon's mouth, to a perpendicular elevation of 6,000 feet,
and La Condamine relates that in 1533 Cotopaxi hurled stones
of eight feet in diameter in an oblique direction to the
distance of seven miles. The lighter scoriæ, carried far
away by the winds, not seldom bury whole provinces under
a deluge of sand and ashes; and their disastrous effects,
spreading over an immense area, are frequently greater than
those of the lava-streams, whose destructive power is necessarily confined to a narrower space. To cite but a few examples, the rain of sand and ashes which in 1812 menaced the
Island of St. Vincent with the fate of Pompeii soon buried
every trace of vegetation, and the affrighted planters and
negroes fled to the town. But here also the black sand,
along with many larger stones, fell rattling like hail upon
the roofs of the houses, while at the same time a tremendous
subterranean thunder increased the horrors of the scene.
Even Barbadoes, though eighty miles from St. Vincent's, was
covered with ashes. A black cloud, approaching from the
sea, brought with it such pitchy darkness that in the
rooms it was impossible to distinguish the windows, and a
white pocket-handkerchief could not be seen at a distance of
five inches.

The fall of ashes caused in April 1815 by the eruption of
the Temboro, in Sumbawa, not only devastated the greater
part of the island, but extended in a westerly direction to
Java, and to the north, as far as Celebes, with such an
intensity that it became perfectly dark at noon. The roofs

of houses at the distance of forty miles were broken in by the weight of the ashes that fell upon them. To the west of Sumatra the surface of the sea was covered two feet deep with a layer of floating pumice or scoriæ, through which ships with difficulty forced their way.

By the terrific eruption of Cosiguina in- the Gulf of Fonseca, in Central America, in 1835, all the ground within a radius of twenty-five miles was loaded with scoriæ to the depth of ten feet and upwards, while the lightest and finest ashes were carried by the winds to places more than 700 miles distant. Eight leagues to the southward of the crater they covered the ground to the depth of three yards and a half, destroying every sign of life. Thousands of cattle perished, their bodies being in many instances one mass of scorched flesh. Deer and other wild animals sought the towns for protection; birds and beasts were found suffocated in the ashes, and the neighbouring streams were polluted with dead fish.

When we consider the amazing quantity of stones and ashes ejected in these and similar instances by volcanic power, we cannot wonder that considerable mountains have frequently been piled up by one single eruption. Thus in the Bay of Baiæ near Naples, Monte Nuovo, a hill 440 feet high, and with a base of more than a mile and a half in circumference, was formed, in less than twelve hours, on September 29, 1538; and near Bronte, on the slopes of Etna, a few days gave birth to Monte Minardo, which rises to the still more considerable height of 700 feet. It would be curious to calculate how many thousands of workmen, and what length of time, man would need to raise mounds like these, produced by an almost instantaneous effort of nature.

In other cases the expansive power of the elastic vapours, which cast up these prodigious masses from the bowels of the earth, is such as to blow to pieces the volcanic cone through which it seeks its vent.

In Quito there is an ancient tradition that Capac Urcu, which means 'the chief,' was once the highest volcano near the equator, being higher than Chimborazo, but at the beginning of the fifteenth century a prodigious eruption took place which broke it down. The fragments of trachyte, says

Mr. Boussingault, which once formed the conical summit of this celebrated mountain, are at this day spread over the plain. On August 11, 1775, the Pepandajan, in Java, formerly one of the highest mountains of the island, broke out in eruption; the inhabitants of the country around prepared for flight, but, before they could escape, the greater part of its summit was shivered to pieces and covered the neighbourhood with its ruins, so that in the upper part of the Gurat valley forty villages were completely buried. During the dreadful eruption of 1815, the Temboro, in Sumbawa, is said to have lost at least one-third of its height from the explosion of its summit, and similar instances are mentioned as having occurred among the volcanoes of Japan.

In the year 1638 a colossal cone called the Peak, in the Isle of Timor, one of the Moluccas, was entirely destroyed by a paroxysmal explosion. The whole mountain, which was before this continually active, and so high that its light was visible, it is said, three hundred miles off, was blown up and replaced by a concavity now containing a lake.

Again, according to M. Moreau de Jonnes, in 1718, on March 6-7, at St. Vincent's, one of the Leeward Isles, the shock of a terrific earthquake was felt, and clouds of ashes were driven into the air, with violent detonations, from a mountain situated at the eastern end of the island. When the eruption had ceased, it was found that the whole mountain had disappeared like the baseless fabric of a dream.

The disastrous effects of the showers of sand, pumice, and lapilli ejected by a volcanic eruption are increased by the transporting power of water. The aqueous vapours which are evolved so copiously from volcanic craters during eruptions, and often for a long time subsequently to the discharge of scoriæ and lava, are condensed as they ascend in the cold atmosphere surrounding the high volcanic peak; and the clouds thus formed, being in a state of high electrical tension, give rise to terrific thunderstorms. The lightning flashes in all directions from the black canopy overhanging the mountain, the perpetually rolling thunder adds its loud voice to the dreadful roar of the labouring volcano, while torrents of rain, sweeping along the light dust and scoriæ which they carry down with them from the air, or meet with on their way, produce currents of mud, often more dreaded than streams of lava, from the far greater velocity with which they move.

It not seldom happens that the eruptions of volcanoes rising above the limits of perpetual snow are preceded or accompanied by the rapid dissolution of the ice which clothes their summits or their sides, owing to the high temperature imparted to the whole mass of the mountain by the vast conflict raging within. Thus in January 1803 one single night sufficed to dissolve or sweep away the enormous bed of snow which in times of rest covers the steep cone of Cotopaxi (18,858 feet high), so that on the following morning the dark mountain, divested of its brilliant robe, gave warning to the affrighted neighbourhood of the terrific scenes that were about to follow. The volcanoes of Iceland, which mostly rise in the midst of vast fields of perpetual ice, frequently exhibit this phenomenon. On October 17, 1758, the eruptive labouring of Kötlingia gave birth to three enormous torrents, which carried along with them such masses of glacier fragments, sand, and stones as to cover a space fifty miles long and twenty-five miles broad. Blocks of ice as large as houses, and partly bearing immense pieces of stone on their backs, were hurried along by the floods; and soon after the eruption took place with a terrific noise.

A very singular phenomenon sometimes occurs in the gigantic volcanoes of the Andes. By the infiltration of water into the crevices of the trachytic rock of which they are composed, the caverns situated at their declivities or at their foot are gradually changed into subterranean lakes or ponds, which frequently communicate by narrow apertures with the Alpine brooks of the highlands of Quito. The fish from these brooks live and multiply in the subterranean reservoirs thus formed, and when the earthquakes which precede every eruption of the Andes chain shake the whole mass of the volcano, the caverns suddenly open and discharge enormous quantities of water, mud, and small fish.

When in the night between the the 19th and 20th of June 1698, the summit of Carguairazo (18,000 feet high) was blown up, so that of the whole crater-rim but two enormous peaks remained, the inundated fields were covered, over a surface of nearly fifty square miles, with fluid tuff and clay-mud enveloping thousands of dead fish. Seven years before, the malignant fever which prevailed in the mountain-town of Ibarra to the north of Quito was attributed to the effluvia arising from the putrid fish ejected by the volcano of Imbaburu.

Amidst all these terrible phenomena—the dreadful noise, the quaking of the earth, the ejection of stones and ashes—which, often continuing for weeks or months, shake the deepest foundations of the volcano, fiery streams of liquid lava gush forth sooner or later as from a vase that is boiling over. Their appearance generally indicates the crisis of the subterranean revolution, for the rage of the elements, which until then had been constantly increasing, diminishes as soon as the torrent has found an outlet. The lava rarely issues from the summit crater of the mountain ; much more frequently it flows from a lateral rent in the volcano's side, which, weakened and dislocated in its texture by repeated shocks, at length gives way to the immense pressure of the lava column boiling within. From the vast size of these eruptive rents, we may form some idea of the gigantic power of the forces which give them birth.

Thus during the great eruption of Etna in 1669, the south-east flank of the mountain was split open by an enormous rent twelve miles long, at the bottom of which incandescent lava was seen. The extreme length of the fissure which gave lateral issue to the lava of Kilauea in 1840 was twenty-five miles, as could distinctly be traced through the disturbance of the surface rocks above; and in the terrific eruption of Skaptar Jökull, which devastated the west coast of Iceland in 1783, lava gushed forth from several vents along a fissure of not less than 100 miles in length. In some cases the whole mass of the volcano has been cleft in two. Vesuvius was thus rent in October 1822 by an enormous fissure broken across its cone in a direction N. W.—S. E.

Here and there along the line of such a rent, cones of eruption are thrown up in succession at points where the gaseous matter obtains the freest access to the surface, and has power to force up lava and scoriæ. Few indeed, if any, of the greater volcanic mountains are unattended by such minor elevations, clustering about its sides like the satellites of a planet. Professor Dana found Mauna Loa covered with numerous parasitic cones, and Mr. Darwin counted several thousands on one of the Gallapagos Islands. . On the flanks of Etna, according to Professor Sartorius von Waltershausen, more than 700 of them are to be seen; almost all possessing craters, and each marking the source of a current of lava. Though they appear but trifling irregularities when viewed from a distance as subordinate parts of so imposing and colossal a mountain, many of them would nevertheless be deemed hills of considerable height in almost any other region. The double hill near Nicolosi, called Monte Rossi, formed in 1669, is 450 feet high and two miles in circumference at its base; and Monte Minardo, near Bronte, on the east of the great volcano, is upwards of 700 feet in height.*

'On looking down from the lower borders of the desert region of Etna,' says Sir Charles Lyell, 'these minor volcanoes, which are most abundant in the woody region, present us with one of the most delightful and characteristic scenes in Europe. They afford every variety of height and size, and are arranged in beautiful and picturesque groups. However uniform they may appear when seen from the sea, or the plains below, nothing can be more diversified than their shape when we look from above into their craters, one side of which, as we have seen, is generally broken down. There are indeed, few objects in nature more picturesque than a wooded volcanic crater. The cones situated in the higher parts of the forest zone are chiefly clothed with lofty pines, while those at a lower elevation are adorned with chestnuts, oaks, and beech-trees.'

As the point where a lava-current finds a vent is often situated at a considerable distance below the surface of the liquid column in the internal chimney of the volcano, the pressure from above not seldom causes the lava to spout forth in a jet, until its level in the crater shaft has been reduced to that of the newly-formed orifice. Thus, when

* See p. 67

Vesuvius was rent by the dreadful poroxysmal eruption of 1794, the lava was seen to shoot up in magnificent fountains as it issued from the openings along the fissure.

Further on, the lava flows down the mountain's side according to the same laws which regulate the movements of any other stream, whether of water, mud, or ice : more rapidly down an abrupt declivity, slower where the slope is more gradual; now accumulating in narrow ravines, then spreading out in plains; sometimes rushing in fiery cascades down precipices, and where insurmountable obstacles oppose its progress, not seldom breaking off into several branches, each of which pursues its independent course.

At the point where it issues, the lava flows in perfect solution, but, as its surface rapidly cools when exposed to the air, it soon gets covered with scoriæ, which are dashed over each other in wild confusion, by successive floods of liquid stone, so as to resemble a stormy sea covered with ice-blocks. But the liquefied stone not only hardens on its external surface; it also becomes solid below, where it touches the colder soil, so that the fluid lava literally moves along in a crust of scoriæ, which lengthens in the same proportion as the stream advances.

The movements of the lava-current are of course considerably retarded by the formation of scoriæ, so that, unless where a greater inclination of the soil gives it a new impulse, it flows slower and slower. Thus the lava-stream which was ejected by Etna during the great eruption of 1669, performed the first thirteen Italian miles of its course in twenty days, or at the average rate of 162 feet per hour, but required no less than twenty-three days for the last two miles. While moving on, its surface was in general a mass of solid rock; and its mode of advancing, as is usual with lava streams, was by the occasional fissuring of the solid walls. Yet in spite of the tardiness of its progress, the inhabitants of Catania watched its advance with dismay, and rushed into the churches to invoke the aid of the Madonna and the Saints. One citizen only, a certain Baron Papalardo, relied more upon his own efforts than upon supernatural assistance, and set out with a party of fifty men, dressed in skins to protect them from the heat, and armed with iron crows and hooks for the purpose of breaking open one of the solid walls of scoriæ that flanked the liquid current, so as to divert it from the menaced city. A passage was thus opened for a rivulet of melted matter, which flowed in the direction of Paterno; but the inhabitants of that town being alarmed for their safety, took up arms against Papalardo, whose fifty workmen would hardly have been able to cope with the powers of nature. Thus, slowly but irresistibly, the lava advanced up to the walls of Catania, which being formed of huge Cyclopean blocks, and no less than sixty feet high, at first stemmed the fiery stream. But the glowing floods, pressing against the rampart, rose higher and higher, and finally reaching its summit, rushed over it in fiery cataracts, and destroying part of the town, at length disgorged themselves into the sea, where they formed a not inconsiderable promontory.

A truly gigantic conflict might naturally be expected from the meeting of two such powerful and hostile bodies as fire and water. This, however, is by no means the case, for as soon as the lava enters the sea, the rapid evaporation of the water that comes into immediate contact with it accelerates the cooling of the surface and thickens the hard external crust to such a degree that very soon all communication is cut off between the water and the fiery mass. While the lava continues to advance from the land, the crust of scoriæ is prolonged in the same proportion, and should it be rent here and there, steam is at once developed with such violence as to prevent all further access of the water into the interior of the fissures. Thus Breislak informs us that, in 1794, the eruption of a lava-stream into the Bay of Naples, near Torre del Greco, took place with the greatest tranquillity, so that he himself was able to observe the advancing of the lava into the sea while seated in a boat immediately near it, without being disturbed by explosions or any other violent phenomenon.

As the crust of scoriæ is so bad a conductor of heat, it occasions a very slow cooling and hardening in the interior of the lava-stream, forming as it were a vessel in which the liquid fire can be retained and preserved for a long time. When Elie de Beaumont visited the lava-stream of Etna, nearly two years after its eruption in 1832, its interior was still so warm that he could not hold his finger in the hot steam issuing from its crevices. It has also been proved, on trustworthy evidence, that after twenty-five and thirty years, many lava-streams of Etna still continued to emit heat and steam; and after twenty-one years it was possible to light a cigar in the crevices of the lava that issued from Jorullo in 1759.

Another extremely curious effect of the scoriæ being such bad conductors of heat is, that masses of snow will remain unmelted though a lava-stream rolls over them. Thus, in 1787, the lava of Etna flowed over a large deposit of snow, which, however, was by no means fully liquefied, but remained for the greatest part entire, and gradually changed into a granular and solid mass of ice. This was traced in 1828, by the geologist Gemellaro, for a distance of several hundred feet under the lava, and most likely still reposes under it as in an ice-cellar. The cliffs which form the vast crater-ring of the Isle of Deception, in the extreme Southern Atlantic, are likewise composed of alternate layers of ice and lava. Probably in both these cases the ice-beds have been covered, before the lava flowed over them, by a rain of scoriæ and volcanic sand, which is so well known among the shepherds in the higher regions of Etna as a bad conductor of caloric, that, to obtain a supply of water for their herds during the summer, they cover some snow a few inches deep with volcanic sand, which entirely prevents the penetration of solar heat.

Most of the recent lava-streams evolve from all their fissures and rents a quantity of vapour, so as to be dotted with innumerable fumaroles, and to exhibit. as they flow along, a smoking surface by day and a luminous one by night. At first these fumaroles are so impetuous that they frequently puff up the lava-crust around their orifices into little cones or hillocks, consisting of blocks of scoriæ irregularly piled up over each other, and from whose summit the vapours continue to ascend. As the mass cools, they are naturally lessened in numbers and in power; but in 1803 Humboldt still saw fumaroles from twenty to thirty feet high, rising from the small cones which covered by thousands the great lava-stream of Jorullo of the year 1759.

The vast dimensions of single lava-streams give proof of

the enormous powers which forced them out of the bowels of the earth. The lava-stream of Vesuvius which destroyed Torre del Greco in 1794, is 17,500 French feet long, and when it reached the town was more than 2,000 feet wide and forty feet deep. While this mighty mass of molten stone, the volume of which has been reckoned at about 457 millions of cubic feet, was descending towards the sea, another stream, whose mass is computed at about one-half of that of the former, was flowing in the direction of Mauro. This single eruption has therefore furnished more than 685 millions of cubic feet of lava, equal to a cube of 882 feet, in which at least a dozen of the largest churches, palaces, and pyramids on earth might conveniently find room. If to the solid lava we add the astonishing quantities of scoriæ, sand, and ashes thrown out by this same eruption, we may form some idea of the masses of matter which were in this one instance ejected from the interior of the earth.

The volume of the lava-stream which flowed from the volcanoe of the Isle of Bourbon in the year 1787 is estimated at 2,526 millions of cubic feet ; but even this astonishing ejection of molten stone is surpassed by that which took place during the eruption of Skaptar Jökull* in 1783, when the lava rolled on to a length of fifty miles, and, on reaching the plain, expanded into broad lakes, twelve and fifteen miles in diameter and a hundred feet deep.

In the great eruption of Mauna Loa, which commenced on the 30th of May, 1840, the lava began to flow from a small pit-crater called Avare, about six miles from Kilauea. The light was seen at a distance, but, as there was no population in that direction, it was supposed to proceed from a jungle on fire. The next day another outbreak was perceived farther towards the coast, and general alarm prevailed among the natives, now aware of the impending catastrophe. Other openings followed, and by Monday the 1st of June the large flow had begun, which formed a continuous stream to the sea, which-it reached on the 3rd. This flood issued from several fissures along its whole course, instead of being an overflow of lava from a single opening ; it started from an

* A detailed account of this eruption, one of the most dreadful on record, is given in 'The Polar World,' chap. vi. p. 81.

elevation of 1,244 feet, as determined by Captain Wilkes, at a point twenty-two miles distant from the first outbreak, and twelve from the shore. The scene of the flowing lava, as we are told by those who saw it, was indescribably magnificent. As it rolled along it swept away forests in its course, at times parting and inclosing islets of earth and shrubbery, and at other times undermining and bearing along masses of rock and vegetation on its surface. Finally, it plunged into the sea with loud detonations, and for three weeks continued to disgorge itself with little abatement.

The light which it emitted converted night into day over all eastern Hawaii. It was distinctly visible for more than one hundred miles at sea, and at the distance of forty miles fine print could be read at midnight. As previous to the eruption, the whole vast pit of Kilauea had been filled to the brim with the lava, which, bursting through the flanks of the mountain, thus found a vent towards the sea, we have some means of estimating the volume of the ejected masses in the actual cubic contents of the emptied pit. The area of the lower pit, as determined by the surveys of the American Exploring Expedition, is equal to 38,500,000 square feet. Multiplying this by 400 feet, the depth of the pit after the eruption, we have 15,400,000,000 cubic feet for the solid contents of the space occupied by lava before the eruption, and therefore the actual amount of the material which flowed from Kilauea. This is equivalent to a triangular range 800 feet high, two miles long, and over a mile wide at base!

Though generally symptoms of violent disturbance, such as shakings of the earth and loud thundering noises, precede the eruption of lava, yet this is not always the case. Thus the craters of Mount Kea have frequently disgorged their masses of molten stone without such accompanying phenomena. In 1843, when the volcano poured out a flood of lava, reaching for twenty-five miles down its side, all took place so quietly that persons at the foot of the mountain were unaware of it, except from the glare of light after the action had begun. Through its progress no sounds were heard below, nor did it cause any perceptible vibrations, except in the region of the outbreak, and there none of much violence.

The lava sometimes cools down with a smooth, solid, undulating surface, marked with rope-like lines and concentric folds, such as are seen on any densely viscid liquid if drawn out as it hardens; but much more frequently it appears as if shattered to a chaos of ruins. The fragments vary from one to hundreds of cubic feet, or from a half-bushel measure to a house of moderate size. They are of all shapes, often in angular blocks, and sometimes in slabs, and are horribly rough, having deep recesses everywhere among them. The traveller shudders as his path leads him over a lava-field, thus bristling with myriads of spikes, where the least false step would precipitate him into the deep cavities, among the jagged surfaces and edges. This scene of horrid confusion often extends for miles in every direction, and, viewed from its central part, the whole horizon around is one wide waste of gray and black desolation, beyond the power of words to describe.

The breaking up of a lava-field into chaotic masses evidently proceeds from a temporary cessation, either complete or partial, and a subsequent flow of a stream of lava. The surface cools and hardens as soon as the stream slackens; afterwards there is another heaving of the lava, and an onward move, owing to a succeeding ejection or the removing of an obstacle, and the motion breaks up the hardened crust, piling the masses together, either in slabs or huge angular fragments, according to the thickness to which the crust had cooled. If the motion of a lava-stream be quite slow, the cooling of the front of it may cause its cessation, thus damming it up and holding it back, till the pressure from gradual accumulation behind sweeps away the barrier. It then flows on again, carrying on its surface masses of the hardened crust—some, it may be, to sink and melt again, but the larger portion to remain as a field of clinkers. The breaking-up of the ice of some streams in spring gives some idea of the manner in which the hardened masses of a lava-field are piled up as it moves along; but to form a just idea of the greatness of the effect, the mind must bring before it a stream, not of the scanty limits of most rivers, but one, not unfrequently, of several miles in breadth : besides, in place of slabs of pure and clear ice, there should be substituted shaggy heaps of-black scoriæ, and a depth or thickness of many yards in place of a few inches.

Where volcanic mud-streams have flooded the land, or a rain of ashes and light scoriæ has descended upon the soil, its fertility may soon be restored under the influence of a sunny sky; but as far as the lava reaches, a stony wilderness often remains for ages, particularly in the colder regions of the earth. Thus, though many of the lava-fields of Iceland have existed long before the first Scandinavian colonists settled in the land, their surface is generally as naked as when they first issued from the volcano; and where signs of vegetation may be seen among their fragments, the eye finds nothing to relieve the horrid monotony of the scene but spare patches of lichen and mosses, or here and there some dwarf herb or shrub that hardly ventures to peep forth from the crevice in which it has found a shelter. But in a milder climate, such as that of Italy, and still more rapidly in the torrid zone, the horrid nakedness of a lava-field undergoes a more rapid transformation, provided a sufficient moisture favours the growth of plants. The rains promote the decomposition of the lava, and a rank vegetation succeeds, which in its turn assists the work of decomposition, and thus hastens the accumulation of soil. Ferns and grasses spring up in the nooks and crevices, and finally the vine or the taro flourish luxuriantly, for nothing can exceed the fertility of a disintegrated lava-field.

Volcanoes have frequently been considered as safety-valves, which, by affording a vent to subterranean vapours, preserve the neighbouring regions from the far more disastrous and wide-spreading effects of earthquakes; and facts are not wanting which seem to justify this opinion. After the soil had trembled for a long time throughout the whole of Syria, in the Cyclades, and in Eubœa, the shocks suddenly ceased when, in the plains near Chalcis, a stream of '*glowing mud*' (lava from a crevice) issued from the bowels of the earth. Strabo, who relates this incident, adds that 'since the craters of Etna have been opened, through which fire ascends, the land on the sea-coast is less subject to earthquakes than at the time when all vents on the surface were stopped up.'

Before the earthquake which destroyed the town of

Riobamba, the smoke of the volcano of Pasto, which is 200 miles distant, disappeared. The Neapolitans and Sicilians consider the eruptions of Vesuvius and Etna, or even a more lively activity of these volcanoes, as a certain preservative against devastating earthquakes, and we meet with the same belief among the inhabitants of Quito and Peru. But in many cases this fancied security has proved to be delusive, as very violent earthquakes have not seldom been found to accompany volcanic eruptions. The great Chilian earthquake of 1835 coincided with an eruption of Antuco ; and the shocks which agitated all Kamtschatka and the long chain of the Kurilian Islands, in 1737, occurred simultaneously with an eruption of Kliutschewskaja Skopa.

Professor Dana doubts whether action so deep-seated as that of the earthquake must be, can often find relief in the narrow channels of a volcano miles in length. He points out the example of Mauna Loa, where lavas are frequently poured out from the summit crater, at an elevation of more than 10,000 feet above Kilauea, so that the latter, notwithstanding its extent, the size of its great lakes of lava, and the freedom of the incessant ebullition, is not a safety-valve that can protect even its own immediate neighbourhood.

In his opinion volcanoes might more fitly be called indexes of danger. They point out those portions of the globe which are most subject to earthquakes, and are results of the same causes that render a country liable to such convulsions.

The phenomena attending an eruption can leave no doubt that below every active volcano a large subterranean cavity must exist in which melted lava accumulates. The partisans of the theory which supposes the earth to consist of a central fluid mass with a solid shell resting upon it, attribute the formation of volcanoes to rents or fissures in this crust through which the lava is cast forth; but the local development of heat by chemical action, or some other unknown cause, is quite sufficient to account for the existence of fiery lakes imbedded in a solid mass, and which, though insignificant when compared with the surface of the globe, may still be large enough to produce volcanic phenomena on the grandest scale.

The cause of the reaction of such a reservoir against the surface of the earth must in all probability be sought for in

the expansive force of steam; for when water, penetrating
through crevices or porous strata, comes in contact with
the heated subterranean mass, it is evident that the steam
thus generated must press upon the lava, and, when formed
in sufficient quantity, ultimately force it up the duct of the
volcano. In other cases, we may suppose a continuous
column of lava mixed with liquid water raised to a red-hot,
or white-hot temperature under the influence of pressure.
A disturbance of equilibrium may first bring on an eruption
near the surface, by the expansion and conversion into gas
of the entangled water, so as to lessen pressure. More
and more steam would then be liberated, bringing up with it
jets of liquid rock, and ultimately ejecting a continuous
stream of lava. Its force being spent, a period of rest succeeds, until the conditions for a new outburst (accumulation
of steam and melted rock) are obtained, and another cycle of
similar changes is renewed. The important part which water
plays in volcanic action is moreover sufficiently proved by the
enormous quantity of steam which is poured forth during
every eruption, or is constantly escaping in the fumaroles of
a crater. The various gases (carbonic, muriatic, sulphurous)
which are likewise exhaled by volcanoes may also have been
rendered liquid by pressure at great depths, and may assist
the action of water in causing eruptive outbursts. The great
number of active volcanoes on sea-coasts and in islands likewise points to the agency of water in volcanic operations ;
and in the few cases where eruptive cones are situated far
inland, their situation on the borders of a lake, or their
cavernous and porous structure, accounts for the absorption
of a quantity of atmospheric water, sufficient for the production of volcanic phenomena.

CHAPTER VII.

DESTRUCTION OF HERCULANEUM AND POMPEII.

State of Vesuvius before the eruption in the year A.D. 79--Spartacus—Premonitory Earthquakes—Letter of Pliny the Younger to Tacitus, relating the death of his Uncle, Pliny the Elder—Benevolence of the Emperor Titus—Herculaneum and Pompeii buried under a muddy alluvium—Herculaneum first discovered in 1713.

OF all the volcanic eruptions recorded in history there is none more celebrated than that which, on the 23rd of August, A.D. 79, buried the towns of Herculaneum and Pompeii under a deluge of mud and ashes. Many other eruptions have no doubt been on a grander scale, or may have spread ruin and desolation over a wider area, but never has a volcano, awakening from the slumber of a thousand years, devastated a more smiling paradise than the fields of happy Campania, or buried more beautiful cities.

Before that terrible catastrophe, Mount Vesuvius, now constantly smoking, even in times of rest, had, ever since the first colonisation of South Italy by the Greeks, exhibited no signs of volcanic activity. Even tradition knew of no previous disturbance. No subterranean thunder, or sulphurous streams, or cast-up ashes, gave token of the fires slumbering beneath its basis; and the real nature of the apparently so peaceful mountain could only be conjectured from the similarity of its structure to other volcanoes, or from the ancient lava-streams that furrowed its abrupt declivities. At that time also its shape was very different from its present form, for instead of two apices, it exhibited, from a distance, the regular outlines of a sharply truncated cone. Plutarch relates that rough rock walls, piled round its summit, and overgrown with wild vines, inclosed the waste of the crater.

When, in 73 B.C., Spartacus, with seventy of his comrades, broke the fetters of an insupportable slavery, he found a secure retreat in this natural stronghold, which could only be scaled by a single narrow and difficult path. By degrees 10,000 fugitive slaves gathered round his standard, and Rome began to tremble for her safety. The prætor Clodius led an army against the rebels, and surrounded the mountain; but Spartacus caused ropes to be made of the branches of wild vines, by means of which he, with the boldest of his followers, was let down from the rocks, where they were supposed to be totally inaccessible, and, falling unawares upon the prætor, put his troops to flight and took his camp. The declivities of the mountain, thus become historically renowned, were covered with the richest fields and vineyards, and at its foot, along the beautiful Bay of Naples, lay the flourishing towns of Herculaneum and Pompeii, the seats of luxury and refinement. Who could, then, have imagined that this charming scene was so soon to be disturbed in so terrible a manner, and that the time was nigh when the ancient volcanic channels, from which, in unknown ages, lava-streams and ashes had so frequently broken forth, were once more to be re-opened? The first sign which announced the awakening energies of the volcano was an earthquake, which, in A.D. 63, devastated the fertile regions of Campania. From that time, to the crowning disaster of 79, slight tremors of the earth frequently occurred, until, finally, the dreadful eruption took place which Pliny the Younger so vividly describes in his celebrated letter to Tacitus.

'My uncle,' says the Roman, 'was at Misenum, where he commanded the fleet. On the 23rd of August, about one o'clock in the afternoon, my mother informed him that a cloud of an uncommon size and form was seen to arise. He had sunned himself (according to the custom of the ancient Romans), and taken his usual cold bath, then dined, and studied. He asked for his sandals, and ascended an eminence, from which the wonderful phenomenon could be plainly seen. The spot from whence the cloud ascended, in a shape like that of an Italian pine-tree, could not be ascertained on account of the distance; its arising from Vesuvius only subsequently became known.

'In some parts it was white, in others black and spotted, from the ashes and stones which it carried along. To my uncle, being a learned man, the phenomenon seemed important, and worthy of a closer investigation. He ordered a light ship to be got ready, and left it to my option to accompany him. I answered that I preferred studying, and by chance he himself had given me something to write. He was on the point of leaving the house when he received a letter from Resina, the inhabitants of which, alarmed at the impending danger—the place lay at the foot of the mountain, and escape was only possible by sea—begged him to help them in their great distress. He now changed his plan, and executed as a hero the undertaking to which he had been prompted as a natural philosopher.

'He ordered the galleys of war to set sail, and embarked to bring help, not only to Resina, but to many other places along the coast, which, on account of its loveliness, was very densely peopled.

'He hastens to the spot from which others are taking flight, and steers in a direct line towards the seat of danger, so unconcerned as to dictate his observations upon all the events and changes of the catastrophe, as they passed before his eyes.

'Already ashes fell upon the ship, hotter and thicker on approaching, and also pumice and other stones blackened and burnt by fire. Suddenly a shallow bottom, and the masses ejected by the eruption, rendered the coast inaccessible. He hesitated for a moment whether he should sail back again, but, soon resolved, said to the steersman who advised him to do so, " Fortune favours the bold ; steer towards the villa of Pomponianus." This friend resided at Stabiæ, on the opposite side of the bay, where the danger, although as yet at some distance, was still within sight, and menacing enough. Pomponianus had therefore caused his effects to be conveyed on board a ship, intent on flight so soon as the contrary wind should have abated. As soon as my uncle, to whom it was very favourable, has landed, he embraces, consoles, encourages his terrified friend, takes a bath to relieve his fears by his own confidence, and dines after the bath with perfect composure, or, what is no less great, with a serene countenance.

'Meanwhile high columns of flame burst forth from Vesuvius in various places, their brilliancy being increased by the darkness of the night. My uncle, with the intention of relieving apprehension, said that they proceeded from the villas which, abandoned by their terrified proprietors and left a prey to the flames, were now burning in solitude. He then retired and slept soundly, for his attendants before the door heard him fetch his breath, which, on account of his corpulence, was deep and loud. But now the court, into which the room opened, became filled to such a height with ashes and pumice that by a longer delay he would not have been able to leave it. They awaken him, he rises, and greets Pomponianus and the others who had watched. They consult together, whether to remain in the house or to flee into the open air, for the ground trembled from the repeated and violent shocks of the earth, and seemed to reel backwards and forwards. On the other hand, they feared in the open air the falling of the pumice-stones and cinders. On comparing these two dangers, flight was chosen : and, as a protection against the shower of stones, they covered their heads with cushions. Everywhere else the day was already far advanced, but the blackest night still reigned at Stabiæ. Provided with torches, they resolved to seek the shore, in order to ascertain whether they could venture to embark, but the sea was found to be too wild and boisterous.

'My uncle now lay down upon a carpet, and asked for some cold water, of which he repeatedly drank. The flames and their sulphurous odour drove away his companions, and forced him to rise. Leaning on two slaves, he tried to move, but immediately sank down again, suffocated as I believe by the dense smoke, and by the closing of his larynx, which was by nature weak, narrow, and subject to frequent spasms. On the third morning after his death the body was found without any marks of violence, covered with the clothes he had worn, and more like a person sleeping than a corpse.'

Thus perished, in his fifty-sixth year, one of the greatest naturalists and noblest characters of ancient Rome, the philosopher to whom we are indebted for the first general description of the world—a work which, in spite of its numerous imperfections and errors, is one of the most interesting monuments of classical literature.

When the rage of the volcanic powers had subsided, the sun, now no longer obscured by clouds of ashes, shone upon a scene of utter desolation, where nature, embellished by art, had, but a few days before, appeared in all her loveliness. The mountain itself had changed its form, and rose with new peaks to the skies; a thick layer of stones and dust had settled with the curse of sterility on the fields; thousands of homeless wretches wandered about disconsolate, and three towns—Pompeii, Herculaneum, Stabiæ—had disappeared, to be brought to light again in a wonderful manner after the lapse of many centuries.

This great catastrophe gave the Emperor Titus a fine opportunity for displaying the benevolence which entitled him to be called 'the delight of mankind.' He immediately hastened to the scene of destruction, appointed guardians of consular rank to distribute among the needy survivors the property of those who had perished without heirs; and encouraged the weak-hearted, assisting them by liberal donations, until a no less terrible misfortune recalled him to Rome, where a fire, which laid almost half the town in ashes, was followed by a plague, which, for some time, daily swept away thousands.

It has often been asked how so many of the relics buried in Herculaneum and Pompeii could have been so perfectly preserved as to form a Museum of the Past for the admiration and instruction of future ages. A stream of lava would undoubtedly have consumed everything on its fiery track, but, fortunately for posterity, it was not a flood of molten stone, but a current of mud, which overwhelmed the devoted cities. We learn from history that a heavy shower of sand, pumice, and lapilli was ejected from Vesuvius for eight successive days and nights, in the year 79, accompanied by violent rains, and thus all these volcanic matters were converted into mud-streams, which, rushing down the sides of the mountain, descended upon Herculaneum and Pompeii. This circumstance satisfactorily explains how the *interior* of the buildings, with all the underground vaults and cellars, was filled up, and how all the objects they contained could be as perfectly moulded as in a plaster cast by the muddy alluvium, which subsequently hardened into pumice tuff. Hence this wonderful preservation of paintings, which, shielded from the destructive influence of the atmosphere, still retained their original freshness of colour when again brought to light by a late generation; these rolls of papyrus which it has been found possible to decipher; this perfect cast of a woman's form, with a child in her arms!

No lava has flowed over Pompeii since that city was buried, but with Herculaneum the case is different. Although the substance which fills the interior of the buildings in that doomed city must have been introduced in a state of mud like that found in similar situations in Pompeii, yet the superincumbent mass differs wholly in composition and thickness.

Herculaneum was situated several miles nearer to the volcano, and has, therefore, been always more liable to be covered, not only by showers of ashes, but by alluvium and streams of lava. Accordingly, masses of both have accumulated on each other above the ancient site of the city, to a depth of nowhere less than 70, and in many places of 112 feet; while the depth of the bed of ashes under which Pompeii lies buried seldom exceeds 12 or 14 feet above the houses, and it is even said that the higher part of the amphitheatre always projected above the surface.

Yet, strange to say, Herculaneum, though far more profoundly hidden, was discovered before Pompeii, by the accidental circumstance that a well sunk in 1713 came right down upon the theatre where the statues of Hercules and Cleopatra were found. Many others were afterwards dug out and sent to France by the Prince of Elboeuf, who, having married a Neapolitan princess, became proprietor of the field under which the theatre lies buried. Further excavations were, however, forbidden by Government, and only resumed in 1736. But the difficulty of removing the large masses of lava accumulated above the city, and the circumstance of its partly lying under the modern towns of Portici and Resina, have confined the exploration of Herculaneum within narrow limits. The large theatre alone is open for inspection, and can be seen only by torchlight, so that its dark galleries, cut through the tuff, are but seldom visited by strangers; while no traveller leaves Naples without having wandered through the ruins of Pompeii, for Italy hardly affords a more interesting sight than that of these streets and forums, these theatres and temples, these houses and villas, which require but the presence of their ancient inhabitants to complete the picture of a Roman town, such as it was eighteen hundred years ago.

CHAPTER VIII.

GAS SPRINGS AND MUD VOLCANOES.

Carbonic Acid Springs— Grotto del Cane—The Valley of Death in Java—Exaggerated Descriptions—Carburetted Hydrogen Springs—The Holy Fires of Baku—Description of the Temple—Mud Volcanoes—The Macaluba in Sicily—Crimean Mud Volcanoes—Volcanic Origin of Mud Volcanoes.

THE numerous gas springs which in many countries are evolved from an unknown depth, afford us a convincing proof that the remarkable chemical transformations of which we find so many traces in the past history of our planet are still perpetually taking place in many of the mysterious crevices and hollows of the earth-rind. In Auvergne, the Vivarrais, the Eifel, and along the whole basaltic range from the Rhine to the Riesengebirge in Silesia, carbonic acid gas is exhaled in incredible quantities from the vast laboratories of the subterranean world.

Professor Bischoff found that a single gas spring near Burgbrohl daily produced 5,650 cubic feet of carbonic acid, a quantity amounting in the course of a year to no less than 262,000 pounds in weight; and, according to Bromeis, the great Artesian spring at Nauheim evolves every minute 71 cubic feet of carbonic acid, equal to a weight of 5,000,000 pounds annually. If from these two instances we judge of the produce of the many carbonic acid gas springs of Germany, and if we further extend our view to the rest of the world, in many parts of which carbonic acid probably escapes in still greater quantities, we can form some idea of the geological importance of these springs, which also exercise no small influence upon the organic world. For the incalculable masses of carbonic acid which are thus constantly pouring from subterranean vents into the atmospheric ocean are again absorbed by millions of plants. They feed the forests and the fields; and thus these chemical changes, which are incessantly but imperceptibly modifying the earth-rind, ultimately tend to the advantage of man.

As a light dipped in carbonic acid gas is immediately extinguished, and every animal inhaling it is liable to instant suffocation, these properties are sometimes made use of for cruel experiments, for which, among others, the insignificant Grotto del Cane, in the kingdom of Naples—a cave or hole in the side of a mountain near the Lake Agnano—has become notorious. Some miserable dogs are thrust into the stratum of fixed air which covers the bottom of the hole, and are alternately almost choked and resuscitated to satisfy the idle curiosity of tourists. Their violent efforts to escape, when about to be plunged into the poisonous vapour, prove the horrible cruelty of the practice.

The carbonic-acid springs in the glen of the Brohl, a small rivulet flowing into the Rhine, near Andernach, are turned to a better purpose, for the manufacture of white lead.

The famous ' Valley of Death,' or Poison Valley, in the Island of Java, is nothing more than a funnel-shaped hollow, measuring about 100 feet in diameter at the top, and with a bare space in its centre fifteen feet broad and long, which is frequently covered with a stratum of carbonic acid gas. The sides of the hollow, and even the bottom, with the exception of the above-mentioned naked spot, are everywhere clothed with shrubbery, or even with forest trees.

The dead bodies of stags, tigers, wild boars, and birds, are said to have been frequently found in the hollow; but Dr. Junghuhn, the author of a classical work on Java, saw in 1838 but one human corpse lying on its back in the centre of the bare spot. It was still there in 1840, and but slightly decomposed. In 1845 it had been removed, most likely by some compassionate wanderers desirous of giving it a decent burial, for not the slightest trace of the skeleton remained. During the years 1838, 1840, and 1845 Junghuhn visited the Valley of Death no less than thirteen times. When he last saw it, the bodies of six wild hogs were lying at the bottom, all more or less in a state of putrefaction. The crows that were feasting upon their remains proved that a descent might be effected without danger, for, on seeing them hopping about on the naked soil, even the Javanese entered the circle without hesitation. Not a single trace of carbonic acid was to be perceived, not even when the bold naturalist stretched himself out upon the ground and drew his breath in the crevices and rents with which it was furrowed. Probably the gas never rises more than three feet above the level of the soil, as at this height a luxuriant vegetation begins.

This simple description of an accurate observer forms a strange contrast to the gross exaggeration of other travellers, whose accounts, copied in many hand-books, have puffed up a phenomenon hardly superior to that of the Grotto del Cane into something like an eighth wonder of the world. Loudon, who in July 1830 visited the Pakamaran (as the natives call the pit), swells its dimensions to a vast crater about half a mile in circumference, thickly strewn with skeletons of men, tigers, game, and birds of all kinds; and another recent traveller goes so far as to give it an extent of twenty miles.

Next to carbonic acid, but of far less general occurrence, carburetted hydrogen, which gives rise to the wonderful phenomenon of fiery springs, is the gas most frequently evolved by volcanic spiracles.

Near Pietra Mala, between Bologna and Florence, on a spot about twelve feet in diameter, several flames rise from the earth, the largest of which ascends to a height of five feet, and is seen burning at night with a pale yellow flame, while its minor satellites around are blue tipped with white. No doubt many a terrible legend is attached to this infernal spectacle. Near Barigazzo, between Modena and Pistoja, near the ruins of Velleji, and in many other parts of the volcanic region of the Apennines, similar flames gush out of the ground. The neat little town of Fredonia, in the State of New York, on the eastern shore of Lake Erie, is lighted by natural springs of carburetted hydrogen, which, being led into a gasometer, feed the seventy or eighty lamps of the town. The thrifty and practical Chinese, who have preceded us in so many useful discoveries, have for centuries made a like use of the many gaseous emanations in the provinces of Yunnan, Szutschuan, Kuangsi, and Schansi, by leading the inflammable air in pipes, wherever they want it for lighting or cooking.

But there is no place in the world more remarkable for its burning springs than Baku, on the western coast of the Caspian Sea, where the holy and eternal fires are worshipped by the pious Parsees as the special symbol of the Almighty.

Like most of the cloisters and convents of the Orient, which are exposed to the incursions of plundering hordes, Aleschga, the temple dedicated to the worship of fire, is a fortified square inclosing a large courtyard, and capable of being defended from the terraced roof. The outer wall forms at the same time the back of the cells, which front the yard. Over the entrance gate, which is situated to the north, rises a high bastion or tower, serving as an additional defence, from the summit of which the visitor enjoys after sunset the fantastic view of the flames which, untarnished by smoke, rise on all sides from rents and crevices in the neighbouring steppe, and wave their bright summits to and fro like tongues of fire. In the centre of the court stands a square tower supported by four columns, and inclosing a basin-like excavation, three or four feet in diameter, into which the gas is conducted by a pipe from sources beyond the walls of the temple. Four chimneys at the four corners of the tower are fed in a similar manner. From the centre of the tower rises a trident, called *Thirsul*. The Parsees relate that the Devil once got possession of the earth, and reigned with despotic fury. But man in his distress prayed to the Almighty, and an angel came down and planted this identical trident in the earth as a token that the dominion of his Satanic Majesty had ceased. Round the court are twenty-two cells, like those of a Catholic convent. They are very small, and, with the exception of a ragged rug, wholly without furniture; but each of them is provided with a gas-pipe, which can be opened or closed at pleasure, and furnishes light and warmth to the inmate. Near the temple a well has been dug fifty feet deep, in which the gas accumulates in larger quantities. Koch ('Wanderungen im Oriente,' 1843-4) tells us that he here enjoyed a sight more wonderful and surprising than any he had ever witnessed before. A carpet was spread over the mouth of the well to prevent the gas from escaping. After a few minutes, a priest seized a bundle of brushwood, in which a piece of burning paper had been stuck, and flung it into the well, after quickly removing the carpet. The strangers had previously been warned to keep at some distance, and the priest and his assistants likewise ran off as fast as they could. About half a minute after the fire-brand had been cast into the pit, a terrific explosion took place, and a vast column of fire, in the shape of an inverted cone (from the gas spreading out as soon as it emerges from the pit), ascended to the skies.

How long the fires of Baku may have been burning is unknown, but it is very probable that they did not exist before the Christian era. No Greek or Roman author mentions them, and it is not before the tenth century that Arab writers take notice of Baku and its wonders. When the Sassanides restored the religion of Zoroaster, the attention of these fire-worshipping princes was naturally directed to a place where fire gushes pure and unbidden from the earth. They raised a temple on the spot, and thousands of pilgrims wandered to the holy fires of Baku. But when the fanatical Arabs overthrew the Persian Empire, times of persecution and distress began for the Parsees; and still later they were almost entirely extirpated by the hordes of Tamerlane. During the last centuries fire-worshipping was again introduced by the Indians, who, after the Sefides had ascended the Persian throne, gradually settled in the Caspian provinces, and whose number must have been considerable, as travellers inform us that in the latter half of the seventeenth century 200 rich Indian merchants were residing in the town of Schemachi. But the anarchical times which followed the usurpation of Nadir Schah forced most of these Indians to leave their adopted country, and since then only solitary pilgrims have found their way to Baku. But the number even of these is constantly diminishing, although the Russians, to whom the sanctuary now belongs, allow them full freedom of access. When Koch was at Baku, he found there only five Indians from Mooltan, whither the majority would gladly have returned, had they but possessed the necessary means. Their squalid appearance and tattered raiment formed the strongest imaginable contrast to the splendour of the element they worshipped. Among them was a Fakir, who had made a vow constantly to remain in the same position absorbed in religious contemplation, and who for sixteen years had never moved from the spot.

The burning springs gush out not only from the ground near the temple and in other parts of the peninsula of Abscheron, but even from the bottom of the neighbouring Caspian Sea : and as Sir Charles Lyell saw carburetted hydrogen rise in countless bubbles through the crystal waters above the falls of the Niagara, and shoot up in bright flames at the approach of a light, so Dr. Abich mentions a spot in the Gulf of Baku where the inflammable gas issues with such force, and in so great a quantity, from the bottom, which is there three fathoms deep, that a small boat is in danger of being overturned when coming too near it.

As gas springs most frequently occur in districts which have been the former seats of volcanic action, and as similar exhalations often arise from still active craters, they are supposed by many geologists to be the last remaining traces of an expiring volcanic energy. Bischoff considers the carbonic acid of the German gas springs to be developed by the decomposition of carbonate of lime by volcanic heat or heated water.

A phenomenon which is sometimes found connected with gas springs is that of the mud volcanoes, which may be described as cones of a ductile, unctuous clay, formed by the continued evolution of a sulphurous and inflammable gas, spurting up waves and lumps of liquid mud. These remarkable cauldrons are found in many parts of the world, in the Island of Milo, in Italy, in Iceland, in India, about 120 miles from the mouths of the Indus, on the coast of Arracan, in Birmah, in Java, Columbia, Nicaragua, and Trinidad; but probably nowhere on a grander scale than at either extremity of the chain of the Caucasus, towards the Caspian on the east and the Sea of Azof on the west, where in the peninsula of Taman, and on the opposite coast of the Crimea, near Kertsch, vast numbers of mud volcanoes are scattered, some of them 250 feet high. Their operations have apparently been going on for countless ages, and have covered a great extent of land with their products.

· The Macaluba, in Sicily, which owes its name to the
Arabs, is the mud volcano most anciently known. It is
mentioned by Plato in his 'Phædon,' and has been described
by Strabo. It is situated five miles to the north of Girgenti,
on a hill of a conical shape, truncated at the top, and 150
feet high. The summit is a plain half a mile round, and the
whole surface is covered with thick mud. The depth of the
mud, which is supposed to be immense, is unknown. There
is not the slightest appearance of vegetation upon it. In the
rainy season the mud is much softened; the surface is even,

MUD VOLCANOES OF TRINIDAD.

and there is a general ebullition over it, which is accompanied with a very sensible rumbling noise. In the dry
season the mud acquires greater consistency, but its motion
still goes on. The plain assumes a form somewhat convex;
a number of little cones are thrown up, which rarely rise to
the height of two feet. Each of them has a crater, where
black mud is seen in constant agitation, and incessantly
emitting bubbles of air. With these the mud insensibly
rises, and as soon as the crater is full of it, it disgorges.
The residue sinks, and the cone has a free crater, until a new
emission takes place.

Such is the ordinary state of the Macaluba; but from

time to time the hill becomes subject to alarming convulsions. Slight earthquake shocks are felt at the distance of
two or three miles, accompanied with internal noises resembling thunder. These increase for several days, and are
followed at last by a prodigious spout of mud, earth, and
stones, which rises two or three hundred feet in the air.

Similar paroxysmal explosions have been observed in the
Caucasian mud volcanoes. In February 1794, the Obu, in
the peninsula of Taman, had an eruption accompanied with
a dreadful noise, and an earthquake which radiated from
the cone, and was felt as far as Ekaterinodor, at a distance
of fifty-five leagues. At the beginning of the eruption
flames were seen, which rose to a prodigious height, and
lasted about half an hour. At the same time dense clouds
of smoke escaped from the crater, and mud and stones were
cast up to the height of 3,000 feet. Six streams of mud,
the largest of which was half a mile long, flowed from the
volcano, and their volume is said to have been equal to
twenty-two millions of cubic feet.

Violent eruptive symptoms accompanied the formation of
a new mud volcano in the vicinity of Baku on the Caspian.
On November 27, 1827, flames blazed up to an extraordinary
height for three hours, and continued for twenty hours more
to rise about three feet above a crater from which mud was
ejected. At another point in the same district, where flames
issued, fragments of rock, of large size, were hurled up into
the air and scattered around.

The phenomena exhibited by the Macaluba and other mud
cauldrons are certainly very distinct from those of true volcanoes, since no scoriæ or lava or heated matters of any kind
are sent forth, the mud being described as cold when
emitted, although the gas, whose violent escape throws it up,
is sometimes ignited. Hence geologists commonly regard
these phenomena as entirely distinct from the volcanic, and
ascribe their origin to chemical action going on at no great
depth beneath the surface, among the constituents of certain
stratified matters ; while other scientific authorities declare
them to be as much connected with internal igneous agency
as any other eruptive phenomena. Their occurrence in dis-

tricts not remote from the sites of vast volcanic disturbance, and their occasional violent paroxysms, certainly afford much support to this view, and show that it is probably the same power, in different degrees of energy, which casts up the mud of the Macaluba and pours forth the lava-streams of Cotopaxi.

CHAPTER IX.

EARTHQUAKES.

Extent of Misery inflicted by great Earthquakes—Earthquake Regions—Earthquakes in England—Great Number of Earthquakes—Vertical and undulatory Shocks—Warnings of Earthquakes—Sounds attending Earthquakes—Remarkable Displacements of Objects—Extent and Force of Seismic Wave Motion—Effects of Earthquakes on the Sea—Enormous Waves on Coasts—Oscillations of the Ocean—Fissures, Landslips, and shattering Falls of Rock caused by Earthquakes—Causes of Earthquakes—Probable Depth of Focus—Opinions of Sir Charles Lyell and Mr. Poulett Scrope—Impressions produced on Man and Animals by Earthquakes.

OF all the destructive agencies of nature there is none to equal the earthquake. The hurricane is comparatively weak in its fury; the volcanic eruption generally confines its rage to the neighbourhood of the labouring mountain, but a great earthquake may cover a whole land with ruins.

The terrible subterranean revolution which convulsed all Asia Minor and Syria, in the reign of Tiberius, destroyed twelve celebrated cities in a single night. The sun, which on setting had gilded their temples and palaces with his parting rays, beheld them prostrate on the following morning.

In A.D. 115 Antioch was the centre of a great commotion. The city was full of soldiers under Trajan; heavy thunder, excessive winds, and subterranean noises were heard; the earth shook, the houses fell; the cries of people buried in the ruins passed unheeded. The Emperor leaped from a window, while mountains were broken and thrown down, and rivers disappeared, and were replaced by others in a new situation. Four centuries later (May 20, 526) the same doomed city was totally subverted by an earthquake, when it is reported that 250,000 persons perished.

Similar catastrophes, in which thousands and thousands of victims were suddenly destroyed, have frequently occurred in

H

Peru and Chili, in the West Indies and Central America, in the Moluccas and Java, in the countries bordering on the Mediterranean and the Red Sea; but a bare mention of the loss of life conveys but a faint idea of the extent of misery inflicted by one of those great earthquakes which mark with an ominous shade many large tracts of the earth's surface.

We must picture to ourselves the slow lingering death which is the fate of many—some buried alive, others burnt in the fire which almost invariably bursts out in a city where hundreds of dwellings have suddenly been laid prostrate—the numbers who escaped with loss of limbs or serious bodily injuries, and the surviving multitude, suddenly reduced to penury and want.

In the Calabrian earthquake of 1783, it is supposed that about a fourth part of the inhabitants of Polistena and of some other towns were buried alive, and might have been saved had there been no want of hands; but in so general a calamity, where each was occupied with his own misfortunes or those of his family, help could seldom be procured. 'It frequently happened,' says Sir Charles Lyell, 'that persons in search of those most dear to them could hear their moans, could recognise their voices, were certain of the exact spot where they lay buried beneath their feet, yet could afford them no succour. The piled mass resisted all their strength, and rendered their efforts of no avail. At Terranova four Augustin monks, who had taken refuge in a vaulted sacristy, the arch of which continued to support a vast pile of ruins, made their cries heard for the space of four days. One only of the brethren of the whole convent was saved, and of what avail was his strength to remove the enormous weight of rubbish which had overwhelmed his companions? He heard their voices die away gradually, and when afterwards their four corpses were disinterred, they were found clasped in each others' arms.

Affecting narratives are preserved of mothers saved after the fifth, sixth, and even seventh day of their interment, when their infants or children had perished with hunger, In his work on the great Neapolitan earthquake of 1857, Mr. Mallet, from innumerable narratives of personal peril and sad adventure, selects the distressing case of a noble family of Monte Murro, as affording a vivid picture of the terrors of an earthquake night. Don Andrea del Fino, the owner of one of the few houses in the city which escaped total destruction, was with his wife in bed, his daughter sleeping in an adjacent chamber on the principal floor. At the first shock his wife, who was awake, leaped from bed, and immediately after, a mass of the vaulting above came down, and buried her sleeping husband. At the same moment, the vault above their daughter's room fell in upon her. From the light and hollow construction of the vaults neither was at once killed. The signora escaped by leaping from the front window, she scarcely knew how. For more than two hours she wandered, unnoticed, amongst the mass of terrified survivors in the streets, before she could obtain aid from her own tenants and dependants, to extricate her husband. They got him out after more than eighteen hours' entombment—alive, indeed, but maimed and lame for life. His daughter was dead. As he lay longing despairingly for release from the rubbish, which a second shock, an hour after the first, had so shaken and closed in around him that he could scarcely breathe, he heard, but a few feet off, her agonising cries and groans grow fainter and fainter, until at last they died away. His wife, to whose devotion his own life was owing, had escaped unhurt.

Unfortunately man too often vies with the brute forces of nature to increase the horrors of a great earthquake. As the arm of the law is paralysed by the general panic, thieves and ruffians are not slow to avail themselves of their opportunity. Thus, in the Calabrian catastrophe of 1783, nothing could be more atrocious than the conduct of the peasants, who abandoned the farms and flocked in great numbers into the towns—not to rescue their countrymen from a lingering death, but to plunder. They dashed through the streets amid tottering walls and clouds of dust, trampling beneath their feet the bodies of the wounded and half buried, and often stripping them, while yet living, of their clothes.

From the vast ruin and misery they entail, it is evident that where earthquakes are frequent, there can never be perfect security of property even under the best government; and as the fruits collected by the labour of many years may be lost in an instant, the progress of civilisation and national wealth must necessarily be retarded.

'Earthquakes alone,' says Mr. Darwin, 'are sufficient to destroy the prosperity of any country. If beneath England the now inert subterranean forces should exert those powers which most assuredly in former geological ages they have exerted, how completely would the entire condition of the land be changed! What would become of the lofty houses, thickly-packed cities, great manufactories, the beautiful public and private edifices? If the new period of disturbance were first to commence by some great earthquake in the dead of the night, how terrific would be the carnage! England would at once be bankrupt; all papers, records, and accounts would from that moment be lost. Government, being unable to collect the taxes, and failing to maintain its authority, the hand of violence and rapine would remain uncontrolled. In every large town famine would go forth, pestilence and death following in its train.'

Fortunately the experience of many ages shows that the regions subject to these terrible catastrophes are confined to a comparatively small part of the surface of the globe. Thus Southern Italy and Sicily; the tract embracing the Canaries, the Azores, Portugal, and Morocco; Asia Minor, Syria, and the Caucasas; the Arabian shore of the Red Sea; the East Indian Archipelago; the West Indies, Nicaragua, Quito, Peru, and Chili, are particularly liable to destructive shocks.

But beyond these limits slighter earthquakes are of far more common occurrence than is generally supposed, and probably they leave no part of the world entirely undisturbed. From the year 1821 to 1830 no less than 115 earthquakes have been felt to the north of the Alps, and since the year 1089, 225 are cited in the annals of England. Some of these earthquakes seem to have but just stopped at the point when a slight increase of their force would have covered the land with ruins. In 1574, on the 26th of February, between five and six in the evening, an earthquake was felt at York, Worcester, Hereford, Gloucester, and Bristol. Norton Chapel was filled with worshippers; they were nearly all overthrown, and fled in terror, thinking that the dead were unearthed or that the chapel was falling. Six years later, on the 6th of April, at 6 p.m., all England was thrown into consternation. The great bell at Westminster began to toll; the students at the Temple started up from table and rushed into the street, knives in hand; a part of the Temple Church fell, and stones dropped from St. Paul's. Two stones fell in Christ's Church, and crushed two persons. In rushing out of the church many were lamed, and there was a shower of chimneys in the streets. At Sandwich, the occurrence was marked by the violence of the sea, which made ships run foul of each other; and at Dover a part of the fortifications fell with the rock which supported it.

On the 6th October 1863, a movement, though gentle when compared to the preceding instances, was felt from the English Channel to the Mersey, and from Hereford to Leamington and Oxford. The Malvern range was about the centre of the area, as it has often been before. Even in alluvial Holland, six or eight slight earthquakes have been felt during the last century. The industrious researches of Kluge show that, during the eight years from 1850 to 1857, no less than 4,620 earthquakes—a great proportion of which (509) fell to the share of Southern Europe—have been noticed in both hemispheres; and when we remember that a very considerable part of the globe is still either totally unknown or removed by the barbarous condition of its inhabitants from all intercourse with the scientific world, and that, consequently, the above list must necessarily be incomplete, it is very probable that not a day passes without some agitation of the surface of the earth in some place or other.

A violent earthquake almost always consists of several shocks following each other in rapid succession. Sometimes they are preceded by slighter vibrations; at other times they suddenly convulse the land without any previous notice. In most instances, each shock lasts but a few seconds; but this is enough to ruin the work of ages. Three violent commotions within five minutes destroyed the town of Caraccas on March 26, 1812; and the earthquake which, in 1692, desolated Jamaica, lasted but three minutes. On January 11, 1839, two shocks within thirty seconds covered Martinique, and the whole range of the Lesser Antilles, with ruins. But a violent earthquake, though itself but of short duration, is generally followed by a series of secondary shocks, which are repeated at gradual widening intervals and with decreasing energy, so that if these subsequent tremors be taken into account, it may often be said to last for weeks or even months. Thus, to cite but one instance, the earthquake of October 21, 1766, destroyed the whole town of Cumana in a few minutes, but during the following fourteen months the earth was in a constant vibratory motion, and scarce an hour passed without a shock being felt.

In countries where earthquakes are comparatively rare (for instance in the south of Europe), the belief is very general that oppressive heat, stillness of the air, and a misty horizon, are always forerunners of the phenomenon. But this popular opinion is not confirmed by the experience of trust-worthy observers, who have lived for years in countries such as Cumana, Quito, Peru, and Chili, where the ground trembles frequently and violently. Humboldt experienced earthquakes during every state of the weather, serene and dry, rainy and stormy.

Brute animals, being more sensitive than men of the slightest movement of the earth, are said to evince extra-ordinary alarm, and it has been often observed that even the dull hog shows symptoms of uneasiness previous to the shock. During the great Neapolitan earthquake of 1857 an unusual halo-light was seen in the sky before, and not long after, the shock. Mr. Mallet was at first inclined to look upon this notion as a superstitious tale; but, finding it widely diffused in a country where communication is bad and news travels slowly, no longer doubted that it was founded on fact.* Conjectures would be useless as to its nature, but future observation directed to the point may determine whether some sort of auroral light may emanate from the vast depths of rock formation under the enormous tensions and compressions that must precede the final crush and rupture; or whether volcanic action, going on in the unseen depths below, may give rise to powerful disturbances of electric equilibrium, and hence to the development of

* Mallet, 'The Great Neapolitan Earthquake of 1857,' vol. i. p. 323.

light; just as from volcanic mountains in eruption lightnings continually flash from the high volumes of steam and floating ashes above the crater. Humboldt is also of opinion that, though in general the revolutions which take place below the surface of the earth are not announced beforehand by any meteorological process, or a peculiar appearance of the sky, it is not improbable that during violent shocks some change may occur in the condition of the atmophere. Thus, during the earthquake in the Piedmontese valleys of Pelis and Clusson, great alterations were observed in the electric tension of the atmosphere without any appearance of a thunder-storm.

Earthquakes are generally attended with sounds, sometimes like the howling of a storm, or the rumbling of subterranean thunder; at others like the clashing of iron chains, or as if a number of heavily laden waggons were rolling rapidly over the pavement, or as if enormous masses of glass were suddenly shivered to pieces. As solid bodies are excellent conductors of sound (burnt clay, for instance, propagating it ten or twelve times more rapidly than the air), the subterranean noise may be heard at a vast distance from the primary seat of the earthquake. In Caraccas, in the grassplains of Calabozo, and on the banks of the Rio Apure, which falls into the Orinoco, a dreadful thunder-like sound was everywhere heard on April 30, 1812, without any simultaneous trembling of the earth, at the time when, at the distance of 158 geographical miles, the volcano of St. Vincent, in the Lesser Antilles, was pouring out of its crater a mighty lava-stream. This was, according to distance, as if an eruption of Vesuvius were heard in the north of France. In the year 1744, during the great eruption of the volcano Cotopaxi, a subterranean noise like the firing of cannon was heard at Honda on the Magdalena river. But the crater of Cotopaxi is 17,000 feet higher than Honda, and both points, situated at a distance of 109 geographical miles, are moreover separated by the colossal mountain masses of Quito, Pasto, and Popayan, and by numberless ravines and deep valleys. The sound was certainly transmitted, not through the air, but through the earth, and must have proceeded from a very considerable depth.

. But noise is not the necessary attendant of an earthquake, for many instances are known in which the most violent shocks have been completely noiseless. No subterranean sounds were heard during the terrific earthquake which destroyed Riobamba on February 4, 1797, and the same circumstance is mentioned in the narratives of many of the Chilian earthquakes.

The phenomenon of sound, when unaccompanied by any perceptible vibration, makes a peculiarly deep impression on the mind, even of those who have long inhabited a country subject to frequent earthquakes. They tremble at the idea of the catastrophe which may follow. A remarkable instance of a long protracted noise without any trembling of the earth occurred in 1784, at the wealthy mining town of Guanaxuato in Mexico, where the rolling of subterranean thunder, with now and then a louder crash, was heard for more than a month, without the slightest shock, either on the surface of the earth, or in the neighbouring silver mines, which are 1,500 feet deep. The noise was confined to a small space, so that a few miles from the town it was no longer audible. Never before had this phenomenon been known to occur in the Mexican highlands, nor has it been repeated since.

Earthquake shocks are either vertical or undulatory. A vertical shock, which is felt immediately above the seat or focus of the subterranean disturbance, causes a movement up and down. Like an exploding mine, it frequently jerks movable bodies high up into the air. Thus, during the great earthquake of Riobamba, the bodies of many of the inhabitants were thrown upon the hill of La Culla, which rises to the height of several hundred feet at the other side of the Lican torrent; and during the earthquake of Chili in 1837, a large mast, planted thirty feet deep in the ground at Fort San Carlos, and propped with iron bars, was thrown upwards, so that a round hole remained behind.

Although to the inhabitants of a shaken district the undulatory wave or vibration of an earthquake appears to radiate horizontally outwards from the spot on the surface where it is first felt, the force does not really operate in a horizontal direction, like a wave caused by a pebble on the

surface of a pond; for at every point, except that immediately above the focus of the shock, it comes up obliquely
from below, causing the ground to move forwards and then
backwards in a more or less horizontal direction. As a ship,
yielding to the oscillatory movements of the waves, alternately inclines to one side or the other, so, during the more
violent undulations of the soil, the objects on its surface are
momentarily moved from their vertical position, and often
considerably inclined towards the horizon. Thus during the
great earthquake which convulsed the valley of the Mississippi in 1811-12, Mr. Bringier, an engineer of New Orleans,
who was on horseback near New Madrid, where some of the
severest shocks were experienced, saw the trees bend as the
wave-motion of the earthquake passed under them, and
immediately afterwards recover their position. The transit
of the wave through the woods was marked by the crash of
countless branches, first heard on one side and then on the
other. It must have been awful to see the giants of the
forest thus move to and fro like a corn-field agitated by the
wind !

Very remarkable displacements of objects are not seldom
caused by earthquakes, such as the rotation of the blocks of
columns or the turning of statues on their pedestals.

At Lima, which, owing to its repeated destructions by
earthquakes, is properly a city of ruins, Professor Dana saw
two obelisks with the upper stone on each displaced and
turned round on its axis about fifteen degrees in a direction
from north to east. These rotations by earthquakes have
been attributed by some authors to an actual rotatory
movement in the earthquake vibration ; but it has lately
been shown by Mr. Mallet that this hypothesis is untenable
and unnecessary, as a simple vibration back and forth is all
that is required to produce a rotatory motion in the stone of
a column, provided that stone be attached below more
strongly on one side of the centre than on the opposite.

The wave-motion of an earthquake sometimes spreads over
enormous spaces. The shocks of the earthquake of New
Granada which took place in the night from the 16th to
the 17th of June 1826, were noticed over a surface of 750,000
square miles. The earthquake of Valdivia (February 20,

1835) was felt southwards on the distant island of Chiloe
to the north as far as Copiapo, in Mendoza to the east of
the Andes, and on the Island of Juan Fernandez, 300 miles
from the coast. Supposing these effects to have taken
place at corresponding distances in Europe, all the land
would have trembled from the North Sea to the Mediterranean, and from Ireland to the centre of France.

It is evident that the extent and force of the wave-motion
of an earthquake must in a great measure depend upon the
nature of the rocks through which it is transmitted. It will
vibrate more easily through solid homogeneous masses, while
in alluvial deposits, or in a soil composed of sand and loose
conglomerate, its undulations will be propagated irregularly
and its effects be far more destructive. This is particularly
the case where the alluvial deposits repose on a substratum
of hard rock. Thus the devastations of the Calabrian earthquake of 1783 were most apparent in the plain of Oppido, in
those parts where the newer tertiary strata rest upon granite.
The earthquake wave generally follows the direction of
mountain-chains, and but rarely crosses them. The great
Chilian earthquakes, which often propagate their vibrations
to distances of many hundred miles along the western foot
of the Andes, remain unfelt on their eastern border; while
the earthquakes along the shores of Venezuela, Caraccas,
and New Granada rarely transmit their vibrations beyond
the high mountain-chains which run parallel with the coast,
This is probably due to the numerous dislocations, rents,
and caverns which are produced by the elevation of the
mountain-chains, and necessarily serve as barriers to the
propagation of the earthquake wave.

Severe earthquakes are not seldom accompanied by a
violent agitation of the sea. First, at the instant of the
shock, the water swells high up on the beach with a gentle
motion, and then as quietly recedes; secondly, some time
afterwards, the whole body of the sea retires from the coast,
and then returns in waves of overwhelming force. The first
movement seems to be an immediate consequence of the
earthquake affecting differently a fluid and a solid, so that
their respective levels are slightly deranged; but the second
is a far more important phenomenon. 'Some authors,' says

Mr. Darwin, 'have attempted to explain it by supposing that the sea retains its level, while the land oscillates upwards; but surely the water close to the land, even on a rather steep coast, would partake of the motion of the bottom; moreover, similar movements of the sea have occurred at islands far distant from the chief line of disturbance. I suspect (but the subject is a very obscure one) that a wave, however produced, *first draws the water from the shore* on which it is advancing to break. I have observed that this happens with the little waves from the paddles of a steamboat. From the great wave not immediately following the earthquake, but sometimes after the interval of even half-an-hour, and from distant islands being affected similarly with the coasts near the focus of the disturbance, it appears that the wave first rises in the offing, and, as this is of general occurrence, the cause must be general. I suspect we must look to the line where the less disturbed waters of the deep ocean join the water nearer the coast which has partaken of the movements of the land, as the place where the great wave is first generated; it would also appear that the wave is larger or smaller according to the extent of shoal water which has been agitated together with the bottom on which it rested.'

The following examples sufficiently prove that no storm, however violent, is capable of raising such prodigious waves as an earthquake.

In the year 1692 the town of Kingston in Jamaica was almost totally destroyed by a huge earthquake wave. A frigate which lay in port was carried forward over the houses, and stranded in the middle of the town. In his 'Principles of Geology,' Sir Charles Lyell relates that, during the Calabrian earthquake of 1783, the Prince of Scilla had persuaded a great part of his vassals to betake themselves to their fishing-boats for safety, and he himself had gone on board. On the night of February 5, when some of the people were sleeping in the boats, and others on a level plain slightly elevated above the sea, the earth rocked and large masses of rock were thrown down with a dreadful crash upon the plain. Immediately afterwards the sea, rising more than twenty feet above the level of this low tract, rolled foaming over it and swept away the multitude. It then retreated, but soon rushed back again with greater violence, bringing back with it some of the bodies it had carried away. At the same time every boat was sunk or dashed against the beach, and some of them were swept far inland. The aged prince was killed, with 1,430 of his people.

After the earthquake which devastated the town of Lima on the 28th of October 1746, the sea rose on the evening of the same day eighty feet above its usual level in the neighbouring Bay of Callao, overwhelmed the town, and destroyed nearly all the inhabitants. Of the twenty-three ships which were lying in the harbour at the time, nineteen immediately sank, while the four others were thrown upon the land at the distance of nearly a league.

Shortly after the shock which desolated Chili on the 20th of February 1835, a great wave was seen from the distance of three or four miles, approaching in the middle of the Bay of Talcahuano with a smooth outline, but tearing up cottages and trees along the shore, as it swept onwards with irresistible force. At the head of the bay it broke in a fearful line of white breakers, which rushed up to a height of twenty-three vertical feet above the highest spring tides. Their force must have been prodigious, for at the Fort a cannon with its carriage, estimated at four tons in weight, was moved fifteen feet inwards. The whole coast was strewed over with timber and furniture, as if a thousand ships had been wrecked. As Mr. Darwin walked along the shore, he observed that numerous fragments of rock, which, from the marine productions adhering to them, must recently have been lying in deep water, had been cast up high on the beach. One of these was six feet long, three broad, and two thick.

During the dreadful earthquake which in 1868 raised the strip of land at the western foot of the Andes from Iburra in Ecuador, to Iquique in Peru, 1,200 miles in length, the receding sea uncovered the bay at Iquique to the depth of four fathoms, and then, returning in an immense wave, a mass of dark blue water, forty feet high, rushed over the already ruined city, and swept away every trace of what had been a town. One spectator, seeing the whole surface of the sea rise like a mountain, ran for his life to the Pampa. The waves overtook him. Fighting with the dark water, amidst wreck and ruin of every kind, carried back into the bay, and again thrown back to the Pampa, wounded and half naked, he crept for safety into a hole of the sand, and waited sadly for the dawn. At Arica, the British Vice-Consul, alarmed at the first shock, rushed out of the house with his family, and made for the high ground, in just terror of the expected sea-wave. Through the ruined town, amidst dead and dying, half stifled with dust, they reached rising ground, and, looking back, saw a dreadful sequel—the sea rushed in and left not a vestige remaining of the lower part of Arica. Six vessels were lost in the bay, or tossed over rocks and houses; an American gunboat was whirled away from her moorings, and laid, without a broken spar or tarnished flag, high and dry on the sand-hills, a quarter of a mile from the sea.

As might be expected from the movable nature of water, the wave-motion of earthquakes is frequently propagated to surprising distances over the sea. The Chilian earthquake of 1835 produced oscillations of the ocean that made themselves felt on the Sandwich Islands, at a distance of 5,000 nautical miles. On Mauai, the sea retreated 120 feet, and then suddenly returned with a tremendous wave that swept away the trees and houses on the beach. In Hawaii, a large congregation had assembled for divine service near Byron's Bay. Suddenly the water began to sink, so that soon a great part of the harbour was laid dry. The spectators hurried to the shore to admire the astonishing spectacle, when a wave, rising twenty feet above the usual tide-mark, inundated the land, destroyed sixty-six huts, and drowned eleven of the islanders, though the best swimmers in the world. So far from its starting-point did the South American earthquake seek its victims. Fifteen hours and a half after the great earthquake of Arica (1868), the water-wave undulating over the vast Pacific was felt at Chatham Islands, a distance of 6,300 miles, and an hour later at New Zealand.

The enormous powers which come into action during a great earthquake show themselves not only in the destruction of edifices and the wide-spread ruin so produced, but in the changes which they effect in the configuration of the soil. Wherever masses of earth rest loosely upon a sloping surface of subjacent rock, or where steep mountain crests overlie wet and unctuous beds of shale, or where the rock itself is composed of incoherent material, or where river-banks are formed of precipitous masses of clays, or where the corroding waters have undermined the ground, the violent commotion caused by an earthquake cannot fail to produce landslips, fissures, and falls of rock. In 1571, on the 17th of February, the ground opened all at once at the 'Wonder,' near Putley, not far from Marcle in Herefordshire; and a large part of the sloping surface of the hill—twenty-six acres, it is said - descended with the trees and sheep-folds, and continued in motion from Saturday to Monday, masses of ground being turned round through half a circle in their descent. This was a great landslip, said to have been occasioned by an earthquake.

Earth-fissures were formerly supposed to be occasioned by a *stretching* of the ground, occasioned by the wavy nature of the shocks; but Mr. Mallet has shown that no earthquake wave can possibly produce any such stretching, and considers them as cases of small and incipient landslips caused by the shaking downwards of a loose mass. His own observations left no doubt in his mind that the descriptions, given by the Neapolitan Academy in their Historical Account of the Earthquake of 1783, of the earth-fissures therein produced, and designated constantly by the pompous term 'voragines,' are gross exaggerations; and that the well-known Jamaica earth-fissures, that were said to have opened and closed with the wave, and to have *bitten people in two*, must be regarded as audacious fables.

'The vulgar mind, filled from infancy with superstitious terrors as to "the things under the earth," is seized at once by the notion of these fissures of profound and fabulous depth with fire and vapour of smoke issuing from within their murky abysses; but they should cease to belong to science.'

Enormous landslips are sometimes occasioned by earthquakes, but their extent depends less upon the power and energy of the shock than upon the conditions of unstable equilibrium presented by great masses of loose material, through the configuration of the country. In consequence of landslips or dislodgements of large masses of rock, alterations in the flow or distribution of the waters frequently take place. Thus, brooks or rivers are not seldom dammed, and temporary ponds or lakes created.

Permanent elevations of the land have been observed after some earthquakes. Thus, after the violent shocks of November 19, 1822, a great part of the coast·of Chili was found to be raised several feet above its previous level; and after the great earthquake which occurred in New Zealand in the night of January 23, 1855, a large tract of land was found to be permanently upraised from one to nine feet. Before the shock there had been no room to pass between the sea and the base of a perpendicular cliff called Muka-Muka, except for a short time at low water, and the herds-men were obliged to wait for low tide in order to drive their cattle past the cliff. But immediately after the upheaval, a gently sloping raised beach, more than 100 feet wide, was laid dry, affording ample space at all states of the tide for the passage of man and beast.

These permanent elevations have often been attributed to the immediate agency of earthquakes; but Mr. Mallet proves this assumption to be a fallacy, as the impulse of the earth-quake wave even right above the focus is utterly incapable of raising the level of the land by a height much more than in-strumentally appreciable, and there is not the least evidence that any part even of this elevation is permanent. That earthquakes occur along with, and as part of, a train of other circumstances which do produce permanent elevation occa-sionally, and that earthquakes are probably always the signals that the forces producing elevation are operative, is another matter, with which that erroneous or loosely expressed view should not be confounded.

The causes of earthquake are still hidden in obscurity, and probably will ever remain so, as the e violent convulsions originate at depths far below the reach of human observa-tion. Mr. Mallet came to the conclusion that the depth of the original Calabrian shock in 1857 did not exceed seven or eight miles, and deduces from all the facts known as to the movements of earthquakes, that the subterranean points where the shocks originate perhaps never exceed thirty geo-graphical miles, so that, even supposing the central nucleus of the earth to be fluid, they cannot possibly be due to the reaction of the internal ocean of molten stone upon the solid shell with which it is enveloped, but must have their seat within the latter. The existence of reservoirs of fused matter at various depths in the solid earth-rind is quite sufficient to account for all seismic and volcanic phenomena; for it is evident that whenever rain-water, or the waters of the sea percolating through rocks, gain access to these subterranean lakes of molten stone, steam must be generated, the pressure of which will in many cases rend and dislocate the incumbent masses.

'During such movements,' says Sir Charles Lyell, 'fissures may be formed and injected with gaseous or fluid matter, which may sometimes fail to reach the surface, while at other times it may be expelled through volcanic vents, stufas, and hot springs. When the strain on the rocks has caused them to split, or the roofs of pre-existing fissures or caverns have been made to fall in, vibratory jars will be produced and propagated in all directions, like waves of sound through the crust of the earth, with varying velocity, according to the violence of the original shock, and the density or elasticity of the substance through which they pass. They will travel, for example, faster through granite than through limestone, and more rapidly through the latter than through wet clay, but the rate will be uniform through the same homogeneous medium.'

According to Mr. Poulett Scrope, the originating cause of the earthquake must be sought in the expansion of some deeply-seated mass of mineral matter, owing to augmentation of temperature or diminution of pressure. By this expansive force the solid rocks above are suddenly rent asunder, and whether below the sea or not, their violent disruption produces a jarring vibration, which is propagated on either side through their continuous masses in undulatory pulsations.

Some geologists are of opinion that earthquakes are frequently the result of the subsiding, sinking in, or cracking of subterranean cavern roofs, in consequence of the pressure of the superincumbent rocks. Small local earthquakes may be explained by this theory; but terrible convulsions which shake a whole continent evidently proceed from a far more formidable cause, and are more satisfactorily explained by the agency of subterranean heat and elastic vapours.

If, even during an ordinary storm, the black clouds, the howling of the wind, the flashes of lightning, and the loud claps of thunder, strike men and brutes with fear, we may naturally expect to see terror carried to its highest pitch by so dreadful a phenomenon as an earthquake. All creatures living or burrowing under the earth—rats, mice, moles, snakes—hastily creep forth from their subterranean abodes, though many no doubt are gripped and suffocated by the suddenly-moved soil before they can effect their escape ; the crocodile, generally silent, like our little lizards, rushes out of the river and runs bellowing into the woods ; the hogs show symptoms of uneasiness ; the horses tremble ; the oxen huddle together ; and the fowls run about with discordant cries. On man, the phenomenon makes a peculiarly deep impression.

'A bad earthquake,' say Mr. Darwin, 'at once destroys our oldest associations. The earth, the very emblem of solidity, has moved beneath our feet like a thin crust over a fluid. One second of time has created in the mind a strange idea of insecurity, which hours of reflection would not have produced.' We can no longer trust the soil on which we stand, and feel ourselves completely at the mercy of some unknown destructive power, which at any moment, without forewarn · ing, can destroy our property or our lives. But as first impressions are always the deepest, so habit renders man callous even to the terrors of an ordinary earthquake. In countries where slight shocks are of frequent occurrence, almost every vestige of fear vanishes from the minds of the natives, or of the strangers whom a long residence has familiarised with the phenomenon.

On the rainless coast of Peru, thunderstorms and hail are unknown. The thunder of the storm is there replaced by the thunder which accompanies the earthquake. But the frequent repetition of this subterranean tumult, and the general belief that dangerous shocks occur only twice or thrice in the course of a century, produce in Lima so great an indifference towards slighter oscillations of the soil, that they hardly attract more attention than a hail-storm in Northern Europe.

I

CHAPTER X.

THE GREAT EARTHQUAKE OF LISBON.

A dreadful All Saints' Day—The Victims of a Minute—Report of an Eye-witness
—Conflagration—Banditti—Pombal brings Order into Chaos—Intrigues of the
Jesuits—Damages caused by the Earthquake in other Places—At Cadiz—In
Barbary—Widespread Alarm—Remarks of Goethe on the Earthquake.

HISTORY exhibits few catastrophes more terrible than
that which was caused by the great earthquake which,
on November 1, 1755, levelled the town of Lisbon to the
dust. On other occasions, such as that of a siege, a famine,
or a plague, calamity approaches by degrees, giving its
victims time to measure its growth and preparing them, as
it were, to sustain an increasing weight of misery; but here
destruction fell upon the devoted city with the rapidity of a
flash of lightning.

A bright sun shone over Lisbon on that fatal morning.
The weather was as mild and beautiful as on a fine summer's
day in England, when, about forty minutes past nine in the
morning, an earthquake shock, followed almost immediately
by another and another, brought down convents, churches,
palaces and houses, in one common ruin, and, at a very
moderate computation, occasioned the loss of 60,000 lives.
'The shocking sight of the dead bodies,' says an eye-witness
of the scene, 'together with the shrieks and cries of those
who were half buried in the ruins, exceeds all description;
for the fear and consternation were so great that the
most resolute person durst not stay a moment to remove a
few stones off the friend he loved most, though many might
have been saved by so doing; but nothing was thought of
but self-preservation. Getting into open places, and into the
middle of streets, was the most probable security. Such as
were in the upper storeys of houses were, in general, more

fortunate than those who attempted to escape by the doors, for they were buried under the ruins with the greatest part of the foot-passengers; such as were in equipages escaped best, though their cattle and drivers suffered severely; but those lost in houses and the streets are very unequal in number to those that were buried in the ruins of churches; for as it was a day of great devotion, and the time of celebrating mass, all the churches in the city were vastly crowded; and the number of churches here exceeds that of both London and Westminster; and as the steeples are built high, they mostly fell with the roof of the church, and the stones are so large that few escaped.' *

Many of those who were not crushed or disabled by the falling buildings fled to the Tagus, vainly hoping that they might find there the safety which they had lost on land. For, soon after the shock, the sea also came rushing in like a torrent, though against wind and tide, and rising in an enormous wave, overflowed its banks, devouring all it met on its destructive path. Many large vessels sank at once; others, torn from their anchors, disappeared in the votex, or, striking against each other, were shattered to pieces. A fine new stone quay, where about three thousand persons had assembled for safety, slipped into the river, and every one was lost; nor did so much as a single body appear afterwards.

Had the misery ended here, it might in some degree have admitted of redress; for, though lives could not be restored, yet a great part of the immense riches that were in the ruins might have been recovered; but a new calamity soon put an end to such hopes; for, in about two hours after the shock, fires broke out in three different parts of the city, caused by the goods and the kitchen fires being all jumbled together. About this time also, a fresh gale suddenly springing up, made the fire rage with such violence that, at the end of three days, the greatest part of the city was reduced to ashes. What the earthquake had spared fell a prey to the fire, and the flames consumed thousands of mutilated victims, who, incapable of flight, lay half buried in the ruins.

According to a popular report, which, true or not, shows

* Philosophical Transactions, vol. xlix. part i. p. 404.

the hatred in which the Holy Office was held, the Inquisition was the first building that fell down, and probably more than one inquisitor, who, in his life-time, had sent scores of Jews or heretics to the stake, was now, in his turn, burnt alive.

As if the unshackled elements were not sufficient agents of destruction, the prisons also cast forth their lawless denizens, and a host of malefactors, rejoicing in the public calamity which paralysed the arm of justice, added rapine and murder to the miseries of the city.

More than 60,000 persons are supposed to have perished in Lisbon from all these various causes. The total loss of property was estimated at fifty millions of dollars—an enormous sum for a small country, and in times, when money was far more valuable than at present. A few shocks sufficed to destroy the treasures accumulated by the savings of many generations.

The royal family was at this time residing in the small palace of Belem, about a league out of town, and thus escaped being buried among the ruins of the capital—a fortunate occurrence in the midst of so many misfortunes: for the anarchy that must have ensued from the destruction of all authority would have filled the cup of misery to the brim. As it was, Government seemed utterly incapable of contending with a disaster of such colossal proportions. 'What is to be done?' said the helpless king to his minister Carvalho, Marquis of Pombal, who, on entering the council-chamber, found his sovereign vainly seeking for advice among his weeping and irresolute courtiers: 'how can we alleviate the chastisement which divine justice has imposed upon us?' 'Sire! by burying the dead and taking care of the living,' was the ready answer of the great statesman, whose noble bearing and confident mien at once restored the king's courage. From that moment José bestowed a boundless confidence upon Pombal. Without losing a single moment, the minister, invested with full powers, threw himself into a carriage, and hastened with all speed to the scene of destruction. Wherever his presence was most needed, there was he sure to be found. For several days and nights he never left his carriage, whence, incessantly active in his efforts to reduce chaos to order, he issued no less than two

hundred decrees, all bearing the stamp of a master-mind.
Troops from the provinces were summoned in all haste, and
concentrated round the capital, which no one was allowed to
leave without permission, so that the robbers who had enriched themselves with the plunder of palaces and churches
were unable to escape with their spoil.

In all his numerous ordinances Carvalho neglected none
of the details necessary for insuring their practical utility,
writing many of them on his knees with a pencil, and sending them, without loss of time, to the various officers charged
with their execution. His wise regulations for ensuring a
speedy supply and a regular distribution of provisions
averted famine. Great fears were entertained of pestilential
disorders in consequence of the putrid exhalations of so
many corpses which it was impossible to bury. To prevent
this additional misfortune, Carvalho induced the Patriarch
to give orders that the bodies of the dead should be cast into
the sea, with only such religious ceremonies as circumstances
permitted.

But the Jesuits, the mortal enemies of the enlightened
minister, did not lose this opportunity of intriguing against
him, and openly ascribed the catastrophe to the wrath of
God against an impious Government. Thus Pombal had not
only to cope with the disastrous effects of the earthquake,
but also with the venomous attacks of hypocritical bigots,
in spite of whose clamours he interdicted all public pro-
cessions and devotional exercises that were calculated still
further to inflame the excited minds of the populace.

Though Lisbon was the chief sufferer from the great
earthquake of 1755, the shocks which destroyed the capital
of Portugal proved disastrous in many other places, and
vibrated far and wide over a considerable portion of the
globe. St. Ubes was nearly swallowed up by the sudden
rising of the sea. At Cadiz the shocks were so violent that
the water in the cisterns washed backwards and forwards so
as to make a great froth upon it. No damage was done, on
account of the excessive strength of the buildings ; but about
an hour after, an immense wave, at least 60 feet higher than
common, was seen approaching from the sea. It broke
against the west part of the town, which is very rocky, and

where, fortunately, the cliffs abated a great deal of its force. At last it burst upon the walls, destroyed part of the fortifications, and swept away huge pieces of cannon. The strong causeway which connects the town with the Island of Leon, was utterly destroyed, and more than fifty people drowned that were on it at the time.

In Seville a number of houses were thrown down, and the bells were set a-ringing in Malaga. In Italy, Germany, and France, in Holland, and in Sweden, in Great Britain and in Ireland, the lakes and rivers were violently agitated. The water in Loch Lomond rose suddenly and violently against its banks, so that a large stone lying at some distance from the shore, in shallow water, was moved from its place and carried to dry land, leaving a deep furrow in the ground along which it had moved. At Kinsale, in Ireland, a great body of water suddenly burst into the harbour, and with such violence that it broke the cables of two vessels, each moored with two anchors, and of several boats which lay near the town. The vessels were whirled round several times by an eddy formed in the water, and then hurried back again with the same rapidity as before. London was shaken, the midland counties disturbed, and one high cliff in Yorkshire threw down its half-separated rocks. At Töplitz, in Bohemia, between eleven and twelve o'clock, the mineral waters increased so much in quantity that all the baths ran over. About half an hour before, the spring grew turbid, and flowed muddy, and having stopped entirely for nearly a minute, broke forth again with prodigious violence, driving before it a considerable quantity of reddish ochre. After this, it became clear, and flowed as pure as before, but supplying more water than usual, and that hotter and more impregnated with its medicinal substances.

In Barbary, the earthquake was felt nearly as severely as in Portugal. Great part of the city of Algiers was destroyed; at Fez, Mequinez, and Morocco, many houses were thrown down, and numbers of persons were buried in the ruins. At Tangiers and Sallee the waters rushed into the streets with great violence, and when they retired they left behind them a great quantity of fish.

Ships sailing on the distant Atlantic received such violent

concussions that it seemed as if they had struck upon a rock, and even America was disturbed.

At the Island of Antigua the sea rose to such a height as had never been known before, and at Barbadoes a tremendous wave overflowed the wharfs and rushed into the streets. The remote Canadian lakes were seen to ebb and flow in an extraordinary manner, and the Red Indian hunter felt the last expiring pulsations of the great terrestrial shock which a few hours before had overthrown the distant capital of Portugal.

Such were the extraordinary effects of this terrible earthquake, which extended over a space of not less than four millions of square miles! Of the enormous sensation it produced over all Europe, as well as of the deep impression it made upon his own youthful mind, Goethe, then about six years old, has given us a masterly account in his autobiography ('Dichtung und Wahrheit').

'For the first time,' says the illustrious poet, 'the boy's peace of mind was disturbed by an extraordinary event. On November 1, 1755, the earthquake of Lisbon took place, and spread consternation over a world which had long been accustomed to tranquility and peace. A large and splendid capital, the seat of wealth and commerce, suddenly falls a prey to the most terrible disaster. The earth shakes, the sea rises, ships are dashed against each other, houses, churches, and towers fall in; the king's palace is partly engulfed by the waves; the bursting earth seems to vomit flames, for smoke and fire appear everywhere among the ruins. Sixty thousand persons, but a moment before in the enjoyment of a comfortable existence, are swept away, and they are the most fortunate who no longer feel or remember their misery. The flames continue to rage, along with a host of criminals whom the catastrophe has set at liberty. The unfortunate survivors are exposed to robbery, to murder to every act of violence; and thus on all sides Nature replaces law by the reign of unfettered anarchy. Swifter than the news could travel, the effects of the earthquake had already spread over a wide extent of land; in many places slighter commotions had been felt; mineral springs had suddenly ceased to flow; and all these circumstances increased

the general alarm when the terrible details of the catastrophe
became known. The pious were now not sparing of moral
reflexions, the philosophers of consolations, the clergy of
admonitions. Thus the attention of the world was for some
time concentrated upon this single topic, and the public,
excited by the misfortunes of strangers, began to feel an
increasing anxiety for its personal safety, as from all sides
intelligence came pouring in of the widely-extended effects of
the earthquake. The demon of fear has indeed, perhaps,
never spread terror so rapidly and so powerfully over the
earth. The boy who heard the subject frequently discussed
was not a little perplexed. God, the Creator and Preserver
of Heaven and Earth, whom the first article of faith represented as supremely wise and merciful, appeared by no
means paternal while thus enveloping the just and the unjust in indiscriminate ruin. It was in vain that his youthful
mind endeavoured to shake off these impressions; nor can
this be wondered at, as even the wise and the learned did
not agree in their opinions on the subject.'

CHAPTER XI.

LANDSLIPS.

Igneous and Aqueous Causes of Landslips—Fall of the Diablerets in 1714 and 1749—Escape of a Peasant from his living Tomb—Vitaliano Donati on the Fall of a Mountain near Sallenches—The Destruction of Goldau in 1806—Wonderful Preservation of a Child—Burial of Vallein and Tauretunum, of Plürs and Scilano—Landslip near Axmouth in Dorsetshire—Falling in of Cavern-roofs—Dollinas and Jamas in Carniola and Dalmatia—Bursting of Bogs—Crateriform Hollows in the Eifel.

LANDSLIPS, or sudden subsidences and displacements of portions of land, result both from igneous and aqueous causes.

Wherever cavities have been formed beneath the surface of the earth, whether in consequence of volcanic eruptions or by the erosive and dissolving action of subterranean waters, the shock of an earthquake or the mere weight of the superincumbent mass may cause the roof to fall in, or the superficial ground, no longer sustained by its undermined foundations, to slide away and sink to a lower level.

In mountainous regions it frequently occurs that the foundations of a rock, undermined by filtering waters, give way, and that huge masses of stone and earth, now no longer reposing on a solid basis, are precipitated into the valley below. More than once, the slipping or falling in of a mountain has brought death and destruction upon the humble dwellings of the Alpine peasants, and added many a mournful page to their simple annals. Thus, in the years 1714 and 1749, large beds of stone were detached from the Diablerets, a mountain stock between the cantons of Vaux and Valais, and burying the meadows of Cheville and Leytron under a mound of rubbish 300 feet deep, killed many herds and shepherds.

In the first of these catastrophes, the life of a peasant was preserved in a wonderful manner. An immense block came toppling down close to his châlet and covered it like a shield, so as to preserve it from being crushed by the following débris, though piled up two hundred feet above it. Thus, immured as it were in a living tomb, the unfortunate man spent miserable weeks and months, subsisting on the stores of cheese hoarded in his hut, without light and air, and in constant fear that the rocks above his head might give way and bury him under their ruins. With all the energy of despair, he endeavoured to find his way out of the mighty mound of rubbish, and at length, after incredible toil, emerged into the open daylight. More like a spectre than a human being, pale and emaciated, with torn clothes, and covered with bruises, he knocked at the door of his house * in the lower valley, where his wife and children, who had already long reckoned him among the dead, were at first terrified at his ghost-like appearance, and called in the village pastor to convince them of his identity, before they ventured to rejoice at his return.

On the road from Sallenches to Servoz, in the Valley of the Arve, well known to all the visitors of Mont Blanc, may be seen the ruins of a high mountain which collapsed in the year 1751, causing so dreadful a crash and raising such clouds of dust that the whole neighbourhood thought the world was at an end. The black dust was taken for smoke ; flames had been seen darting about in the murky clouds, and the report spread to Turin that a new Vesuvius had suddenly opened its subterranean furnaces among the highest of the Alpine mountains. The king, alarmed or interested at the news, immediately sent the famous geologist Vitaliano Donati to gather accurate information on the spot. Donati, travelling night and day, with all the eagerness of a zealous naturalist, arrived while the appalling phenomenon was still in full activity.

'The peasants,' writes Donati to a friend, 'had all fled from the neighbourhood, and did not venture to approach

---

* It is almost superfluous to mention that in the Alps many of the peasants lead a migratory existence. During the summer they ascend, with their herds, into the higher valleys, where they remain, separated from their families, until the first night-frosts force them to return to their homes on a lower level.

the crashing mountain within a distance of two Italian miles. The country around was covered with dust, which closely resembled ashes, and had been carried by the wind to a distance of five miles. I examined the dust, and found it to consist of pulverised marble. I also attentively observed the smoke, but could see no flames, nor could I perceive a sulphurous smell; the water also of the rivulets and sources showed no trace of sulphurous matter. This convinced me at once that no volcanic eruption was taking place, and penetrating into the cloud of dust which enveloped the mountain, I advanced close to the scene of the commotion. I there saw enormous rocks tumbling piecemeal into an abyss with a dreadful noise, louder than the rolling of thunder or the roar of heavy artillery, and distinctly saw that the smoke was nothing but the dust rising from their fall.

'Further investigations also showed me the cause of the phenomenon, for I found the mountain to consist of horizontal strata, the lowest being composed of a loose stone of a slaty texture, while the upper ones, though of a more compact nature, were rent with numerous crevices. On the back of the mountain were three small lakes, the water of which, penetrating through the fissures of the strata, had gradually loosened their foundations. The snow, which had fallen during the previous winter more abundantly than had ever been known within the memory of man, hastened the progress of destruction, and caused the fall of six hundred million cubic feet of stone, which alone would have sufficed to form a great mountain. Six shepherds, as many houses, and a great number of cows and goats, have been buried under the ruins.

'In my report to the king I have accurately described the causes and effects of the catastrophe, and foretold its speedy termination—a prediction which has been fully verified by the event—and thus the new volcano has become extinct almost as soon as its formation was announced.'

Fortunately, this grand convulsion of nature, which spread consternation far and wide, caused the death of but a few victims. The landslip of the Rufi or Rossberg, which, on September 2, 1806, devastated the lovely Vale of Goldau, and overwhelmed four villages, with their rich pasture-grounds and gardens, was far more disastrous. The preceding years had been unusually wet, the filtering waters had loosened the Nagelflühe, or coarse conglomerate of which the mountain is composed, and the rains having latterly been almost continuous, a great part of the mountain, undermined by the subterranean action of the waters, at length gave way and was hurled into the valley below.

Early in the morning the shepherds who were tending their herds on the mountains perceived fresh crevices in the ground and on the rock-walls. In many parts the turf appeared as if turned up by a ploughshare, and a cracking noise as if roots were violently snapped asunder was heard in the neighbouring forest. From hour to hour, the rents, the cracking, the rolling down of single stones increased, until finally, at about five in the afternoon, a large chasm opened in the flanks of the mountain, growing every instant deeper, longer, and broader. Then from the opposite Righi the forest might be seen to wave to and fro like a storm-tossed sea, and the whole flank of the mountain to slide down with a constantly increasing velocity, until finally hundreds of millions of cubic feet of rock came sweeping down into the valley with a noise as if the foundations of the earth were giving way. The friction or clash of the huge stones, hurled against each other in their fall, produced so intense a heat that flames were seen to flash forth from the avalanche, and the moisture with which they were saturated, being suddenly changed into steam, caused explosions like those from the crater of a volcano. Dense clouds of dust veiled the scene of destruction, and it was not before they slowly rolled away that the whole extent of the disaster became visible. Where, but a few hours since, four prosperous villages—Goldau, Busingen, Upper and Lower Röthen, and Lowerz—had been gilded- by the sun, and numerous herds had been grazing on the rich pastures along the borders of the lake of Lowerz, nothing was now to be seen but a desolate chaos of rocks, beneath which 457 persons lay buried. From this terrible disaster some wonderful escapes are recorded. High on the slope of the Rossberg, lived Bläsi Mettler, with his young wife Agatha. When, in the morning, the first premonitory signs of the disaster appeared, and the labouring mountain began to raise its warning voice, the superstitious peasant, fancying he heard the jubilee of demons, hastened down to Arth, on the bank of the lake of Zug, and begged the parish priest, with tears and lamentations, to accompany him, and exorcise the evil spirits with a copious sprinkling of holy water. While he was still speaking, the catastrophe took place, and he now rushed back again to his hut, where beyond all doubt his beloved wife and his only child, which was but four weeks old, had found a premature grave.

Meanwhile Agatha had spent several anxious hours. She was preparing her humble evening meal when the thundering uproar and the shaking of the hut filled her with the terrors of death. Seizing the infant, which lay awake in its cradle, she crossed the threshold, while the soil under her feet slid down into the valley. Escaping into the open air, she looked back and saw her hut and a sea of huge stone blocks roll down into the vale below, while the spot on which she stood remained unmoved. In this situation she was found by Bläsi, who, though a poor and ruined man, still thanked God for the wonderful preservation of his family.

About a thousand feet lower down the mountain lived Bläsi's brother Bastian, who, when the mountain slipped, was tending his herd on the opposite Righi. But his wife and her two little children were in his hut when it was buried beneath the stony avalanche. After the terrible commotion had subsided, the relations of Frau Mettler, anxious to ascertain her fate, hastened to the scene of desolation. The hut had disappeared, the green Alpine meadow was covered with a heap of ruins, but, not far from the former site of the humble cottage, the youngest child lay quietly sleeping. At the peril of his life, one of the infant's relations clambered over the ruins and rescued the little sleeper, who, unhurt amidst the falling rafters of the hut and the ruins of the crumbling mountain, had been carried away with the bed on which he was reposing. On my last visit to Switzerland, I was informed that ·Sebastian Meinhardt Mettler, the child thus wonderfully saved, died in the year 1867, at the age of sixtyone.

Some of the victims who had been buried in the ruins of the villages were dug out and restored to daylight; others, less fortunate, may have slowly perished, immured in a living grave; but by far the greater number were no doubt suddenly killed. The total number of those who were saved, either by the assistance of their friends, or by a timely flight, or by absence from their homes at the time of the disaster, amounted to 220; but more than double that number perished, and probably there was not one among the survivors who had not to lament the loss of friends and kinsfolk.

This dreadful catastrophe also levied its tribute among the strangers whom the beauties of Alpine scenery annually attract to Switzerland. A party of tourists had left Arth in the afternoon, with the intention of spending the night in Schwyz. Part of the company had already entered the ill-fated village of Goldau, and the others were about to follow, when suddenly the thundering roar of the sliding mountain caused them to stop. Looking up and seeing rocks, forests, huts, all rushing down in horrible confusion, they instinctively ran back for their lives. The warning came not one instant too soon, for close behind the spot where they stopped panting for breath, the stones still fell like hail. But their unfortunate companions, the wife and daughter of Baron Diesbach, Colonel Victor von Steiger, and some boys, whose tutor had been slowly following them with the Baron, were buried beneath the ruins.

From the Righi the traveller still looks down upon the avalanche of stones, and the flank of the Rossberg still plainly shows the spot where, more than half a century ago, the masses of rock now reposing in the valley detached themselves from the mountain. But the beautifying hand of vegetation has already done much to adorn the scene of ruin. Green mosses have woven their soft carpet over the naked stones, while grasses and flowers, and in some places even shrubs and trees, have sprung up between them. The tears also which once were shed over the victims of the great catastrophe have long since been dried, and its last witnesses have passed away to make room for a new generation, who remember the mountain-slip which buried their fathers only as a legend of the past.

This terrible disaster, however appalling through the far-spread desolation it entailed, has yet been equalled or surpassed by others of a like nature. In the fifth century, the old Roman town of Velleja was buried under the ruins of the Ravinazzo Mountain, and the bones and coins dug out of its ancient site prove that no time was left to the inhabitants for flight. Tauretunum was once a flourishing Roman town, situated on the south bank of the lake of Geneva, at the foot of the Dent d'Oche. In 563 it was utterly destroyed by a disruption of the overhanging mountain. The avalanche of stones which at that time was hurled down upon the devoted city is still visible as a promontory projecting far into the lake, which is here at least 500 feet deep. The immense wave caused by the rocky mass as it plunged into the water inundated the opposite shore from Morges to Vevay, and swept away every homestead that lay on its path.

In the night of September 4, 1618, the falling of the Monte Conto, in the Vale of Chiavenna, so completely buried the small town of Plürs and the village of Scilano, that of their 2,430 inhabitants but three remained alive, and but one single house escaped the universal destruction. At present, magnificent chestnut-trees grow upon the mound of ruins and cast their shade over the graves of the long-forgotten victims. Three villages, with their whole population, were covered in the district of Treviso when the Piz mountain fell in 1772; and the enormous masses of rock which in 1248 detached themselves from Mount Grenier, south of Chambery in Savoy, buried five parishes, including the town and church of Saint André, the ruin occupying an extent of about nine square miles.

Sometimes the same village has been repeatedly destroyed by mountain-slips. Thus excavations have shown that Brienz, a hamlet built on the borders of the lake of the same name, on a mound of accumulated ruins, has been twice overwhelmed by a deluge of stones and mud, and twice reconstructed.

It would be useless to multiply examples of the undermining power of water. I will merely add that it is impossible to wander through the valleys of Switzerland without being struck by the sight of the sloping hillocks of rubbish piled up against the foot of every gigantic rock-wall, which in many cases can only be attributed to that cause. Some are entirely overgrown with large firs, thus showing that the last stony avalanche took place at a remote period; others are desolate heaps of rubbish, which evidently prove that the work of destruction is constantly going on, and that the highest peaks will ultimately be levelled with the plain. Over many a hamlet the sword of Damocles is continually suspended in the shape of a precipitous rock-wall, or of a forest-crowned mountain-brow. For years the undermining waters are slowly and secretly at work, and then suddenly the crisis takes place.

AXMOUTH LANDSLIP.

Were the history of the Andes or of the Himalayas as familiar to us as that of the Alps, we should be able to relate many like instances of disastrous mountain-slips. But the high places of the earth do not alone bear witness to the power of aqueous erosion, for wherever the soil is undermined it may be precipitated to a lower level. Thus the pheno-menon is by no means uncommon in England, though rarely occuring on so large a scale as in the landslip which took place at Axmouth in Dorsetshire, on December 24, 1839.

' The tract of downs ranging there along the coast,' says Sir Charles Lyell ('Principles of Geology'), 'is capped by chalk, which rests on sandstone, beneath which is more than 100

feet of loose sand, the whole of these masses reposing on
retentive beds of clay shelving towards the sea. Numerous
springs, issuing from the loose sand, have gradually removed
portions of it, and thus undermined the superstratum. In
1839, an excessively wet season had saturated all the rocks
with moisture, so as to increase the weight of the incumbent
mass, from which the support had already been withdrawn
by the action of springs. Thus, the superstrata were precipitated into hollows prepared for them, and the adjacent
masses of partially undermined rock, to which the motion was
communicated, were made to slide down, on a slippery basis
of watery sand, towards the sea. These causes gave rise to
a convulsion, which began on the morning of December 24,
with a crashing noise : and, on the evening of the same day,
fissures were seen opening in the ground, and the walls of
tenements rending and sinking, until a deep chasm or ravine
was formed, extending nearly three-quarters of a mile in
length, with a depth of from 100 to 500 feet, and a breadth
exceeding 240 feet. At the bottom of this deep gulf lie fragments of the original surface, thrown together in the wildest
confusion. In consequence of lateral movements, the tract
intervening between the new fissure and the sea, including
the ancient undercliff, was fractured, and the whole line of
sea-cliff carried bodily forwards for many yards. This motion
of the sea-cliff produced a further effect, which may rank
among the most striking phenomena of this catastrophe.
The lateral pressure of the descending rocks urged the neighbouring strata, extending beneath the shingle of the shore,
by their state of unnatural condensation, to burst upwards in a
line parallel to the coast, so that an elevated ridge, more than
a mile in length and rising more than forty feet, covered by a
confused assemblage of broken strata and immense blocks of
rock, invested with seaweed and corallines, and scattered over
with shells and starfish, and other productions of the deep,
forms an extended reef in front of the present range of cliffs.'

Landslips caused by the falling in of cavern roofs are nowhere more common than in the cretaceous strata, which are
more liable than others to be undermined by the action of running waters. In the vast chalk-range extending from Carinthia
to the Morea, they occur of all sizes, from a diameter of a few

fathoms to one of many thousand feet, and are not seldom of considerable depth. They are generally funnel-shaped, sometimes elongated; and the bottom of the larger ones is generally covered with villages, orchards, vineyards, or considerable tracts of arable land. In Dalmatia, Carinthia, Carniola, and Istria, where the country consists chiefly of arid plateaux or mountain-chains, exposed to the dry north-easterly winds, the cultivation of the soil is almost exclusively confined to these depressions or *dollinas*, which, as a further protection against the cutting blasts, are inclosed with walls of loose stones.

Besides the funnel-shaped landslips or *dollinas*, there are others with perpendicular sides like walls or shafts, which are called *jamas* or mouths. One of these (near Breschiak) descends to a depth of 384 feet. The hares seek a winter refuge in the *dollinas*, and the *jamas*, as the favourite resort of pigeons, are also called pigeon-holes or *golubinas*. Many a pedestrian has lost his life by falling into a *jama*, particularly in former times, when fewer precautions were taken to protect the stranger against these treacherous precipices.

In the Jura Mountains there are also whole rows of cauldron-shaped depressions; and in North Jutland, where the chalk formation is likewise very extensive, a recent landslip suddenly emptied the Norr Lake, which lost itself in subterranean channels.

Effects very similar to those of an ordinary landslip are sometimes produced by the bursting of a bog. On the western confines of England and Scotland, the Solway Moss occupies a flat area about seven miles in circumference. Its surface is covered with grass and rushes, presenting a dry crust and a fair appearance; but it shakes under the least pressure, the bottom being unsound and semi-fluid. The adventurous passenger therefore, who sometimes, in dry seasons, traverses this treacherous waste, must pick his way over the rushy tussocks as they appear before him, for here the soil is firmest. If his foot slip, or if he venture to move in any other part, it is possible he may sink never to rise again.

On December 16, 1772, this quagmire, having been filled, like a great sponge, with water, during heavy rains, swelled to an unusual height above the surrounding country, and then burst. The turfy covering seemed for a time to act like the skin of a bladder retaining the fluid within, till it forced a passage for itself, when a stream of black half-consolidated mud began at first to creep over the plain, resembling in the slow rate of its progress an ordinary lava-current. No lives were lost, but the deluge totally overwhelmed some cottages, and covered 400 acres with a mass of mud and vegetable matter, which in the lowest parts of the submerged area was at least fifteen feet deep.

It may easily be imagined that in Ireland, the classical land of bogs, such phenomena are not uncommon. In the peat of Donegal an ancient log-cabin was found, in 1833, at the depth of fourteen feet. The cabin was filled with peat, and was surrounded by other huts, which were not examined. Trunks and roots of trees, preserved in their natural position, lay around these huts. There can be little doubt that we have here one instance out of many in which villages have been overwhelmed by the bursting of a moss.

In many volcanic regions we find circular cauldron-shaped depressions in the earth's surface, which might easily be mistaken for landslips, but which have in reality been formed by explosive discharges of confined vapours. When vents or fissures are produced by a paroxysm of volcanic energy, we can easily understand how in some cases the pent-up gases, finding a sudden outlet through some weaker part of the surface, must act like a powder mine, and scattering the rocks that surrounded the orifice, leave a deep hollow behind as a memento of their fury. The depressions thus caused bear a great resemblance to real craters, from which they are, however, distinguished by the absence of a cone of scoriæ and from their never having ejected lava.

These curious crateriform hollows are very common in the Eifel, a volcanic region in Rhenish Prussia, where, probably owing to the clayey nature of the soil, they have become reservoirs of water, or Maare, as they are called by the natives. Most of them still have small lakes at their bottom, while others have been drained for the sake of cultivation, or by the spontaneous rupture or erosion of their banks. Some of them are of considerable dimensions, such as that of

Meerfeld, the diameter of which falls very little short of a mile ; or the Pulvermaar of Gillenfeld, remarkable for the extreme regularity of its magnificent oval basin.

Similar lakes or Maare occur in Auvergne, in Java, in the Canary Islands, in New Zealand, and in the volcanic districts of Italy. The beautiful lakes of Albano and Nemi, which have been so often sung by ancient and modern poets, belong to this class; but Fr. Hoffman, a celebrated German geologist, ascribes the origin of the former to a landslip caused by the falling in of the roof of a vast subterranean cavern.

CHAPTER XII.

ON CAVES IN GENERAL.

Their various Forms—Natural Tunnels—The Ventanillas of Gualgayoc—Eimeo—Torgatten—Hole in the Mürtschenstock—The Trebich Cave—Grotto of Antiparos—Vast Dimensions of the Cave of Adelsberg and of the Mammoth Cave—Discovery of Baumann's Cave—Limestone Caves—Causes of their Excavation—Stalactites and Stalagmites—Their Origin—Variety of Forms—Marine Caves—Shetland—Fingal's Cave—The Azure Cave—Cave under Bonifacio—Grotta di Netuno, near Syracuse—The Bufador of Papa Luna—Volcanic Caves—The Fossa della Pulomba—Caves of San Miguel—The Surtshellir.

THE natural excavations which abound in many mountain chains, or on rocky shores washed by the stormy sea, are extremely various in their forms. Many are mere rents or crevices in the disruptured rocks; others wide vaults, frequently of hall or dome-like dimensions, or long and narrow passages branching out in numerous ramifications. Not seldom the same cave alternately expands into spacious chambers, and then again contracts into narrow tunnels or galleries. The walls of many are smooth and nearly parallel; the sides of others are irregular and rugged. Many have narrow entrances and swell at greater depths into majestic proportions; while others open with wide portals, and gradually diminish in size as they penetrate into the rock. Sometimes an excavation pierces a mountain from side to side like a natural tunnel, so as to allow a passage to the light of day. Such, among others, are the numerous perforations or *windows* (ventanillas) in the serrated bastions of the rich silver mountain Gualgayoc in the Peruvian Andes, or the opening through one of the high peaks of the romantic island of Eimeo which rises within sight of Tahiti out of the dark blue ocean. According to a popular tradition, this hole owes its origin to Oro, the powerful god of war, who, having one day quarrelled with the minor god of Eimeo, hurled his mighty spear at him over the sea. As even gods, when losing their temper, are apt to miss their aim, the puny delinquent escaped unhurt, while the dreadful lance flew like a thunderbolt through the mountain, leaving the perforation as a lasting memorial of its passage. In Europe we likewise meet with several remarkable instances of such natural tunnels. One of the most celebrated is the grotto of Torgatten in Norway, which perforates a huge rock, 400 feet above the level of the sea. Its proportions are truly colossal, as it is no less than 900 feet long by from 80 to 100 feet broad; and the arches of its vast portals measure respectively 200 and 120 feet. Its floor is nearly horizontal, and covered with fine sand; its sides are smooth, as if they had been chiselled by the hand of man. The sea, with its numberless cliffs and white-crested breakers, appears though the immense gallery as through the tube of a gigantic telescope, and in fine sunny weather affords a spectacle of incomparable beauty.

Whoever has visited the romantic lake of Wallenstädt, in Switzerland, will have had his attention directed to a tunnel near the summit of the almost inaccessible Mürtschenstock, a favourite resort of the chamois. It is visible from the lake near the hamlet of Mühlehorn, and, though of considerable dimensions, appears to the eye like a mere speck of snow on the huge grey rock-wall, which towers to a height of 7,517 feet. From the 1st to the 3rd of February, at two o'clock in the afternoon, the inhabitants of Mühlehorn see through this aperture the disk of the sun for the first time after a long winter.

In the structure of some caves a vertical direction predominates; as, for instance, in the Trebich Cave, three leagues from Trieste, which consists of several perpendicular shafts, connected by narrow transversal passages, and descending one after another, until finally, at a depth of more than a thousand feet, the cavern terminates in a wide vaulted space spanning a subterranean river. Such, also, is the renowned Grotto of Antiparos, into which the visitor is let down by a rope to a depth of about twenty fathoms. After reaching a tolerably even platform, he is obliged to descend another precipice, and then to proceed over slippery rocks until he finally reaches the terminal vault.

In most caverns, however, the chief direction is horizontal, either on several planes, separated from each other by more or less steep passages, or on a single level. The dimensions of caves are as various as their forms. Many are small and of inconsiderable depth—mere holes worn in the rock; while others are of a truly astonishing size, and fatigue the wondering spectator as he wanders through their lofty halls or endless galleries. The famous Cave of Adelsberg in Carniola has been explored to a distance of 1,243 fathoms from the chief entrance; and in the Mammoth Cave in Kentucky no less than 226 avenues branch out to the right and left from the main gallery, so as to form a network of subterranean passages and halls of various dimensions, whose total length has been computed at about 160 miles!

As many caves are without any visible communication with the external world, and the entrance of others is frequently narrow, and concealed behind rocks in solitary ravines on wild hill slopes or steep sea shores, far from the busy haunts of man, we cannot wonder that chance has frequently been instrumental in their discovery. Sometimes a hunter pursuing a wild animal has been led to the hidden cave in which it sought a refuge, or the workmen in a quarry have been suddenly surprised at meeting with a hollow in the rock, which opened an unexpected passage into the bowels of the mountain. The digging of wells, of cellars, of foundations, the boring for mines or Artesian wells, has often revealed the existence of unknown subterranean chambers; and so recently as 1868, one of the finest known caverns, which already attracts a number of delighted visitors, was discovered in the neighbourhood of the thriving manufacturing town of Iserlohn in Westphalia, on blasting a rock for the making of a railway. We may thus infer that a vast number of caves must still be totally unknown; many so situated that chance may one day lead to their discovery; while others are hollowed out at such vast depths in the earth-rind as to be for ever inaccessible to man.

Even of those caves which have been objects of curiosity

for centuries, many have still been by no means thoroughly explored. In the year 1848 an American gentleman persuaded the guides of Baumann's Cave in the Hartz Mountains to accompany him on a voyage of discovery through parts of the cavern hitherto untrodden by man. It was no easy task to clamber over slippery rocks and deep chasms yawning into black abysses; but curiosity and the spirit of adventure kept leading them on from passage to passage and vault to vault, when suddenly the lights began to burn more dimly; and the glass of the guiding compass having been accidentally broken warned them to retrace their steps. They had been wandering for twenty-four hours in the subterranean labyrinth, and after so long an absence from the light of day joyfully hailed the green hill slope which decks the mysterious palace of the gnomes. Franz Baumann, the first discoverer of the cavern, was less fortunate. Its tortuous windings confused the expert and intrepid miner, who lost his way in the recesses of the cave. While seeking in vain for an outlet, his sparing light went out. Three days he groped about in darkness, until at length, worn out and exhausted, he was led by a wonderful chance to the mouth of the cave. Before he died he had yet sufficient strength briefly to mention the wonders he had seen during his fatal expedition. His descendants still enjoy the privilege of serving as guides to the visitors of the cave, and never fail to relate the melancholy end of their ill-fated forefather.

Grottoes and caves occur in every kind of rock; in lavas, basalt, slate, and granite, as well as in limestone, dolomite, and gypsum; for the volcanic powers are capable of rending the hardest stone, and the foaming breakers of a turbulent ocean meet with no cliff that is able ultimately to resist their never-tiring assaults.

But, owing to their great fragility and to the solubility of limestone (carbonate of lime) in water containing carbonic acid, calcareous rocks are more liable than any others to be shattered and undermined, both by volcanic and aqueous causes. Its water readily absorbs carbonic acid gas. Every drop of rain that falls upon the ground necessarily contains some small portion of this gas, which, as we all know, is constantly mixed with the atmosphere, and thus becomes a solvent for chalk; more particularly if the latter, as, for instance, in the Karst Mountains of Carniola, contains some proto-carbonate of iron, which, changing into an oxide when

ENTRANCE TO THE CAVE OF ADELSBERG.

in contact with water, yields its carbonic acid to the percolating fluid, and consequently increases its solvent powers. Hence every shower of rain that filters through the crevices of a limestone rock wears away some part of its mass; and if we consider the vast number of years over which these

operatious have extended, and add to their effects the transporting powers of the waters on their progress through the
subterrauean channels which they have excavated or enlarged,
we can easily comprehend how in the course of ages whole
mountains may be hollowed out.

As the streams that flow on the surface of the earth are
constantly altering their course, thus also the subterranean
waters are ever active in excavating new channels in the
bosom of the rock. Finding at length new outlets on a
lower level, they abandon their ancient beds, and the explorer now wanders dry-footed where once a foaming river
gushed along. The Cave of Adelsberg is a remarkable
example of the changes which the subterranean waters,
aided by time or by the disrupting power of earthquakes,
may thus bring about; for the Poik now flows beneath its
galleries in the same north-easterly direction, in a channel
which is for the greatest part unknown and inexplorable, so
that the dry cave of the present day must evidently have
been the old river-bed. But nowhere can be found such
perfect, unequivocal, and abundant proofs of the action of
running water in corroding and excavating new passages
in a soluble rock as in the huge Mammoth Cave. The
rough-hewn block in the quarry does not bear more distinct
proof of the hammer and the chisel of the workman than
these interminable galleries afford of its denuding and
dissolving power. At Niagara we see a vast chasm evidently
cut by water for seven miles, and still in progress; but we
cannot see beneath the cataract the water-worn surface, nor
the rounded angles of the precipice; while the frosts and
rains of countless winters have reduced the walls of the
chasm itself to a talus of crumbling and moss-grown rocks.
But in the Mammoth Cave we see a freshness and perfection
of surface such as can be found only where the destructive
agencies of meteoric causes are wholly absent. Here we
have the dry beds of subterranean rivers exactly as they
were left thousands of years ago by the streams which
flowed through them when Niagara was young. No angle
is less sharp, no groove or excavation less perfect, than it
was originally left when the waters were suddenly drained
off by cutting their way to some lower level. The very sand

and rounded pebbles, which now pave the galleries and which anciently formed the bed of the stream, have remained in many of the more distant galleries untrodden even by the foot of man. 'The rush of ideas was strange and overpowering,' says Professor Silliman, 'as I stood in one of these before unvisited avenues, in which the glow of a lamp had never before shone, and considered the complex chain of phenomena which were before me. There were the delicate silicious forms of cyathophylli and eucrinites protruding from the softer limestone, which had yielded to the dissolving power of the water; these carried me back to that vast and desolate ocean in which they flourished, and were entombed as the crystalline matrix was slowly cast around them, mute chroniclers of a distant epoch. Then succeeded the long periods of the upper secondary, and, these past, the slow but resistless force of the contracting sphere elevated and drained the rocky beds of the ancient ocean. The action of the meteorological causes commenced, and the dissolving power of fresh water, following the almost invisible lines of structure in the rocks, began to hollow out these winding paths slowly and yet surely.' What a lesson for the thoughtful spectator, and how vast a prospect into the dark abysses of the past here unrolls itself before him!

After abandoning the vaults where they once collected and formed a running stream, the waters, filtering through the porous limestone, begin to ornament them with lustrous petrifactions; for whether below or above the surface of the earth, Nature ever loves to decorate her works. The moisture, charged with carbonate of lime, evaporates or parts with its free carbonic acid in coming into contact with the air of the cave; the carbonate, now no longer held in solution, precipitates and forms calcareous incrustations or excrescences, which in course of time assume every variety of fantastic shape, either hanging like icicles from the vault (stalactites), or rising in columns (stalagmites) from the floor of the cave where the dripping water deposited its spar. Sometimes stalactities and stalagmites join as they continue to grow in opposite directions, and ultimately form pillars which appear to sustain the roof.

On considering the simple physical and chemical agencies

which are at work in the formation of these beautiful productions—solution, mechanical dripping, evaporation, and precipitation—a great similarity might naturally be expected in their forms; but here also Nature shows herself as a consummate artist, and with the simplest means brings forth an astonishing variety of effects. As among the leaves of a forest there are not two perfectly alike, thus also every stalactite differs from another; and the celebrated traveller Kohl affirms that every stalactital cave has its peculiar style or character of decoration. The causes to which stalactites owe their existence are indeed everywhere the same, but the circumstances under which the drops fall and evaporate are so various that in each case some new shape is produced. Thus all the infinite diversity of form which we admire in the corals and sponges of the seas, is wonderfully repeated in the dark vaults of the subterranean world.

The variety and beauty of their colouring likewise contribute to adorn these formations. They are generally white, sometimes rivalling the purity of snow, and translucent, even when of considerable thickness, but often also green, brown, yellow, red, orange—a variety of tints which produces the most pleasing effects, and is chiefly owing to the metallic salts with which the water has been impregnated while filtering through the calcareous rock.

All these wonderful plays of Nature, in which form and colour contribute to delight the eye or to charm the fancy, are, however, still less interesting than the reflections suggested by the slow growth of stalactites in general, and the enormous size which some of them attain.

Inscriptions seventy or eighty years old appear covered only with a thin translucent coat of sinter, and in the Cave of Adelsberg names scratched in the walls more than six centuries ago are still perfectly legible. How many ages must, then, have passed before such colossal stalagmites could have been formed as, for instance, in the Australian cavern explored by Mr. Woods,* or in the Cave of Corncale, near Trieste, where we find one of these formations measuring fifty feet in circumference, and another rising thirty-five feet above the ground, with a trunk as massive as that

* 'The Geology of South Australia.'

of an old oak.   The ruins of Thebes, or the rock-temples of
Ipsamboul, appear almost as works of the present day when
compared with those amazing monuments of time.   But,
while meditating on their colossal dimensions, the mind
is necessarily carried still further back, and wanders through
the countless ages which the filtering waters, collecting into

STALACTITAL CAVERN IN AUSTRALIA.

subterranean streams, required for hollowing out the vast
cavities on whose floor those gigantic stalagmites were subsequently deposited.   An epoch of still older date presents
itself when the limestone rocks, now pierced with vast stalactital caverns, were first slowly forming at the bottom of
the primeval sea by the accumulation of countless exuviæ
of zoophytes, star-fishes, and foraminifera, and after growing

into strata many hundred feet thick, were then forced upwards by plutonic powers, and became portions of the dry
land. Nor is this the end of the vast perspective, for
changes still more remote loom in the fathomless distance.
The mind grows giddy while thus plunging into the abyss
of time, and, in spite of the ideas of sublimity awakened
by such meditations, feels a painful sense of its incapacity
to conceive a plan of such infinite extent.

While on land the running or filtering waters restlessly
pursue their work of excavation, the tumultuous waves of
the ocean impress on every rocky shore the seal of their
tremendous power. As, day after day, and year after year,
the billows strike against the cliffs that oppose their
progress, they undermine their foundations, scoop out wide
portals in their projecting headlands, and hollow out deep
caverns. Here also water appears as the beautifying
element, decorating inanimate nature with picturesque
forms; and the sea nowhere exhibits more romantic scenes
than on the bold coasts against which her waves have been
beating for many a millennium. During the calm ebb tide
seals are often seen sunning themselves at the entrance of
the oceanic grottoes, while cormorants stand before them
as guardians of the dark galleries beyond; the waves
murmur in softer strains, and the screeching sea-mew glides
with his silvery pinions through the tranquil air; but when
the stormy flood batters against the coast, the billows rush
into the caverns, scaring all animal life away, and no voice
is heard but that of the tumultuous ocean.

Our coasts abound in beauties such as these, particularly
on the wild shores of Shetland or the stormy Hebrides—

> ' Where rise no groves and where no gardens blow,
> Where even the hardy heath scarce dares to grow;
> But rocks on rocks, in mist and storm arrayed,
> Stretch far to sea their giant colonnade.'—Scott.

Along the coast of the mainland of Shetland and the
neighbouring islets of Bressay and Noss, cape follows upon
cape, consisting of bold cliffs hollowed into caverns, or
divided into pillars and arches of fantastic appearance, by
the constant action of the waves. As the voyager passes

the most northerly of these headlands, and turns into the open sea, the scenes become yet more sublime. Rocks, upwards of three or four hundred feet in height, present themselves in colossal succession, sinking perpendicularly into the sea, which is very deep, even within a few fathoms of their base. All these huge precipices abound with caves, many of which run much farther into the rock than the boldest islander has ever ventured to penetrate. One of these marine excavations, called 'The Orkneyman's Harbour,' is remarkable for the circumstance of an Orkney vessel having once run in there to escape a French privateer. Sir Walter Scott, who visited this interesting spot, found the entrance lofty enough to admit his six-oared boat without striking the mast, but a sudden turn in the direction of the cave would have consigned him to utter darkness if he had gone in further. The dropping of the sea fowl and cormorants into the water from the sides of the cavern, when disturbed by his approach, had something in it wild and terrible.

The shores of Caithness and of Sutherland, and of many of the islets in the Highland seas, likewise exhibit many wonderful specimens of the fantastic architecture of the ocean; but pre-eminent above all in grandeur and renown is Fingal's Cave.

Sir Walter Scott, who twice visited this celebrated grotto (in 1810 and 1814), pronounced it above all description sublime. 'The stupendous columns and side-walls, the depth and strength of the ocean with which the cavern is filled, the variety of tints formed by stalactites dropping and petrifying between the pillars, and resembling a sort of chasing of yellow or cream-coloured marble, filling the interstices of the roof—the corresponding variety below, where the ocean rolls over a red, and in some places a violet-coloured rock, the basis of the basaltic pillars—the dreadful noise of those august billows, so well corresponding with the grandeur of the scene—are all circumstances elsewhere unparalleled.'

In the Azure Cave of Capri, the Mediterranean possesses a marine grotto rivalling Fingal's Cave in celebrity, and no less wonderful in its peculiar style of beauty. As the roof of its narrow entrance rises only a few feet above the level of the sea, it is probable that no human eye had ever been delighted with its charms before 1826, when it was accidentally discovered by two Prussian artists who were swimming in the neighbourhood. After passing the low portal the cave widens to grand proportions, 125 feet long and 145 feet broad, and except a small landing-place on a projecting rock at the further end, its precipitous walls are on all sides bathed by the influx of the waters, which, in that sea, are so clear that the smallest objects may be distinctly seen on the bottom of the deep basin, the most beautiful bathing-place a mermaid might wish for. All the light that enters the grotto must first penetrate the whole depth of the waters before it can be reflected into the cave, and it thus acquires so blue a tinge, from the large body of clear water through which it has passed, that the walls of the cavern are illumined by a radiance of the purest azure. Had Byron known of the existence of this magic cave, Childe Harold would surely have devoted some of his most brilliant stanzas to its praise.

In many other parts of the Mediterranean the limestone rocks that fringe its shores have been worn into magnificent caverns, less singular, indeed, than the fairy grot of Capri, but still of rare and wonderful beauty. Such, among others, is the Antro di Nettuno, in the island of Sardinia, about twelve miles from the small seaport of Alghero.*

Exceedingly picturesque caverns have also been worn by the chafing waters in the chalk cliffs under Bonifacio, in the island of Corsica. Their entrances festooned with hanging boughs, they penetrate far into the interior of the rocks, and the water percolating through their vaulted roofs has formed stalactites of fantastic shapes. The boat glides through the arched entrance, and the glaring sunshine without is replaced by cool and grateful shade. Fishes are flitting in the clear water; limpid streams oozing through the rocks form crystal basins with pebbly bottoms; and the channels from the blue sea, flowing over the chalk, become cerulean. Poetic fancy has never pictured anything more enchanting than these lovely caves.

' The Sea and its Living Wonders,' 3rd edit. p. 49.

The rocky coast of Sicily is likewise hollowed out with numerous marine grottoes, which, though rarely noticed by travellers, may well be ranked amongst the greatest natural beauties of the island. One of the most remarkable is the Grotta di Nettuno, near Syracuse, which, in calm weather, admits a boat to a considerable distance. Its rugged vaults rise to a height of about twenty or thirty feet, and are covered with stalactites wherever the water does not reach. There is no landing-place, and throughout the whole cave the water is as deep as in the open sea beyond. Nothing

CAVE UNDER BONIFACIO.

can be more charming than to look back from the dark recesses of the grotto upon the bright sunshine without, and to listen to the soothing murmurs of the clear waves as they ripple against the rocky walls. The atmosphere is so pure in this delicious climate that not a trace of fog or mist obscures even the remotest parts of the cave, and the serene daylight falling through the entrance renders even its deepest shadows translucent. Here a lover of nature might linger for hours enjoying the most delicious coolness, and watching the charming effects of light and shade in their ever-varying play.

L

On many rocky shores the ocean has worn out subterranean channels in the cliffs, against which it has been beating for ages, and then frequently emerges in water-spouts, or fountains, from the opposite end. Thus in the Skerries, one of the Shetland Islands, a deep chasm or inlet, which is open overhead, is continued underground, and then again opens to the sky in the middle of the island. When the tide is high, the waves rise up through this inland aperture, with a noise like the blowing of a whale.

Similar phenomena occur on the south side of the Mauritius,[*] on the north coast of Newfoundland, near Huatulco on the Mexican coast of the Pacific, and near Peniscola in Spain, where a cave, through whose roof at storm tides the sea bursts with a terrific noise, has received the name of the 'Bufador, or the water-spout of Pope Luna,' the family name of Benedict XIII., who, having been deposed by the Councils of Pisa and Constance, retired to the small Spanish town where he was born. As the chief occupation of the holy father in exile was to vent a continuous torrent of curses and excommunications upon his numerous enemies, it is probable that this circumstance caused his name to be given to the noisy but harmless Bufador.

Though water, aided by time, is probably the chief excavating power, there can be no doubt that the action of subterranean fire has likewise produced, and still produces, many hollows in the hard crust of the earth. Wherever a volcano has been piled up to the skies, the matter ejected from its vents must necessarily have left a void behind, and given rise to corresponding cavities in the space beneath. The shock of an earthquake must frequently rend asunder deep-seated rocks, and the slow upheaval of considerable tracts of land can hardly be supposed to take place without the formation of hollows and crevices.

When the lavas poured forth during an eruption are in a liquid state, they do not form on cooling a compact homogeneous mass, but generally exhibit a porous, spongy texture, due to the bubbles of the vapour generated through, or entangled in, their mass. These bubbles frequently unite in

---

[*] 'The Sea and its Living Wonders,' 3rd edit. p. 52.

larger volumes, which, influenced by their elasticity and inferior specific gravity, rise towards the surface of the lava as it flows on, and, when sufficiently powerful, raise its crust in dome-like or conical protuberances, which not seldom burst open at the summit, or crack at the sides. The hollows thus formed are often so large as to entitle them to the name of caves.

According to Sir Charles Lyell, the sudden conversion into steam of lakes or streams of water, overwhelmed by a fiery current, may perhaps explain the formation of many of the extensive underground passages or caverns which form a common feature in the structure of a volcano. Great volumes of vapour, thus produced, may force their way through liquid lava, already coated over externally with a solid crust, and may cause the sides of such passages, as they harden, to assume a very irregular outline. The famous cave on Etna, called the Fossa della Palomba, which opens near Nicolosi, not far from Monte Rossi, has not improbably been thus formed. After reaching the bottom of a hollow 625 feet in circumference at its mouth, and 78 feet deep, the explorer enters another dark cavity, and then others in succession, sometimes descending precipices by means of ladders. At length the vaults terminate in a great gallery ninety feet long, and from fifteen to fifty feet broad, beyond which there is still a passage never yet visited, so that the whole extent of the Fossa remains unknown.

The volcanic caves of Punta Delgada in the Island of San Miguel, one of the Azores, are still more grand. Their entrance is through a narrow crevice, which soon, however, expands into an enormous hall, whose vault even the strongest torch-light is unable to illumine. In one spot, an opening in the floor shows that the lava, which is here but a foot thick, forms the roof of a second cave, situated below the first, into which even the boldest explorer has never ventured, but the noise of stones cast into the abyss proves it to be of considerable size. The first-mentioned cave leads into another, the width of which is estimated by Webster * at 120 feet; but the height he was unable to measure. Gradually this cave becomes narrower and lower, until, about

' Description of the Island of Saint Michael.'

L 2

45C feet from the entrance, it terminates in a low vault.
Black lava stalactites everywhere hang down from the roof,
and the floor is so covered with sharp-sided blocks of the
same volcanic material that walking among them becomes a
task of the greatest difficulty.

While a lava-stream is flowing along, a slag-crust forms
on its surface, inclosing the internal fluid matter as in a
canal. But when the supply of fresh lava from the vent
diminishes or entirely ceases, the still liquid interior at the
central part of the current continuing for some time to flow
on, often leaves behind hollow gutters, arched over by a
thin and brittle roof—so thin sometimes as to yield to the
weight of a person stepping on it. Such vaulted roofs have
pseudo-stalactitic projections left by the subsidence of the
liquid, and are coated with a glossy varnish. Sometimes
very large caverns are thus formed beneath the surface of a
lava-stream, and even rival in their extent and windings the
caves worn by water in limestone rocks. The famous Surtshellir,* situated near Kalmanstunga in Iceland, is a remarkable instance of a vast lava excavation owing its origin to
this cause. It has very appropriately been named after Surt,
the prince of darkness and fire, of the ancient Scandinavian
mythology; for this gloomy deity could not possibly have
chosen a fitter residence than its vast and dismal halls, once
glowing with subterranean fires, and now the seat of perpetual darkness.

* ‘The Polar World,’ p. 58.

CHAPTER XIII.

CAVE RIVERS.

The Fountain of Vaucluse—The Fontaine-sans-fond—The Katabothra in Morea—Subterranean Rivers in Carniola—Subterranean Navigation of the Poik in the Cave of Planina—'The Stalactital Paradise'—The Piuka Jama.

WHEREVER large bodies of water gush forth in a rapid stream from the bowels of the earth, they must either have flowed through wide underground channels or they must come from extensive lake-like reservoirs, for the mere drainage of a porous stratum is evidently incapable of accounting for their production.

Thus the celebrated fountain of Vaucluse, near Avignon, which has the volume and power of a river at its very source, is undoubtedly fed by a subterranean sheet of water of considerable extent. Even when least abundant, it pours forth upwards of 13,000 cubic feet of water in a minute; and after the country has been flooded with abundant rains, this volume is increased fourfold. The environs of the fountain are extremely picturesque, and justify the praises which have been lavished upon them in the immortal strains of Petrarch. It fills a large oval basin, vaulted by a spacious cave, and its waters, which, when low, escape through subterranean channels into the deep bed of the Sorgue, rise, when high, over the rock-wall at the mouth of the grotto, and form a broad cataract, rushing down with a dreadful noise.

Near Sable in Anjou, a source, or rather a pit from eighteen to twenty-four feet in diameter, well known in the country under the name of the Bottomless Fountain (*Fontaine-sans-fond*) sometimes overflows its brink, and then casts forth a large quantity of fish, so that it is evidently a mere aperture in the vault of a large subterranean pool.

In the department of the Haute Saône, another pit, called

the Frais Puits, presents a similar phenomenon. After abundant rains, the water gushing forth from its mouth inundates the neighbourhood, and on its retiring pikes and other fishes are not seldom found scattered over the surface of the flooded meadows.

Boswell mentions a visit which he paid with Dr. Johnson to Islam, formerly the seat of the Congreves, where two rivers burst near each other from the rock, after having run for many miles underground. Plott, in his 'History of Staffordshire,' gives an account of this natural curiosity, but Johnson would not believe it, in spite of the gardener's attestation, who said he had put in corks where the river *Manyfold* sinks into the ground and had caught them in a net placed before one of the openings where the water bursts out.

In the vast limestone formation which, under various names, extends through Carinthia, Carniola, Istria, Dalmatia, Albania, and Greece, the whole country is perforated like a sponge by an intricate system of subterranean water-courses.

In the more elevated districts of the Morea there are many deep land-locked valleys or basins inclosed on all sides by mountains of cavernous limestone. When the torrents are swollen by the rains, they rush from the surrounding heights into these basins; but, instead of forming temporary lakes as would be the case in most other countries, they are swallowed by chasms, which are sometimes situated in the middle of the plain, constituting the bottom of the closed basin, but more commonly at the foot of the surrounding escarpment of limestone. During the dry season, which in Greece alternates almost as distinctly as between the tropics with a period of rain, these chasms are the favourite retreats of wild animals.

Sometimes, in the limestone formation, the same stream repeatedly gushes forth from some cavernous recess, and then again disappears. The caves of Adelsberg, Planina, and Upper Laibach in Carniola, are traversed by the same river, which, losing its name every time it plunges into a new subterranean channel, is called, first, the Poik, then the Unz, and finally the Laibach. In the same manner the Temenitz, an affluent of the Save, thrice disappears under the earth, and thrice emerges as a new-born river with another name.

As far as these subterranean streams have been explored, their course exhibits a wonderful variety of interesting underground scenery. Sometimes they form high cataracts, leaping over rocks so picturesquely grouped that, were they illumined by the sun, and of more easy access, they would be admired by numberless tourists; and not seldom they expand into dark and melancholy lakes. Sometimes they with difficulty force a passage through a chaos of rocks, and then again they flow gently in a deep and even channel, so as to be navigable to a considerable distance. Generally, not the least breath of air sweeps over their placid waters, but sometimes their surface is rippled by the wind pouring in through some unseen chasm.

Among the bold explorers who have launched forth their barques on unknown subterranean rivers, the late Adolph Schmidl, of Vienna, holds a conspicuous rank. In a canoe specially constructed for the purpose he trusted himself to the dark streams of Carniola, which rewarded his adventurous zeal with many a scene of incomparable beauty, where the water-spirits and the gnomes seemed to have rivalled each other in the work of decoration. To give an idea of the difficulties and of the enjoyments of these subterranean explorations, we will follow the intrepid naturalist on his voyages of discovery through the famous Cave of Planina, through which flows the Poik, a river which is at all times deep enough to carry a boat. The course of the navigation is stream-upwards, and consequently much safer than would otherwise have been the case; but in many places the rapidity of the current calls for great caution, and considerable strength is needed to overcome its violence; while at the same time great care must be taken to avoid striking against the rocks that lie hidden under the water. As far as the end of a magnificent dome, situated above 600 feet from its entrance, the cave can be traversed on foot; but here the sullen stream, completely filling its whole width, compels the explorer to trust to his canoe. When he has passed a portal about eight fathoms high and half as broad, with proportions as symmetrical as if it had been sculptured by the hand of man, the thundering roar of a distant cataract announces still grander scenes. The portal widens, and

the astonished explorer suddenly emerges on a lake 250 feet
long and 150 feet broad, beyond which the cave is seen to
divide into two arms, giving passage to two streams whose
confluent waters form the lake. This broad sheet of water
affords an imposing but melancholy sight. The walls of the
cave rise everywhere abruptly out of the water, with the
exception of one small landing-place opposite to the portal at
the foot of a projecting rock or promontory. Here and there
large masses of stalactite hang like petrified cascades from
the rocks, which are generally naked and black. The vault
is so high that the light of a few torches fails to pierce its
gloom, which is rendered still more impressive by the roar
of the waterfall in the left branch of the cavern.

As far as the lake, the cave is of comparatively easy access,
and has been repeatedly visited, but the subterranean course
of the two brooks beyond was first explored by Dr. Schmidl.
In the left or western branch of the cave, into which he
penetrated to a distance of more than a mile, his boat had to
be unloaded no less than eleven times on account of the reefs
that obstructed its passage, while the explorers, wading
through the water, dragged it over the shallows. Once even,
where the navigation was interrupted by large masses of
rock, under which the tumultuous waters disappear with a
dreadful roar, they were obliged to take the little shallop to
pieces, and to reconstruct it on the opposite side of the
mound. The navigable part of this western branch ends in
a circular dome, the floor of which is entirely filled with a
lake 180 feet long, and from 40 to 45 feet deep. On the
western bank of this lake a chasm opens at the top of a
mound of rubbish, the only place where it is possible to land.
A violent gust of wind descends from this chasm, which,
sloping upwards, soon narrows to a small crevice, through
which the current of air sets in.

To a lateral gallery, opening beyond the mouth of the
chasm, Dr. Schmidl gave the name of 'The Stalactital
Paradise,' on account of the uncommon beauty of the spar-
crystals with which its walls were incrusted. It was the first
time that the foot of man had ever penetrated into this charming laboratory of nature ; no torch had ever soiled its brilliant
decorations ; no profane hand had ever damaged its gem-like

tapestry. Here whole groups of stalagmitic cones, of all
shapes and sizes, some like tiny icicles, others six feet high
and as thick as a man's waist, rose from the ground, while
further on the brown wall formed a dark background, from
which projected in bold relief the colossal statue of a scep-
tered king. Near the entrance stood a magnificent white
figure, which fancy might have supposed to be a cherub with
a flaming sword, menacing all those who should dare to
injure the wonders which he guarded.

'"The Stalactical Paradise" remained intact,' says Dr.
Schmidl.* 'I begged my companions not to strike off the
smallest piece of spar as a memorial of our visit, and they
all willingly consented. Our feet carefully avoided trampling
down any of its delicate ornaments; we left it with no other
memorial than our admiration of its beauty. The nymphs
of the grot will no doubt have pardoned us for having
intruded upon the sanctuary, where for countless centuries
they had reigned in undisturbed solitude and peace.'

The eastern branch of the cave, through which the main
stream flows, is much larger than the branch above described;
it is also easily navigated, as it contains but two reefs and a
small number of cliffs. On first ascending the stream, the
continually increasing roar of waters announces a considerable waterfall. Enormous masses of stone, piled up by the
falling in of the roof, have blocked up and narrowed the bed
of the river to fifteen feet, and cause the stream to shoot down
in a broad sheet ten feet high. The cataract, madly rushing
over the jet-black rocks and casting up flakes of milk-white
foam, is very beautiful, and, when brightly illuminated, must
produce a truly magical effect.

Beyond the cataract the river flows for a short space in an
invisible channel, as its waters are completely hidden under
rocks. It was no easy task to carry the planks of the dis-
membered boat over these rugged blocks of stone; but after
reconstructing it on the opposite side of the mound, and overcoming the minor obstacles of a couple of reefs, the river was
found to flow in a deep channel between steep walls, and a
free navigation opened to a distance of at least a league and
a half.

* 'Die Höhlenkunde des Karstes.' Wien, 1854.

'No description,' said Dr. Schmidl, 'can do justice to the fascination of this subterranean voyage. In some parts the roof is adorned with coral-shaped draperies of snow-white stalactites, but generally the walls are mere black, naked stone. Here and there sources gurgle down their sides, and, along with the melancholy trickling of single drops of water from the vault, alone break the silence of the dark interminable cave. The breathless attention we bestowed on the guidance of our boat and on the wonders that surrounded us sealed our lips, and we glided silently along through the dark waters, that now, for the first time since they began to flow, reflected the glare of a torch.'

Throughout the whole distance of 1,140 fathoms beyond the second reef, there is but one landing-place; everywhere else the walls rise precipitously from the water. In some parts the roof descends so low that the explorers were obliged to lie down in the boat and to shove it along by holding to the projections of the vault, which finally left but a few inches' space above the water, and thus opposed an invincible obstacle to all further progress.

In another grotto—called the Piuka Jama—the Poik again flows in the midst of the grandest subterranean scenery. About a league to the north of Adelsberg, the wanderer, after traversing a thicket of underwood, suddenly finds himself on the brink of a yawning precipice, from the bottom of which is heard distinctly the noise of a rushing stream. The walls of the chasm are almost perpendicular, except where a small ravine, overgrown with shrubs, leads to an enormous rock, on which it is possible to stand, and, if perfectly free from giddiness, to look down into the gulf below, where the huge portal of a cave is seen to open.

From this rock, which projects over the abyss, the only descent is by means of a rope. The bottom of the pit is covered with large blocks of stone irregularly piled up, and here one first sees the river rushing through the cave from right to left. The Piuka Jama may thus be compared to a window pierced through a vault overspanning a subterranean stream. Clambering down a heap of rubbish, the explorer at length stands upon the floor of the cave, and reaches the bank of the Poik. Stream-upwards, about 300 fathoms from the aperture, he meets with a rock gate, through which the river rushes so violently that a boat can master the current only when the water is unusually low.

After crossing this broad portal, the last faint traces of daylight glimmering from the distant aperture in the Piuka entirely disappear, and the scene suddenly changes. The expanding cavern assumes the proportions of an imposing dome. On its left side a mound has been formed by the falling in of the roof; but every block of stone is completely covered with calcareous incrustations of the purest white. From the floor to the centre of the vault millions upon millions of brilliant spars reflect the light: every hollow in the walls is a cabinet of gems. The background of the dome completes the beauty of the scene, and exhibits one of the most imposing cavern decorations it is possible to conceive. A monstrous pillar rises from its centre, forming two colossal ogival portals. The larger one is on the left, and at its entrance a mighty stalagmite, above twelve feet high, seems to forbid intrusion. The pillar itself and the vaults of both the portals are ornamented with the richest stalatical drapery.

When the river is swollen it rushes tumultuously through both the gates, where now Dr. Schmidl found but a scanty rill whispering and babbling among the stones.

CHAPTER XIV.

SUBTERRANEAN LIFE.

Subterranean Vegetation—Fungi—Enormous Fungus in a Tunnel near Doncaster
—Artificial Mushroom-beds near Paris—Subterranean Animals—The Guacharo
—Wholesale Slaughter—Insects in the Cave of Adelsberg—The Leptodirus and
the Blothrus—The Stalita tænaria—The Olm or Proteus—The Lake of Cirknitz—The Archduke Ferdinand and Charon—The Blind Rat and the Blind Fish
of the Mammoth Cave.

OF all the phenomena which attract the naturalist's attention as he wanders over the surface of the earth, there
is none which makes a deeper impression on his mind than
the omnipresence of life.   On the snow-clad cone of Chimborazo, 18,000 feet above the level of the sea, Humboldt found
butterflies and other winged insects, while, high over his
head, the condor was soaring in solitary majesty.   At the
still greater elevation of 18,460 feet, at the Doonkiah Pass
in the Himalaya Mountains, Dr. Hooker plucked flowering
plants, and saw large flocks of wild geese winging their flight
above Kunchinjinga (22,750 feet) towards the unknown regions of Central Asia.   Thus man meets with life as far as
he is able to ascend, or as far as his sight plunges into the
atmospheric ocean.   Besides the objects visible to his eye,
innumerable microscopical organisms pervade the realms of
air.   According to Ehrenberg's brilliant discovery, the im-
palpably fine dust which, wafted by the Harmattan, often
falls on ships when hundreds of miles from the coast of
Africa, consists of agglomerations of silica-coated diatoms,
individually so small as to be invisible to the naked eye;
and everywhere numberless minute germs of future life—eggs
of insects and sporules of cryptogamic plants—well fitted by
cilia and feathery crowns for an aërial journey, float up and
down in the atmosphere; while the waters of ocean are
found, in like manner, filled with myriads of animated atoms.

But organic life not only occupies those parts of our globe which are accessible to solar light; it also dives profoundly into the subterranean world, wherever rain, or the melted snow, filtering through the porous earth, or through vents and crevices, is able to penetrate into natural caverns or artificial mines. For the combination of moisture, warmth, and air is able to develop organic life even thousands of feet below the surface of the earth; while light, though indispensable to most creatures, would blight and destroy the inhabitants of the subterranean vaults.

On surveying the flora of these dismal recesses, we find it consisting exclusively of mushrooms or fungi, the lowest forms of vegetation, which, shunning the light, love darkness and damp. Their appearance in the caves is, as everywhere else, dependent upon the existence of an organic basis, and thus they are most commonly found germinating on pieces of wood, particularly when in a state of decomposition, which have been conveyed into the caverns either through the agency of man or by the influx of water. Species of a peculiarly luxuriant growth are sometimes seen to spread over the neighbouring stones, or apparently to spring from the rocky ground, where, however, on closer inspection, vestiges of decayed organic substances will generally be detected.

Thus vegetation in caves most commonly keeps pace with the quantity of mouldering wood which they contain, and flourishes not only near their entrance but in their deepest recesses; as, for instance, in the cave of Adelsberg, where, at a distance of more than a thousand fathoms from its entrance, the pegs which have been driven into the stalactital walls for the purpose of measuring its length are covered with a rich coat of fungi. Nothing can be more curious than to see these plants, thriving and luxuriating in deep stillness and gloom, under circumstances so alien to the ordinary conditions of life. Among the fungi found in caves, many also vegetate upon the surface of the earth exposed to the influence of light, and not seldom degenerate into monstrous forms in their less congenial subterranean abodes; but many are the exclusive children of darkness. The Austrian naturalist Scopoli published in 1772 the first exact description of more than seventy subterranean fungi, collected chiefly in the mines of Schemnitz and Idria; and about twenty years later Humboldt wrote his celebrated treatise on the same subject.[*] Since then G. F. Hoffmann has described the subterranean flora of the Hartz Mountains;[†] and latterly the botanists Welwitsch and Pokorny have examined the caves of Carinthia, where they discovered no less than eighteen species of fungi, among others the mouse-tail mushroom (*Agaricus myurus*, Hoffm.), which is also found in the Hartz, and bears on a slender hairy stalk, more than a foot long, a small hat, scarcely a quarter of an inch in diameter. Some of these fungi are remarkable for their size (*Telephora. rubiginosa, sanguinolenta*), others for their elegance (*Diderma nigripes*).

Some years ago a gigantic fungus, found growing from the woodwork of a tunnel near Doncaster, afforded a striking proof of the luxuriancy of subterranean vegetation. It measured no less than fifteen feet in diameter, and was, in its way, as great a curiosity as one of the colossal trees of California.

Even the plants that flourish in the darkness of caves have been rendered subservient to our use. The cultivation of the edible mushroom in spacious caverns or ancient quarries is practised to a great extent in the environs of Paris, at Arcueil, Moulin de la Roche, and St. Germain, but particularly at Montrouge, on the southern side of the city. The mushroom-beds are entirely underground, seventy or eighty feet below the surface, at a depth where the temperature is nearly uniform all the year round. These extensive catacombs, formed by long burrowing galleries, have no opening but by a circular shaft, to be descended by clambering down a perpendicular pole or mast, into the sides of which large wooden pegs are fixed, at intervals of ten or twelve inches, to rest the feet upon.

The baskets containing the ripe mushrooms are hoisted from below by a pulley and rope. The compost in which they grow consists of a white gritty earth, mixed with good stable manure, and is moulded into narrow beds about twenty inches high, ranged along the sides of the passages or gal-

---

[*] 'Flora Fribergensis Plantas Cryptogamicas præsertim subterraneas exhibens.'
[†] 'Vegetabilia in Hercyniæ Subterraneis collecta Norinbergæ.' 1811.

leries, and kept exquisitely neat and smooth. The mushroom-sporules are introduced to the beds either to flakes of earth taken from an old bed, or else from a heap of decomposing stable manure in which mushrooms have naturally been engendered. The beds are covered with a layer of earth an inch thick, the earth being merely the white rubbish left by the stone-cutters above. They must be well watered, and removed after two or three months, when their bearing qualities are exhausted. In one of the caves at Montrouge alone there are six or seven miles of mushroom-bedding—a proof that this branch of industry is by no means unimportant.

While subterranean vegetation is exclusively confined to mushrooms, animal life of almost every class has far more abundant representatives, for plants are in general much more dependent on the vivifying influence of light.

The various animals which are found dwelling in caves may be subdivided into two groups; one, which, though preferring darkness, and spending a great part of its existence under the earth, yet often voluntarily seeks the light of day, or at least wanders forth at night; while the other is *exclusively* subterranean, and is never seen above the surface of the earth, unless by chance or when driven out by violence.

To the first group belong most of the insectivorous and rodent quadrupeds that dwell in self-made burrows, or pursue a subterranean prey, such as the armadilloes and the moles. The large family of the bats likewise love to sleep by day, or to hibernate in warm and solitary caves, where they are sometimes found in numbers as countless as the sea-birds which flock round some rocky island of the north. When Professor Silliman visited the Mammoth Cave (October 16, 1822), he everywhere saw them suspended in dense clusters from the roofs, though a large number had not yet retired into winter-quarters. In a small space scarcely four or five inches square, he counted no less than forty bats, and convinced himself that at least one hundred and twenty find room on a square foot, as they held not only by the surface of the walls of their retreat, but by each other, one closely crowding over another. Such clusters are found in the interior of the cavern, which branches out in many directions

as far as two miles from the entrance, so that a very superficial survey allows them to be counted by millions. Who,
in these dismal regions, where no change of temperature or
of light announces the various seasons, tells them that the
reign of winter is past? who awakes them at the proper
time out of the deep sleep in which they remain plunged
for months? The same mysterious voice of instinct which
regulates the migrations of the birds and the wanderings of
the fishes, and which in this case, as in every other, is
equally wonderful and incomprehensible.

In the class of birds we find many cave-haunting species.
The pigeons like to nestle in grottoes, which also serve as
welcome retreats to the moping owl; and various swallows
and swifts breed chiefly in the darkness of caverns. One
of the most remarkable of these troglodytic birds is the
Guacharo, which inhabits a large cave in the Valley of Caripe,
near the town of Cumana, and of which an interesting
account has been given by Humboldt, who first introduced
it to the notice of Europe.

The *Cueva del Guacharo* is pierced in the vertical profile of
a rock, and the entrance is towards the south, forming a
noble vault eighty feet broad and seventy-two feet high. The
rock surmounting the cavern is covered with trees of gigantic
growth, and all the luxuriant profusion of an inter-tropical
vegetation. Plantain-leaved heliconias and wondrous orchids,
the Praga palm and tree arums, grow along the banks of
a river that flows out of the cave, while lianas, and a
variety of creeping plants, rocked to and fro by the wind,
form elegant festoons before its entrance. What a contrast
between this magnificently decorated portal and the gloomy
mouth of the Surtshellir, imbedded in the lava wildernesses
of Iceland? As the cave at first penetrates into the mountain in a straight direction, the light of day does not disappear for a considerable distance from the entrance, so that
visitors are able to go forward for about four hundred and
thirty feet without being obliged to light their torches; and
here, where light begins to fail, the hoarse cries of the nocturnal birds are heard from afar.

The guacharo is of the size of the common fowl. Its hooked
bill is wide, like that of the goat-sucker, and furnished at the

base with stiff hairs directed forwards. The plumage, like that of most nocturnal birds, is brownish grey, mixed with black stripes and large white spots. The eyes are incapable of bearing the light of day, and the wings are disproportionately large, measuring not less than four feet and a half from tip to tip. It quits the cavern only at nightfall, especially when there is moonlight; and Humboldt remarks that it is almost the only frugivorous nocturnal bird yet known, for it does not prey upon insects like the goatsucker, but feeds on very hard fruits, which its strong hooked beak is well fitted to crack. The horrible noise made by thousands of these birds in the dark recesses of the cavern can be compared only to the wild shrieks of the sea-mews round a solitary bird mountain, or to the deafening uproar of the crows when assembled in vast flocks in the dark fir-forests of the North. The clamour increases on advancing deeper into the cave, the birds being disturbed by the torch-light; and as those nestling in the side avenues of the cave begin to utter their mournful cries when the first sink into silence, it seems as if their troops were alternately complaining to each other of

Such as wand'ring near the secret bower
Molest their ancient solitary verge.

By fixing torches to the end of long poles, the Indians, who serve as guides into the cavern, show the nests of these birds, fifty or sixty feet above the heads of the explorers, in funnel-shaped holes with which the cavern roof is pierced like a sieve. Once a year, about midsummer, the Guacharo Cavern is entered by the Indians. Armed with poles they ransack the greater part of the nests, while the old birds, uttering lamentable cries, hover over the heads of the robbers. The young which fall down are opened on the spot. The peritoneum is found loaded with fat, and a layer of the same substance reaches from the abdomen to the vent, forming a kind of cushion between the birds' legs. The European nocturnal birds are meagre, as, instead of feasting on fruits and oily kernels, they live upon the scanty produce of the chase; while in the guacharo, as in our fattened geese, the accumulation of fat is promoted by darkness and abundant food. At the period above mentioned, which is known at Caripe as

M

the 'oil harvest,' huts are erected by the Indians with palm leaves near the entrance, and even in the very porch of the cavern. There the fat of the young birds just killed is melted in clay pots over a brushwood fire, and is said to be very pure and of a good taste. Its small quantity, however, is quite out of proportion to the numbers killed, as not more than 150 or 160 jars of perfectly clear oil are collected from the massacre of thousands.

The way into the interior of the cavern leads along the banks of the small river which flows through its dark recesses; but sometimes large masses of stalactites obstruct the passage, and force the visitor to wade through the water, which is, however, not more than two feet deep. As far as 1,458 feet from the entrance the cave maintains the same direction, width, and height of sixty or seventy feet, so that it would be difficult to find another mountain cavern of so regular a formation. Humboldt had great difficulty in persuading the natives to pass beyond the part of the cave which they usually visit to collect the oil, as they believed its deeper penetralia to be the abode of their ancestors' spirits; but since the great naturalist's visit, they seem to have abandoned their ancient superstitions, or to have acquired a greater courage in facing the mysteries of the grotto, for, while they would only accompany Humboldt as far as 236 fathoms into the interior of the cave, later travellers, such as Codazzi and Beaupertuis, have advanced with their guides to double the distance, though without reaching its end. They found that beyond the furthest point explored by Humboldt the cave loses its regularity, and has its walls covered with stalactites. In the embranchments of the grotto Codazzi found innumerable birds. It was formerly supposed that the guacharo was exclusively confined to this cave; latterly, however, it has also been found in the province of Bogota.

The discovery of animals destined to a life of perpetual darkness is but of modern date, and as the vast majority of caves have not yet been thoroughly explored by zoologists, the number of genera and species already known gives us reason to believe that future investigations will add considerably to their number. In the Adelsberg, Lueg, and Magdalena grottoes, which form but an inconsiderable part of the extensive cavernous regions of Carniola, seven exclusively subterranean insects—one spider, two scorpionides, one millepede, two crustaceans, one snail, and one reptile ; in all fifteen different species of animals, belonging to no less than six different classes—have been found.

Among these dwellers of the dark, warfare ·is as rife as in the regions of light. Thus, in the recesses of the Grotto of Adelsberg, the cavern beetle (*Leptodirus Hochenwartii*) is persecuted and devoured by the scorpioniform *Blothrus spelæus*, and by the eyeless spider (*Stalita tænaria*). The black and brown Leptodirus discovered in the

LEPTODIRUS HOCHENWARTII.

Grotto of Adelsberg in 1831 by Count Hochenwart, is distinguished by long and delicate antennæ and legs, and comparatively small translucent and smooth elytra. The unique specimen found at the time was unfortunately lost ; and although twenty-five florins were offered to the cavern guides for one of these beetles, fourteen years passed before it was re-discovered in the same cave. Since then other collectors have been more fortunate, particularly Prince Robert Khevenhüller, who, during his repeated visits to the Cave of Adelsberg, captured no less than twenty specimens of the Leptodirus.

Cautiously feeling its way with its long antennæ, the beetle slowly ascends the damp stalactital columns, and accelerates its movements at the approach of a light. The greater number were found in the evening, thus giving reason for supposing that the Leptodirus is a nocturnal beetle, although it is hardly possible to conceive how the alternating influence of night and day can still be felt in these regions of darkness. The manner in which it is pursued by the eyeless Blothrus (discovered in 1833 by Mr. F. Schmidt), has been several times observed by Prince Khevenhüller. He once saw one of these cavern scorpions slowly crawling along, stretching out its palpi in all directions, and evidently on the search. He immediately guessed that the animal was engaged in a hunting expedition, and soon found that he was not mistaken, for a fine Leptodirus was crawling about four feet higher on the opposite wall. For a long time the Prince left the two insects undisturbed, until he had thoroughly convinced himself that the movements of the Blothrus were evidently regulated by those of the Leptodirus, and that the former was, beyond all doubt, in pursuit of the beetle. A Leptodirus having been thrown along with a Blothrus into a phial, was immediately cut to pieces and devoured.

The eyeless cavern spider (*Stalita tonaria*), with brownish palpæ and a snow-white abdomen, is not seldom found in the hollows of the stalactites, lying in wait for the unfortunate Leptodirus. On the surface of the earth spiders are frequently obliged to fast for a very long time; but in caverns, where life is so sparingly distributed, the patience of the Stalita must be exemplary, even among spiders. Her appearance on the snow-white stalactital columns, where she only becomes visible when illumined by the full light of a taper,* is very striking. Like a vision, she sweeps away in her ivory robe, accompanied by her increasing shadow, until she finally disappears in the darkness.

But the largest and most interesting of all the European cave animals is undoubtedly the Olm (*Proteus anguinus; Hypochthon*). This enigmatic reptile was first found in the famous Lake of Cirknitz, which, communicating with numerous subterranean caves, alternately receives and loses its waters through openings in the rock. After long and heavy rains the floods, which the hidden vaults are no longer able to contain, gush forth in foaming cataracts, and the lake, which generally forms but a long and narrow channel, then swells to at least three times its ordinary width. Sometimes, after a long drought, the contrary takes place, and the whole lake disappears under ground. Thus, from December 1833

---

* Torches are not allowed to be carried in the Grotto of Adelsberg, that the whiteness of the stalactites may not be tarnished by the smoke.

to October 1834, not a trace of it was visible, so thoroughly
had it concealed itself in its subterranean reservoirs, where its
fishes, secure from the persecutions of man, multiplied in a
remarkable manner.   The Olm, which only casually comes to
the light of day, along with the overflowing waters of the Cirknitz Lake, was first discovered in 1814, in one of its perma-
ment subterranean abodes. The Magdalena or ' Black Grotto,'
situated about a league to the north of Adelsberg, slants
abruptly into the bowels of the mountain.   After a long and
difficult passage over blocks of stone, or through soft mud, a
tranquil pool is at length reached, which rises or falls simultaneously with the waters of the Poik, and proves, by this
reciprocal action, that, in all probability, all the numerous
grottoes and subterranean river channels of this so strangely
undermined country form but one vast and intricate network.
It was in this pool, which no light illumines and no wind ever

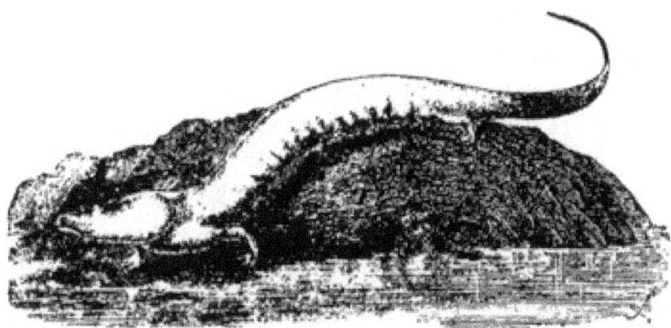

THE PROTEUS ANGUINUS.

stirs, that numerous Protei were first discovered ; but as hundreds of specimens have since found their way to the cabinets
of naturalists, to be observed, dissected, or bottled up in
spirits, their number has very much decreased, and the time is
perhaps not far distant when they will be entirely extirpated
in the grotto, where from time immemorial they had enjoyed
an undisturbed security.   The Proteus is one of those remarkable reptiles which breathe at the same time through lungs
and gills, having on each side of the neck three rose-red
branchiæ, which it retains through life, as its lungs are but
imperfectly developed.   It has a long, eel-like body, with an
elongated head, a compressed tail, and four very short and

thin legs. The skin is flesh-coloured, and so translucent
that the liver and the heart, which beats about fifty times in
a minute, can be distinctly seen underneath. In spite of its
apparent weakness, it is able to glide rapidly through the
water. Its four little legs remain immovable while swimming; they are only used for creeping, and then in a very
imperfect manner. During rapid movements the gills swell
and assume a lively scarlet colour; when quiet, they collapse,
and become white like the rest of the body. Sometimes the
animal raises its head above the water to breathe, but pul-
monary respiration evidently plays but a secondary part in
its economy, as it can only live a very short time out of the
water. The skeleton consists almost entirely of cartilage.
The eyes (two little black spots) lie buried under the skin,
and, as may well be imagined, are very imperfectly developed.
Although more than a thousand specimens have been observed, yet but little is known about its mode of life, nor has
it been ascertained whether it is oviparous or brings forth live
young. In a captive state the Proteus is able to live for
several years without any apparent food; but on fastening a
small worm to the extremity of a thin stick, and holding it
under the water close to the head of the reptile, it shoots
rapidly towards it, swallows it with the same velocity, then
ejects it again, and repeats this manœuvre several times,
until it finally retains the morsel. The untiring zeal of the
German naturalists has discovered the Proteus in thirty-one
different caverns, and ascertained seven distinct species,
varying by their size, the form of the head, the position of
the eyes, and the colour of the skin. Six of these species
belong to the caverns of Carniola, and the seventh to those
of Dalmatia. Two different species never inhabit the same
cavern.

During the visit which the Archduke Ferdinand paid, in
1819, to the Magdalena Grotto, the most remarkable parts
of the cave were brilliantly illuminated, so as to produce a
magical effect. Charon's boat, issuing from a dark recess,
came gliding along over the black surface of the pool. The
grim ferryman drew up his net before the august visitors, and
presented them with six Protei that had been entangled in
its meshes. Dr. Schmidl mentions part of the subterranean

river in the Planina Cave, 1,715 fathoms from the entrance, as the spot where the Protei are most abundant. Near to a small cascade which the rivulet here forms over a reef, the waters absolutely swarm with them, and the light-coloured animals, darting about in all directions in the dark stream, afford a strange and picturesque spectacle. As' the cavern is of most difficult access, they here enjoy a tranquillity rarely disturbed, and no doubt they have many other still more hidden retreats, to which man is incapable of penetrating. The best method for transporting the Proteus is now perfectly understood, and living specimens have been conveyed as far as Russia, Hungary, and Scotland. All that they need is a frequent supply of fresh water, and a careful removal of all light. Their food need cause no trouble, as the water contains all they require. It is recommended to lay a piece of stalactite from their native grotto in the vase in which they are transported. When resting or sleeping, they then coil themselves round the stone, as if tenderly embracing it. In this manner they have already been kept above five years out of their caverns. The guides to the Grotto of Adelsberg have always got a supply on hand, and sell them for about two florins a-piece.

On turning our attention from the grottoes of Carniola to those of the New World, we find, in the vast Mammoth Cave in Kentucky, a no less interesting animal creation, which, though different from that of the Austrian caverns, still shows a certain family resemblance, and affords another proof that a similarity of external circumstances always produces analogous forms of organic life. Thus, the two blind beetles which are found in the Mammoth Cave belong to the same genera (*Anophthalmus* and *Adelops*) that have also their representatives in the Grotto of Adelsberg. The largest insect is here a species of cricket, with enormously long antennæ; there are also two small white eyeless spiders and a few crustaceans. The Mammoth Cave has no proteiform reptile to boast of, but a peculiar blind rat and a peculiar blind fish.

The cavern rat, which is tolerably numerous, but which, on account of its remarkable timidity, seldom shows itself, differs from the common or Norway rat, by its bluish colour, its white abdomen, neck, and feet, and its soft hair. It has

large black eyes, like those of a rabbit, but entirely destitute
of an iris, and uncommonly long whiskers, as if Nature had
wished to indemnify it for the loss of sight by a more perfect
development of the sense of touch.   Although the eyes of this
rat are large and brilliant, yet Professor Silliman convinced
himself of their perfect insensibility to light.   All proof is
wanting that it ever visits the upper world.

The blind fish (*Amblyopsis spelæus*) is now become tolerably rare from its having been so frequently fished out of the
Lethe stream, as the subterranean river of the Mammoth Cave
is called.   Many physiologists have already made it the subject of their observations, and are generally of opinion that the

BLIND FISH (AMBLYOPSIS SPELÆUS).

Amblyopsis was not originally blind, but that, having found
its way into the cave, it gradually lost its powers of vision.
The celebrated naturalist Agassiz, however, being perfectly
convinced that all animals existing in a wild state have been
created within their actual bounds with all the peculiarities
of structure which distinguish them at the present day, is of
opinion that the blind fish and all the other blind animals of
the Mammoth Cave are the aboriginal children of darkness,
and have at no time been connected with the world of light.

CHAPTER XV.

CAVES AS PLACES OF REFUGE.

The Cave of Adullam—Mahomet in the Cave of Thaur—The Cave of Longara—The Cave of Egg—The Caves of Rathlin—The Cave of Yeormalik—The Caves of Granada—Aben Aboo, the Morisco king—The Caves of Gortyna and Melidoni—Atrocities of French Warfare in Algeria—The Caves of the Dahra—The Cave of Shelas—St. Arnaud.

IN time of war or persecution, caverns have often served as places of concealment. It was in the cave of Adullam that David hid himself to escape from the fury of Saul; and on the flight from Mecca to Medina, Mahomet and his disciple Abu Bekr took refuge in a cave in Mount Thaur. They left Mecca while it was yet dark, making their way on foot by the light of the stars, and the day dawned as they found themselves at the foot of the hill, about an hour's distance from the holy city. Scarcely were they within the cave when they heard the sound of pursuit. Abu Bekr, though a brave man, quaked with fear. 'Our pursuers,' said he 'are many, and we are but two.' 'Nay,' replied Mahomet, 'there is a third : God is with us !'

And here the Moslem writers relate a miracle dear to the minds of true believers. By the time, say they, that his pursuers, the Koreishites, reached the mouth of the cave, an acacia tree had sprung up before it, in the spreading branches of which a pigeon had made its nest and laid its eggs, and over the whole a spider had woven its web. When the Koreishites beheld these signs of undisturbed quiet, they concluded that no one could recently have entered the cavern, so they turned away and pursued their search in another direction.

But caverns have not always proved safe places of refuge, and a barbarous enemy has often used them for the destruction of those who there vainly sought safety. Thus, in Palestine, the Jews, who hid themselves with their wives and children in deep caverns hollowed in the flanks of a precipitous mountain, could not escape the satellites of Herod, who, let down from above in large baskets or tubs, put these defenceless fugitives to the sword. During the Gallic war Cæsar ordered his lieutenant Crassus to wall up the mouths of the caves in which the inhabitants of Aquitaine had sought a refuge, and many of them were thus immured alive.

In the year 1510, when the French army, on its retreat from Italy, was traversing the defiles of Piedmont, the rearguard, commanded by the famous Chevalier Bayard, the good knight 'without fear or reproach,' having halted at Longara, the mercenaries, who formed a considerable part of his troops, spread over the country, pillaging and destroying wherever they went.

To escape from these savage bands, the nobles of the district persuaded about two thousand of the peasantry to accompany them, with their families and an abundant supply of provisions, into the Cave of Longara, which forms a vast though low vaulted hall, about 1,200 feet long and 300 feet broad, but with an entrance so narrow that only a single person can pass at a time. The mercenaries, having discovered the secret of the cave, rushed to the spot, eager for pillage. The unfortunate refugees vainly strove to soften the hearts of these barbarians; but, finding all supplications vain, they took courage from despair, and, favoured by the natural strength of the cave, repelled the attack of the first banditti who attempted to force an entrance.

The ruffians now returned to the charge in greater numbers; but being still unable to accomplish their object, they formed the diabolical plan of setting fire to a heap of hay, straw, and greenwood, which they piled up before the mouth of the grotto. The smoke penetrated into the cave and in a short time the two thousand wretches it contained— mostly women and children—were suffocated. Bayard, enraged at this barbarous act, which sullied his own honour, ordered the ringleaders to be seized and hanged before the entrance of the cave.

While these malefactors were in the hands of the

executioner, a lad of fourteen, the only survivor of the catastrophe, was seen to crawl out of the grotto. Bayard ordered every aid to be rendered him which his state required, and could not refrain from tears while listening to his lamentable tale. The boy related that when the smoke began to spread in the cavern, the nobles, resolving to die at least like soldiers, wanted to sally forth, sword in hand, but were prevented by the peasants, who fell upon them, and disarmed them, saying, 'You have led us hither, and here with us you shall die!' Thus, a few moments before a common doom was to destroy both nobles and serfs, a horrible strife had arisen between them in the darkness of the cave.

'And thou, my friend,' asked Bayard, 'by what miracle hast thou escaped death?' 'I had remarked,' answered the lad, 'a feeble ray of daylight in a corner of the grotto, and applied my mouth to the crevice through which it passed. I soon fainted, but this small portion of fresh air preserved my life. When I recovered my senses, I remembered all that had passed, but I was alone, and it took me a long time to crawl out of the grotto.' 'All thy companions,' answered Bayard, 'have been buried, by my orders, in consecrated ground; and behold! there hang their assassins!'

Unfortunately the history of our land is tarnished with similar deeds of atrocity.

A cave in the Isle of Egg, one of the Hebrides, has a very narrow entrance, through which one can hardly creep on knees and hands, but it rises steep and lofty within, and runs into the bowels of the rock to the depth of 255 measured feet. The rude and stony bottom of this cave is strewed with the bones of men, women, and children, the sad relics of the ancient inhabitants of the island, 200 in number, of whose destruction the following story is related. 'The Macdonalds, of the Isle of Egg, a people dependent on Clanranald, had done some injury to the Laird of Macleod. The tradition of the isle says that it was by a personal attack on the chieftain, in which his back was broken; but that of the other isles bears that the injury was offered to two or three of the Macleods, who, landing upon Egg, and behaving insolently towards the islanders, were bound hand and foot, and turned adrift in a boat, which the winds and waves safely conducted to Skye.

To avenge the offence given, Macleod sailed with such a body of men as rendered resistance hopeless. The natives, fearing his vengeance, concealed themselves in the cavern; and, after strict search, the Macleods went on board their galleys, after doing what mischief they could, concluding the inhabitants had left the isle. But next morning they espied from their vessels a man upon the island, and, immediately landing again, they traced his retreat, by means of a light snow on the ground, to the cavern. Macleod then summoned the subterraneous garrison, and demanded that the inhabitants who had offended him should be delivered up. This was peremptorily refused. The chieftain thereupon caused his people to divert the course of a rill of water, which, falling over the mouth of the cave, would have prevented his purposed vengeance. He then kindled at the entrance of the cavern a huge fire, and maintained it until all within were destroyed by suffocation.' *

In the reign of Queen Elizabeth, a no less horrible deed occurred during the campaign of Essex against the Irish rebels. When the English forces entered Antrim, the Scots of that county had sent their wives and children, their aged and their sick, to the island of Rathlin for safety. Sir John Norris was directed by the Earl to cross over and kill all that he could find. The run up the Antrim coast was rapidly and quietly accomplished. Before an alarm could be given the English had landed close to the church which bears Columba's name. The castle was taken by storm, and every soul in it—about two hundred—put to the sword. It was then discovered that the greater part of the fugitives, chiefly mothers and their little ones, were hidden in the caves about the shores. There was no remorse,' says Froude, 'not even the faintest shadow of perception that the occasion called for it. They were hunted out as if they had been seals or otters, and all destroyed.'

When the barbarian Genghis Khan invaded Koondooz in Central Asia, 700 men took refuge, with their wives and families, in the cave of Yeermalik, and defended themselves so valiantly that, after trying in vain to destroy them by fire,

---

* Voyages in the Lighthouse Yacht, published in Lockhart's 'Life of Sir Walter Scott.'

the invader built them in with huge natural blocks of stone, and left them to die of hunger. In the year 1840, the cave was visited by Captain Burslem and Lieutenant Sturt, probably the only Europeans that ever entered its sepulchral recesses, as the people in the neighbourhood believe it to be the abode of Sheitan (the devil), and are as reluctant to guide a stranger as to explore it themselves. The entrance is half way up a hill, and is fifty feet high, with about the same breadth. Not far from the entrance the travellers found a passage between two jagged rocks, possibly the remains of Genghis Khan's fatal wall, so narrow that they had some difficulty in squeezing through, and then before long came to a drop of sixteen feet, down which they were lowered by ropes. Here they left two men to haul them up on their return, and bade farewell to the light of day. The narrow path, which led by the edge of a black abyss, sometimes over a flooring of smooth ice for a few feet, widened gradually till they reached a damp and dripping hall, so vast in size that the light of their torches did not enable them to form any idea of its size. In this colossal hall, or rather tomb, they found the remains of the victims of Genghis Khan: hundreds of skeletons, in a perfectly undisturbed state—one, for instance, still holding the skeletons of two infants in its long arms—while some of the bodies had been preserved, and lay shrivelled, like the mummies reposing in the sepulchral vault of the Great St. Bernard.

In the dark history of Philip II. of Spain, one of the darkest passages is that of the rebellion and final destruction of the Moors of Granada. Driven to despair by an intolerable tyranny, the unfortunate people at length rose in arms against their oppressors; but all their bravery, aided by the natural strength of their mountain fastnesses, failed to defend them against the superior arms of their pitiless enemy. Defeated in every encounter, driven from every stronghold, thousands perished by famine or the sword, and those who submitted were either condemned to a cruel death or exiled from their native soil. Many were driven to seek a refuge in the caves of the Alpujarras, south-east of Granada, and of the bold sierras that stretch along the southern shores of Spain. Their pursuers followed up the chase with the fierce

glee of the hunter tracking the wild beast of the forest to his
lair. There they were huddled together, one or two hundred
frequently in the same cavern. It was not easy to detect
the hiding-place amidst the rocks and thickets which covered
up and concealed the entrance. But when it was detected,
it was no difficult matter to destroy the inmates. The green
bushes furnished the material for a smouldering fire, and
those within were soon suffocated by the smoke, or, rushing
out, threw themselves on the mercy of their pursuers. Some
were butchered on the spot; others were sent to the gibbet
or the galleys; while the greater part, with a fate scarcely
less terrible, were given up as the booty of the soldiers, and
sold into slavery.

Aben Aboo, the last chief of the insurgents, who had
hitherto eluded every attempt to sieze him, but whose capture
was of more importance than that of any other of his nation,
had a narrow escape in one of these caverns not far from
Berchal, where he lay hid with a wife and two of his
daughters. The women were suffocated, with about seventy
other persons; but the Morisco chief succeeded in making his
escape through an aperture at the further end, which was
unknown to his enemies.

Unfortunately, the little king of the Alpujarras, as he was
contemptuously called by the Spaniards, was soon after
killed in another cavern by a traitor's hand, and with him
fell the last hope of the Moriscos. His corpse, set astride on a
mule, and supported erect in the saddle by a wooden frame,
concealed beneath ample robes, was led in triumphal procession through the streets of Granada, and then decapitated.
The body was given to the rabble, who, after dragging it
through the streets with scoffs and imprecations, committed
it to the flames, while the head, inclosed in a cage, was set
up over the gate which opened on the Alpujarras. There it
remained for many a year, no one venturing to remove it, for
on the cage was inscribed, 'This is the head of the traitor
Aben Aboo. Let no one take it down, under pain of
death.'

The neighbourhood of Gortyna, in the island of Crete, has
become celebrated in modern times for a mountain-labyrinth,
with numerous and intricate passages, which exists in a valley
near it, and in which the myth of the Dædalean labyrinth

was probably localised. During the revolutionary war against the Turkish yoke (1822-1828), the Christian inhabitants of the adjacent villages, for months together, lived in this cavern, merely sallying out by day to till their lands, or to gather their crops, when it was safe to do so. Though the dark recesses of the cavern were not very inviting abodes for human beings for any long period, yet the sense of safety gave it, doubtless, a peculiar charm ; for no one could approach within range of the numerous muskets pointed from masked loopholes at its entrance, without being immediately shot down ; nor could either fire or smoke suffocate or dislodge its inmates, as the entrance is in the side of a steep hill, 500 or 600 feet above the bed of the wild valley in which it is situated, and thus is safe from attack in every direction. History as well as tradition states that in all troubled times in Crete, the labyrinth of Gortyna has been the retreat of the inhabitants of the neighbourhood ; but when Captain Spratt visited it in 1852, its only inhabitants were bats, who, by their mode of hooking on to each other, were hanging from the ceiling like clusters of bees. Under good native guides, he spent nearly two hours in threading its tortuous passages, which turn in so many ways and have so many branches as to justify the conclusion that a master hand must have directed the excavation. The mark of the tool is seen upon every side of the avenues and chambers, indicative of its artificial character.

Less fortunate than their brethren of Gortyna were the unfortunate Cretans who, during the same war, took refuge in the cave of Melidoni. In 1822, when Hussein Bey marched against the neighbouring village, the inhabitants, to the number of three hundred, repaired to the cave, taking with them their valuables and provisions sufficient for six months. The entrance is so narrow and steep that they were perfectly secured against an attack, and the Turks in their first attempt lost twenty-five men. Finding that they refused submission on any terms, Hussein Bey ordered a quantity of combustibles to be brought to the entrance and set on fire. The smoke, rolling into the cavern in immense volumes, drove the miserable fugitives into the remoter chambers, where they lingered a little while longer, but were all eventually suffocated. The Turks waited some days, but still did not dare to enter, and a Greek captive was finally sent down on the promise of his life being spared. The Turks then descended and plundered the bodies. A week afterwards three natives of the village stole into the cavern to see what had become of their friends and relatives. It is said that they were so overcome by the terrible spectacle that two of them died within a few days. Years afterwards, when the last vestiges of the insurrection had been suppressed, the Archbishop of Crete blessed the cavern, making it consecrated ground; and the bones of the victims were gathered together, and partially covered up, in the outer chamber—a vast elliptical hall, about eighty feet in height, and propped in the centre by an enormous stalactitic pillar. On all sides the stalactites hang like fluted curtains from the roof, here in broad-sheeted masses, there dropping into single sharp folds, but all on a scale of Titanic grandeur. In this imposing and silent hall, under the black banners of eternal Night, lay heaped the mouldering skulls and bones of the poor Christians. They could not have had a more appropriate sepulchre.

Such have been the atrocities of Turkish warfare within the memory of living man; but French officers have in our days emulated the cruelty of Ottoman commanders, and shown that the nation which boasts of marching at the head of civilisation has still retained much of its ancient Gallic barbarism. When Marshal Pelissier filled with smoke the crowded caves of the Dahra in 1844, and destroyed many hundreds of Kabyls, whose great crime it was to defend their country against the French hordes, it has been stated, as an excuse for this atrocity, that he left open some of the entrances to the caves, and that he only resorted to the smoke as a means of compelling the fugitives to come out and surrender; but no such excuse can be pleaded in favour of his successor, St. Arnaud. In the summer of 1845,* this French commander received private information that a body of Arabs had taken refuge in the cave of Shelas. Thither he marched a body of troops. Eleven of the fugitives came out and surrendered; but it was known to St. Arnaud, though not to any other French-

* Kinglake.

man, that five hundred men remained in the cave.   All these people Colonel St. Arnaud determined to kill, and at the same time to keep the deed secret even from the troops engaged in the operation, as the smoking of the caves of the Dahra had not greatly tended to raise France in the public opinion of Europe.   Except his brother and Marshal Bugeaud, whose approval was the prize he sought, no one was to know what he did.   He contrived to execute both his purposes.   Thus he writes to his brother :—'I had all the apertures hermetically stopped up.   I made one vast sepulchre.   No one went into the caverns ; no one but myself knew that under these there were 500 brigands who will never again slaughter Frenchmen.   A confidential report has told all to the marshal, without terrible poetry or imagery.   Brother, no one is so good as I am by taste and by nature.   From the 8th to the 12th I have been ill, but my conscience does not reproach me.   I have done my duty as a commander, and to-morrow I would do the same over again; but I have taken a disgust to Africa.'   With such nauseous sentiment wrote the man, '*good* by taste and nature,' who seven years later was to attach the memory of his name to the bloody days of December, and, to deal with many a French republican as he had dealt with the Arabs.

CHAPTER XVI.

HERMIT CAVES—ROCK-TEMPLES—ROCK-CHURCHES.

St. Paul of Thebes—St. Anthony—His visit to Alexandria, and death—Numerous Cave Hermits in the East—St. Benedict in the Cave of Subiaco—St. Cuthbert—St. Beatus—Rock-temples of Kanara—The Wonders of Ellora—Ipsamboul—Rock-churches of Lalibala in Abyssinia—The Cave of Trophonios—The Grotto of St. Rosolia near Palermo—The Chapel of Agios Niketas in Greece—The Chapel of Oberstein on the Nahe—The Repentant Fratricide.

THE dim twilight of a forest, its leafy vaults, its majestic silence, or its foliage moaning in the wind, are all apt to strike the mind with a religious awe. But the solitude and stillness of caverns is equally well adapted to awaken feelings of devotion; and thus we find that contemplative minds in every age, and of every creed, have found in them congenial retreats. The Indian fakir and the Mahometan dervish love the seclusion of the silent grotto, and here also the Hebrew prophets not seldom enjoyed their ecstatic visions.

There can be no doubt that during the first ages of Christianity many an unknown anchorite retired to some solitary cave, as to a harbour of refuge from the rude contact of the world; but the first hermit mentioned in ecclesiastical history is St. Paul of Thebes, in Egypt, who, during the persecution of Decius, in the middle of the third century, retreated to the desert, where, dwelling in a cave, and living on the fruits of trees, he reached his hundredth year. His friend and disciple, St. Anthony, who first roused among his contemporaries a wide-spread inclination for hermit seclusion, plays a far more conspicuous part in the annals of the Church. Born of wealthy Coptic peasants, this remarkable man, at the age of twenty, divided his whole property among the poor, and thenceforth devoted himself to a life of the strictest asceticism. He retired first to a rock-cave in the neighbourhood of his native village, and then to the more distant ruins of a deserted castle, where he spent twenty years as a hermit. Meanwhile his reputation for sanctity had spread throughout all Egypt, and numerous candidates for a hermit life besought him to take them under his spiritual care. He yielded to their entreaties, and soon the neighbouring desert was crowded with the huts of zealous anchorites, who revered him as their model. But he was surrounded not only by these pious disciples; the worldly-minded also came flocking to his cave for advice or assistance ; for the belief was general that, like the first Apostles, he was gifted with the power of casting out devils and foretelling future events. Anthony, thus disturbed in his solitary meditations, resolved to bury himself still deeper in the desert, and fled to a cave in the furthest parts of Egypt, near a well shaded by a few date-palms. Here he hoped to be able to live entirely for prayer and contemplation, but his hopes proved vain ; for, after a long search, his disciples discovered his retreat, and again anchorites and worldlings broke in upon his solitude. In his hundredth year he was prevailed upon by St. Athanasius to visit Alexandria, where, whenever he appeared, crowds gathered round him to kiss the hem of his garment and to implore his blessing. Even the Emperor Constantine the Great wrote to him ; yet so indifferent had he become to all worldly distinctions that he could with difficulty be prevailed upon to have the letter read to him. Thus, honoured by high and low, and yet avoiding all honour, Anthony reached the patriarchal age of 105. At the approach of death he begged two of his most beloved disciples to conduct him to the wildest part of the desert. Here he died in their arms, after having first made them promise to keep the place of his burial secret, as he feared that an undue reverence might be paid to his bones.

Anthony's example was followed far and wide over the Eastern world. Whole colonies of hermits settled in the desert of Thebes, near Lake Moeris, in Southern Palestine, in Armenia, and Pontus. Their numbers amounted to thousands, many living in rude huts, which they erected with their own hands, while others found a congenial retreat in the grottoes and rock-tombs which abound in many of the countries where they dwelt.

From the East the spirit of monastic seclusion soon spread

to Western Europe. St. Benedict, the founder of the order which has rendered such signal service to learning during the Middle Ages, spent three years in an almost inaccessible cave near Subiaco, five leagues from Tivoli. Romanus, a monk in a neighbouring convent, alone knew of his retreat, and daily let down by a rope, from the top of the rock in which the cave was situated, the small quantity of bread which he needed for his subsistence. Here he was at length discovered by some shepherds, who at first sight took him for a wild beast, as he was clothed in skins, but soon discovered that he was a saint by the wise lessons he gave them.

A similar longing for a life of pious seclusion induced St. Cuthbert to quit the Convent of Lindisfarne, of which he had been prior, and to seek a retreat in a grotto excavated by his own hands, on one of the Farne Islands, on the coast of Northumberland. An ox-hide, which he hung before its entrance and turned towards the side whence the wind blew, afforded a scanty shelter against the rigours of a northern winter. But the fame of his sanctity spread over all England, and numerous pilgrims resorted to his cave, to profit by his advice, or to seek consolation in their troubles. One day, when he had spent eight years in seclusion, the king of Northumbria, attended by his principal nobles, landed on Cuthbert's island-rock to beg him to accept the episcopal dignity of Durham, to which he had been elected. The holy anchorite yielded, with many tears, and after an obstinate resistance, for he was loth to accept duties which tore him from his solitude. After two years he resigned his bishopric, and returned to his beloved cave, where he shortly after died. According to a popular legend, the Entrochi, or calcareous joints of the petrified Lily-Encrinites, which are found among the rocks of the Farne Islands, are forged by his spirit, and pass there by the name of ' St. Cuthbert's beads.' While at this task, he is supposed to sit during the night upon a certain rock, and to use another as his anvil.

> ' Such tales had Whitby's fishers told,
>   And said they might his shape behold,
>   And hear his anvil sound ;
>   A deadened clang—a huge dim form,
>   Seen but, and heard, when gathering storm ~
>   And night was closing round.'--*Marmion.*

The Beatenberg, situated on the northern side of the lake of Thun, is named after a celebrated cave in which St. Beatus, originally a British noble, who had come to preach the Gospel to the wild men of the district, dwelt for many years, and died at the advanced age of ninety. His relics remaining there, his fête-day attracted such crowds of pilgrims that reforming Berne sent two deputies in 1528 to carry off the saint's skull and bury it between the lakes of Thun and Brienz ; but still the pilgrimages continued, and at length, in 1566, the Protestant zeal of Berne went to the expense of a wall, and thus effectually shut out the pilgrims, who, in more modern times, have been profitably replaced by crowds of tourists.

Both in the heathen and the Christian world, grottoes, particularly such as had been hallowed by the lives of sainted anchorites, have frequently been consecrated to divine service ; and to render them still more worthy of their destination, the rude excavations of nature have not seldom been enlarged and beautified with all the resources of art.

Among these subterranean places of worship, those of India are deservedly renowned for their colossal size, and for the vast labour bestowed upon the sculptures with which they are adorned. A description of the famous rock-temples of Kanara, in the island of Salsette, near Bombay, will give the reader some idea of their magnificence.

The way leads over a narrow mountain-path, through a jungle so dense that the traveller is obliged to quit his palanquin, and to ascend on foot the steep acclivity from which, at some distance from the summit, the large temple overlooks the country. This colossal work is hewn out of the solid rock, ninety feet long and thirty-eight feet broad, with a corresponding height, and forms an oblong square with a vaulted roof. Two colossal rows of columns divide the hall into three naves or avenues, and give it the form of an ancient basilica.

As the Temple of Kanara served the Portuguese for some time as a church, during their occupation of the small archipelago of Bombay, the heathen sculptures which decorated the interior have naturally been mostly destroyed. This is the more to be regretted, as the well-preserved and masterly-executed capitals of the mighty columns justify the belief that their artistic merit must have been worthy of the grand dimensions of the hall. The beautiful portico, however, is still richly decorated. On each side a recess contains a colossal well-executed statue, and long inscriptions in unknown characters are carved on the square pillars of the entrance. The charms of a mysterious past thus add to the interest of this beautiful monument, the work of an astonishing patience and perseverance. The outer face of the portico, as well as the vestibule extending before it, twenty-eight feet

ROCK-TEMPLES OF AJUNTA.

deep, have been considerably injured by the ravages of time: many stones have started from their joints, and a multitude of creeping plants cling to the mouldering statues. Thus the efforts of man to rear eternal monuments are vain; they must necessarily yield to the living powers of nature.

> ' The cloud-capped towers, the gorgeous palaces,
> The solemn temples, the great globe itself,
> Yea, all which it inherit, shall dissolve,
> And, like this unsubstantial pageant faded,
> Leave not a rack behind.' *

Steps are hewn in the rock to the summit of the mountain, and various intricate paths lead to smaller excavations, con-

* Shakespeare, ' Tempest,' iv. 1.

sisting mostly of two cells and a portico. Near each of them is a well or basin, likewise hewn out of the rock, in which the rain-water collects, affording a grateful beverage to the tired wanderer. Many of these caves are larger and more perfect than the others, and some in their general effect resemble the great temple, although far inferior in size and decoration.

The whole aspect of this perforated mountain shows that a complete cave-town, capable of containing several thousand inhabitants, has been hollowed out in its flanks. The largest excavation was, beyond a doubt, the chief temple. The smaller caves, arranged according to the same plan, likewise served for devotional purposes ; and the rest were dwellings more or less commodious and large according to the rank or means of their possessors, or, what is still more likely, the abode of pious Brahmans and their scholars at the time when India was the cradle of arts and sciences, while the nations of Europe were still plunged in barbarism.

From the summit of this wonderful mountain the spectator enjoys a beautiful prospect. The island of Salsette lies before him as if spread out on a map, affording a most agreeable variety of rice-fields and cocoa-nut groves, of villages and meadows, of woody hills and fruitful vales. The surrounding mountains form a foreground of grey rocks, dotted with trees, or excavated into dark grottoes, once the abode of fakirs, but now the retreats of tigers, snakes, huge bats, and enormous swarms of bees; while towards the south the horizon is bounded by the island of Bombay, with its forest of masts, towards the east by the mainland, towards the north by Bassein and the neighbouring mountains, and towards the south by the ocean. The enjoyment of the picturesque scene is marred only by the many tigers which infest the mountains, and frequently descend into the plain, where they not only carry away sheep and oxen, but also tear many a poor Hindoo to pieces.

The rock-temples of Kanara are rivalled by those of Elephanta, Karli, and Ajunta, and far-surpassed in magnificence and extent by the excavations of Ellora, near the town and fort of Dowlatabad, where a whole mountain of hard red granite has been hollowed out into an immense range of highly ornamented grottoes and temples, fit for the residence of a whole pantheon of deities, and for the reception of a whole nation of pilgrims.

About three miles to the north of Madras, where the rock touches the sea, navigators had long remarked some pillars of stone rising from the water and covered with rude sculptures. From these the spot received the name of the Seven Pagodas. Most of them have since been destroyed by the tides, and one only is still standing, though tottering to its fall. These, however, were but the advanced posts of the colossal excavations in the rocky wall behind; for here also are seen large grottoes, porticoes, and temples, as at Ellora, though of somewhat smaller proportions, and of less beautiful execution. They are dedicated to the worship of Vishnu and Siva, and covered with inscriptions. A whole rock-town, or at least a vast sanctuary, thus lies concealed on this solitary coast.

Similar cave-temples are met with in Cochin China, Birmah, Malwah, and Ceylon, where the spacious rock-temple of Dambool is deservedly celebrated for its antiquity, and for its numerous statues of Budda, in the varied attitudes of exhortation and repose.

On the banks of the mysterious Nile we find rock-temples rivalling those of India in colossal grandeur, and among these Ipsamboul is pre-eminent in splendour.

'After sailing for some hours,' says Warburton,[*] 'through a country quite level on the eastern bank, we come upon a precipitous rocky mountain, starting up so suddenly from the river's edge, that its very summits are reflected in the water. We moored under a sand-bank, and, accompanied by half-a-dozen of the crew with torches, approached this isolated and stupendous rock. Yet even here the daring genius of Ethiopian architecture ventured to enter into rivalry with nature's greatness, and found her material in the very mountains that seemed to bid defiance to her efforts.

'On the face of the vertical cliff a recess is excavated to the extent of about a hundred feet in width. From this four gigantic figures stand out in very bold relief. Between the two central stony giants, a lofty doorway opens into a

[*] 'The Crescent and the Cross.'

vast hall, supported by square pillars, each the size of a tower, and covered with hieroglyphics. Just enough painting still glimmers faintly on these columns to show that they were formerly covered with it; and the walls are carved into historic figures in slight relief; these, as our torches threw an uncertain glare over them, seemed to move and become instinct with life.

'This temple was dedicated to Athor, the lady of Aboccis (the ancient name of Ipsamboul), who is represented within under the form of the sacred bow. This was, however, a mere "chapel of ease" to the great temple, excavated from a loftier rock, about fifty yards distant. Between these two a deep gorge once ran to the river, but this is now choked up with sand, in whose burning waves we waded knee-deep to the Temple of Osiris.

'Here a space of about 100 feet in height is hewn from the mountain; smooth, except for the reliefs. Along the summit runs a frieze of little monkeys in long array, as if the architect felt the absurdity of the whole business, or as Byron sometimes finishes off a sublime sentence with a scoff. Then succeeds a line of hieroglyphics and some faintly-carved figures also in relief, and then four colossal giants that seem to guard the portal. They are seated on thrones (which form, with themselves, part of the living rock), and are about sixty feet high. One is quite perfect, admirably cut, and the proportions admirably preserved; the second is defaced as far as the knee; the third is buried in sand to the waist; and the fourth has only the face and neck visible above the desert's sandy avalanche.

'The doorway stands between the two central statues, and is surmounted by the statue of Isis wearing the moon as a turban.

'On entering, the traveller finds himself in a temple, which a few days' work might restore to the state in which it was left, just finished, three thousand years ago. The dry climate and its extreme solitude have preserved its most delicate details from injury; besides which it was hermetically sealed by the desert for thousands of years, until Burckhardt discovered it, and Mr. Hay cleared away its protecting sands.

'A vast and gloomy hall, such as Eblis might have given

Vathek audience in, receives you in passing from the flaming sunshine into that shadowy portal. It is some time before the eye can ascertain its dimensions through the imposing gloom, but gradually there reveals itself, around and above you, a vast aisle, with pillars formed of eight colossal giants, upon whom the light of heaven has never shone. These images of Osiris are backed by enormous pillars, behind which run two great galleries; and in these torchlight alone enabled us to peruse a series of sculptures in relief representing the triumphs of Rameses the Second, or Sesostris. The painting, which once enhanced the effect of these spirited representations, is not dimmed, but crumbled away; where it exists the colours are as vivid as ever.

'This unequalled hall is one hundred feet in length, and from it eight lesser chambers, all sculptured, open to the right and left. Straight on is a low doorway opening into a second hall of similar height, supported by four square pillars, and within all is the adytum, wherein stands a simple altar of the living rock in front of four large figures seated on rocky thrones. This inner shrine is hewn at least one hundred yards into the rock; and here, in the silent depth of that great mountain, these awful idols, with their mysterious altar of human sacrifice, looked very preadamitic and imposing. They seemed to sit there waiting for some great summons which should awaken and reanimate these " kings of the earth, who lie in glory, every one in his own house."

'We wandered through many chambers, in which the air is so calm and undisturbed that the very smell of the torches of the last explorers of these caverns was perceptible.'

In Abyssinia the rock-churches of Lalibala likewise give proof of an ancient state of civilisation, strongly contrasting with the barbarism of the present times.

Like the temples of Ellora, some of these curious structures have been hollowed out of single blocks of stone left standing in the centre of open courts excavated in the bosom of the rock, while others are completely subterranean. Though far inferior in magnificence and extent to those wonderful edifices, they are yet very remarkable. The courts, in which the three principal monolithic churches are respec-

tively dedicated to our Saviour, to the Holy Virgin, and to
St. Emmanuel, communicate with each other by narrow
passages, the whole thus forming a continued series of excavations. The church of St. Emmanuel is forty-eight feet
long, thirty-two feet broad, and forty feet high; but it is
surpassed in size by the Church of the Holy Virgin, where
the rock-walls of the court are moreover perforated with
sepulchral vaults and with cells for the habitation of monks.
The town of Lalibala is situated in a beautiful country,
7,000 feet above the level of the sea, on the slope of the
mighty Ascheten mountain, and commands a prospect of
Alpine magnificence. Though it is now reduced to about
2,000 inhabitants, its eight rock-churches (five monolithic
and submërial, three subterranean), prove that it must once
have been a place of considerable importance. Divine service
is still performed in all these churches, which are the resort
of numerous pilgrims, and to whose service above 500 priests,
monks, and nuns are attached.

Though ancient Greece has no such huge rock-temples to
boast of as India or Egypt, yet caverns played an important
part in her ancient religious history. 'Before the old tribes
of Hellas erected temples to the divinities,' says Porphyry, in
his treatise *De Antro Nympharum*, 'they consecrated caves
and grottoes to their service; in the island of Crete to Zeus,
in Arcadia to Artemis and to Pan; in the isle of Naxos to
Dionysos.'

Caves were the site of some of the most celebrated
Grecian oracles. The tripod of the Delphian pythoness stood
over a subterranean hollow, from which the divine inspiration
was supposed to ascend; and pilgrims from all parts of
Hellas resorted to a cave in the neighbourhood of Lebadeia,
a city of Beotia, and named after Trophonios, a mythical
personage who was supposed to have lived there for many
years, and was subsequently deified as an oracular god. Those
who repaired to this cave for information were required, after
passing some preparatory days in a chapel dedicated to
Fortune and to the 'good genius,' to anoint themselves with
oil, to bathe in a certain river, and to drink of the water of two
neighbouring springs called Léthé and Mnémosyné, the first
of which made them forgetful of the past, while the second

fixed in their memory all they heard and saw in the cavern. They were then clothed in a linen robe, took a honeyed cake in their hands, and, after praying before an ancient statue of Trophonios, descended into the subterranean chamber by a narrow passage. Here it was that the future was unfolded to them, either by visions or extraordinary sounds. The return from the cave was by the same passage, but the persons consulting were obliged to walk backwards. They generally came out astonished, melancholy, and dejected. The priests on their return placed them on an elevated seat, called the seat of Mnêmosynê or remembrance, and the broken sentences they uttered in their confused state of mind were considered as the answer of the oracle. They were then conducted to the chapel of the 'good genius,' where by degrees they recovered their usual composure and cheerfulness. There can be no doubt that the priests introduced themselves into the cave by secret passages, and worked upon the excited imagination of their dupes by terrible sounds and apparitions. During the palmy days of the oracle, the neighbourhood of the Cave of Trophonios was decorated with temples and statues; at present its very site is uncertain.

Like ancient paganism, Christianity not seldom celebrates her rites in caves hallowed by the memory of saints and anchorites. A stately church rises over the Grotto of the Nativity, at Bethlehem, and a magnificent pile has been constructed at Jerusalem over the rock-tomb in which our Saviour was buried. The grotto on Mount Carmel, to which the prophet Elijah retreated from the world, is now dedicated to divine worship, in the convent which bears his name; and the cave in which John the Evangelist is said to have written the Apocalypse during his exile in the island of Patmos has also been converted into a chapel.

One of the most celebrated rock-churches is the grotto of St. Rosolia, the patroness of Palermo. This illustrious lady was niece to King William the Good, and, as the legends inform us, no less remarkable for her beauty than for her virtues, which made her the admiration of all Sicily. Never was a princess more fitted to adorn society; but the world had so few attractions for a spirit that could only breathe in the

pure regions of piety, that, at the age of fifteen, she retired
to the solitary mountains, and from the date of her disappearance, in 1159, was never more heard of for about five hundred
years. The people thought she had been taken up to heaven,
as the fitting abode for her more than human perfections;
but in the year 1624, during the time of a dreadful plague, a
holy man had a vision that the saint's bones were lying in a
cave near to the top of the Monte Pellegrino, and that if these
were taken up with due reverence and carried in procession
thrice round the walls of the city, they should immediately
be delivered from the scourge. The bones were accordingly
sought and found, thrice carried round the town, as the vision
had described, and the plague suddenly ceased. From that
time St. Rosolia was revered as the patron saint of Palermo,
and the remote cave where she probably spent many years of
her solitary life became one of the most renowned sanctuaries
of the Catholic Church, and the resort of innumerable pilgrims.
The mountain is extremely high, and so steep that before the
discovery of St. Rosolia it was looked upon as almost inaccessible; but a fine road, very properly termed La Scala, or the
stair, has been cut out in the rock, and leads from terrace to
terrace, over almost perpendicular precipices, to the entrance
of the holy grotto, which is situated near the very top of the
mountain, and commands a magnificent prospect. Within
two miles of the foot of the mountain, the eye discerns the city
of Palermo, with its beautiful villas and luxuriant gardens,
and then, taking a wider range, glances to the north, over
the dark blue sea bounded by the Lipari Islands and the ever-
fuming cone of Stromboli; while to the east a large portion
of Etna, although at the distance of almost the whole length
of Sicily, towers like a giant above the minor mountain chains.
A church and other buildings, forming a kind of courtyard, where some priests reside, appointed to watch over the
treasures of the place, and to receive the offerings of pilgrims
that visit them, have been erected round the grotto.

As may easily be imagined, the history of rock-chapels
has frequently been embellished with legendary tales. The
chapel of Agios Niketas (St. Nicholas) in Crete, is at present
merely a smoky-looking cave beneath a large detached mass
of rock, lying on the slope of an abrupt mountain; but there

are still the remains of a building which once extended far
beyond the present limits. The roof of the cavern, although
very uneven, is also elaborately ornamented with paintings,
representing the remarkable events in the life of the Saviour
and of St. Nicholas, and showing that considerable cost and
artistic care have been bestowed upon it. Though it is now
abandoned, an event that is said to have happened about four
or five centuries ago gives this cave a special interest with the
natives. The church was crowded with Christians from the
adjacent villages on the eve of the festival of their patron
saint Agios Niketas, so as to be ready (as is usual with the
Greeks) for the matin service at daybreak. But the fires
which the assembled party had lighted near it had been
observed at sea by a Barbary corsair then cruising off the
island, and guided him to the spot, where, under the darkness of the night, he landed his crew in a neighbouring cove.
Thus unobserved, they stole up to the church, and, finding it
full of the natives, closed the door and windows upon them,
and waited for day, the better to secure their captives for
embarkation. In this dreadful plight the unfortunate Cretans raised their voices in a general prayer to Saint Niketas.
Their supplications were heard, for the priest soon after informed them that the saint had shown him a way of escape
—through the back part of the cavern, by opening a small
aperture communicating with another cavern that led finally
out upon the mountain slope over the rock. Through this
aperture they all silently crept, unseen and unheard by the
corsairs.

Another interesting legend is attached to a small rock-
chapel situated beneath the ruins of the ancient Castle of
Oberstein, on the Nahe. The Baron of Oberstein, having, in
a fit of jealousy, hurled his younger brother from the balcony
of the castle, fled from the scene of his crime. For years he
wandered, a wretched outcast, from land to land; but wherever he went the curse of Cain was upon him, and left him
no rest by night or day. At length he came to Rome, to confess his fratricide at the feet of the Sovereign Pontiff, who
comforted him with the assurance that he would recover his
lost peace by returning to Oberstein and excavating with his
own hands a rock-chapel for the interment of his brother on
the spot where he fell.

Soon after the self-banished lord made his appearance at Oberstein in a hermit's garb, and set to work upon the hard rock with indefatigable zeal. Never was labour performed with better will; and such, consequently, was the progress of the excavation that it seemed as if he were assisted by the angels in his penitential task. At the expiration of four years, the rock-chapel was completed, and the bones of the murdered man were conveyed with great ceremony to the tomb which had been prepared for their reception at the foot of the altar. As soon as they were lowered into the grave, the murderer bent over them; a smile of ineffable happiness was seen to illumine his emaciated features, and he dropped down dead upon the remains of his brother.

CHAPTER XVII.

ICE-CAVES AND WIND-HOLES.

Ice-caves of St. Georges and St. Livres—Beautiful Ice-stalagmites in the Cave of La Baume—The Schafloch—Ice-cataract in the Upper Glacière of St. Livres—Ice-cavern of Eisenerz—The Cave of Yeermalik—Volcanic Ice-caves—Æolian Caverns of Terni—Causes of the low temperature of Ice-caves.

SOME caves, remarkable for an extremely low temperature even in summer, form natural ice-cellars, though unconnected with glaciers or snow mountains, and in latitudes and at altitudes where ice could not under ordinary circumstances be supposed to exist. Besides the interest attaching to these natural curiosities, these ice-caves are sometimes lucrative sources of revenue to their owners, or answer various purposes of use or comfort.

In hot summers, when the supplies of the artificial ice-houses fail in Geneva and Lausanne, the hotel-keepers have recourse to the stores laid up for them by nature in the ice-caves or glacières of St. Georges and St. Livres, situated in the canton of Vaud, on the slope of the Jura. Other ice-caves are made use of as dairies, or as storehouses for cheese; and the quarries of Niedermendig, near the small town of Andernach, on the Rhine, which are likewise remarkable for a glacial temperature, serve as excellent beer-cellars.

To Mr. Browne, who has made them his special study, we are indebted for an interesting account of the ice-caves of France and Switzerland,* which, as may naturally be expected in halls and galleries hung with drapery of transparent silver, not seldom offer scenes of beauty rivalling the most renowned stalactital grottoes.

In the Glacière of Grâce-Dieu, or La Baume, near Besançon,

* 'Ice-caves of France and Switzerland: a Narrative of Subterranean Exploration.' By the Rev. G. F. Browne. Longmans, 1865.

Mr. Browne was particularly struck with three large stalagmites of ice rising in a line across the middle of the cave.
The central mass was remarkable only for its size, the girth
being sixty-six and a half feet at some distance from the ice-
floor, with which it blended; but nothing could be more strik-
ingly lovely than the stalagmite to its right, owing to the

LOWER GLACIÈRE OF ST. LIVRES.

good taste of some one, who had found that much ice was
wont to accumulate on that spot, and had accordingly fixed
the trunk of a small fir-tree, with the upper branches complete, to receive the water from the corresponding fissure in
the roof. The consequence was that, while the actual tree
had vanished from sight under its crystal covering, the ice
with which it was incrusted showed every elegance of form

o

which a mould so graceful could suggest, each twig of the
different boughs becoming, to all appearance, a solid bar of
frosted ice, from which complicated groups of icicles streamed
down.   But the mass to the left was the grandest and most
beautiful of all.   It consisted of two lofty heads, like weeping
willows in Carrara marble, with three or four others less
lofty, resembling a family group of lions' heads in a subdued
attitude of grief, richly decked with icy manes.   Similar
heads seemed to grow out here and there from the solid
sides of the huge mass, which measured seventy-six and a half
feet in girth about two feet from the floor.   When this
column was looked at from the side removed from the entrance to the cave, so that it stood in the centre of the light
which poured down the long slope from the outer world, the
transparency of the ice made the whole look as if it were set
in a narrow frame of impalpable liquid blue—the effect of
light penetrating through the mass at its extreme edges.

Other and no less striking beauties rewarded our subterranean explorer on his visit to the Schafloch or Trou-aux-
Moutons, a vast ice-cave on the Rothhorn, in the canton of
Berne, which takes its name from the fact that, when a
sudden storm comes on, the sheep and goats make their way
to it for shelter, though never going so far as the spot
where the ice begins.   On entering the cave the way lies
over a wild confusion of loose masses of stone, which soon,
however, begin to be intermingled with ice, until the latter
entirely hides the naked rock under a crystal mantle.

'On either side of the cave was a grand column of ice,
forming the portal, as it were, through which we must pass
to further beauties.   The ice-floor rose to meet these columns
in a graceful swelling curve, perfectly continuous, so that
the general effect was that of two columns whose roots expanded and met in the middle of the cave.   Convinced that
internal investigations would prove interesting, I began to
chop a hole in one of the pillars about two feet from the
ground, and having made an entrance sufficiently large, proceeded to get into the cavity which presented itself.   The
flooring of the dome-shaped grotto in which I found myself
was loose rock, at a level of about two feet below the surface
of the ice-floor, on which my guide Christian still stood.   The

dome itself was not high enough to allow me to stand upright, and from the roof, principally from the central part,
a complex mass of delicate icicles passed down to the floor,
leaving a narrow burrowing passage round, which was itself
invaded by icicles from the lower part of the sloping roof,
and by stubborn stalagmites of ice rising from the floor.
The details of this central cluster of icicles, and, in fact, of
every portion of the interior of the strange grotto, were exceedingly lovely, and I crushed with much regret, on hands
and knees, through fair crystal forests and frozen dreams of
beauty. In making the tour of this grotto, contorting my
body like a snake, to get in and out among the ice-pillars,
and do as little damage as might be, I yet, with all my
care, was accompanied by the incessant shiver and clatter of
breaking and falling ice. Having squeezed myself out again
through the narrow hole, 1 now passed between the two
gigantic columns, and found that the sea of ice became still
broader and bolder, until we came to the edge of a glorious
ice-fall, round and smooth and perfectly unbroken, passing
down like the rapids of some river too deep for its surface to
be disturbed, and plunging majestically into a dark gulf, of
which we could see neither the roof nor the end.'

We will now follow Mr. Browne to the Upper Glacière of
St. Livres, where the interesting discovery was made that
the ice-stream which filled the cave, instead of terminating
with the wall of rock at its end, turned off to the right, and
was lost in darkness. By tying a candle to a long stick,
and thrusting it down the slope of ice, it was further found
that the stream passed down at a very steep incline, and
poured under a narrow and low arch in the wall of the cave,
beyond which nothing could be seen. Steps were now cut
down the slope by one of the party, who was carefully let
down, and his work being completed, the others followed
him through the arch—a rather awkward undertaking, for,
on pushing through, their breasts were pressed on to the
ice, while their backs scraped against the rock which formed
the roof.

'As soon as this trough was passed,' says Mr. Browne,
'the ice spread out like a fan, and finally landed us in a
second cavern, 72 feet long by 36 feet broad, to which this

was the only entrance. The breadth of the fan at the
bottom was 27 feet, and near the archway a very striking
column poured forth from a vertical fissure in the wall, and
joined the main stream. The fissure was partially open to the
cave, and showed the solid round column within the rock:
this column measured 18½ feet in circumference, a little
below the point where it became free of the fissure, and it
had a stream of ice 22 feet long pouring from its base.
The peculiar structure of the ice gave the whole mass the

ICE-STREAMS IN THE UPPER GLACIÈRE OF MT. LIVHEN.

appearance of coursing down very rapidly, as if the water
had been frozen while thus moving, and had not therefore
ceased so to move. . . . At the farthest end of the cave,
a lofty dome opened up in the roof, beneath which a very
lovely cluster of columns had grouped itself, formed of clear
porcelain-like ice, and fretted and festooned with the utmost
delicacy, as if Andersen's Ice Maiden had been there in one
of her amiable moods and had built herself a palace.'

In Upper Styria, the Frauenmauer Mountain, which overlooks the mining town of Eisenerz, contains a remarkable ice-chamber, consisting of a grotto from thirty to forty fathoms long, decked with ice-crystals, pillars of ice, and cascades of the same material, the floor being composed of ice as smooth as glass. In the summer pleasure parties assemble in the cave, and amuse themselves with sliding down its sloping ice-floors.

In his work on the Natural Wonders of the Austrian Empire, the naturalist Sartori describes his visit to an ice-cave on the Brandstein, a peak situated in the same district, which thus appears to be rich in glacières. He found crimpons necessary for descending the frozen snow, which led from the entrance to the floor of the cave, where he discovered pillars and capitals and pyramids of ice of every possible shape and variety, as if the cave had contained the ruins of a Gothic church or a fairy palace. At the further end, after passing large cascades of ice, his party reached a dark grey hole, which lighted up into blue and green under the influence of the torches; they could not discover the end of this hole, and the stones which they rolled down into it seemed to go on for ever.

Other natural glacières are also mentioned as occurring in Bohemia, Hungary, the Hartz, in several places in North America, and probably there are few mountainous regions without them.

The Cave of Yermalik, already mentioned among the silent retreats of nature which have been rendered infamous by the cruelty of man, is likewise highly interesting as a natural glacière. After leaving the roomy dome in which they found the skeletons of the victims of Genghis Khan, Captain Burslem and Lieutenant Sturt proceeded through several low arches and smaller caves, and reached at length a vast hall, in the centre of which was an enormous mass of clear ice, smooth and polished as a mirror, and in the form of a gigantic bee-hive, with its dome-shaped top just touching the long icicles which depended from the jagged surface of the rock. A small aperture led to the interior of this wonderful congelation, which was divided into several compartments of every fantastic shape. In some the glittering icicles hung like curtains from the roof; in others, the vault was smooth as glass. Beautifully brilliant were the prismatic colours reflected from the varied surface of the ice when the torches flashed suddenly upon them as they passed from cave to cave. Around, above, beneath, everything was of solid ice, and being unable to stand on account of its slippery nature, they slid, or rather glided mysteriously, along the glassy surface of this hall of spells. In one of the largest compartments the icicles had reached the floor, and gave the idea of pillars supporting the roof.[*]

Rocks of volcanic formation seem to afford favourable opportunities for the congelation of water. Ice-caves are found in Mount Etna, on the Peak of Teneriffe, and among the lava-currents of Iceland.[†] Scrope visited one of these natural glacières near the village of Roth, in the neighbourhood of Andernach, on the Rhine. It formed the mouth of a deep fissure in a current of basalt derived from an ancient volcanic cone above it, and its floor was covered with a crust of ice at the time of his visit, about noon on a very hot day in August.

The phenomenon of wind-grottoes is analogous to that of ice-caves, and not seldom associated with them. Here cold currents of air, increasing in violence as the day is hotter, are found to blow from the interstices of rocks. One of the most celebrated of these Æolian caverns is found near Terni in Italy. The entrance is closed by an old gate, through the crevices of which the wind issues with a rustling noise, while in the grotto itself the current is sufficiently strong to extinguish a torch. The proprietors of some neighbouring villas have put the phenomenon to an ingenious use. Leaden pipes, branching out from the grotto, convey on sultry summer days an agreeable coolness through masks of gypsum with wide distended mouths, which are fixed in the walls of the apartments.

The small town of Roquefort in France has been renowned ever since the time of the Romans for the delicious flavour of its cheese, which is said to owe its excellence to the cool cellars in which it is matured. These are excavated on the northern slope of a great chalk plateau, and communicate with numerous fissures in the rock, from which air-currents

---

[*] Burslem, 'A Peep into Toorkistan.   [†] The Cave of Surtshellir.

stream forth of so low a temperature as to cause a thermometer marking $+23°$ R. in the shade, and in the external atmosphere, to fall to $+4°$ R. when exposed to their influence. The cellars are so valuable that one, which cost 12,000 francs in construction, sold for 215,000 francs.

In times of ignorance, superstition could not fail to attach its fables to the phenomenon of wind-grottoes. A cave near Eisenach was supposed to be the seat of purgatory, and popular credulity or terror willingly transformed the sounds produced by the rushing air-currents into the wailings of tormented souls.

Fortunately, modern science affords us a more satisfactory explanation of the phenomenon. Pictet represents the case of a cave with cold currents of air to be much the same as that of a mine with a vertical shaft ending in a horizontal gallery, of which one extremity is in communication with the open air at a point much lower, of course, than the upper extremity of the shaft. The cave or wind-hole corresponds to the horizontal gallery, and the various fissures in the rock take the place of the vertical shaft, and communicate freely with the external air. In summer the columns of air contained in these fissures assume nearly the temperature of the rock in which they rest—that is to say, the mean temperature of the district; and therefore they are heavier than the corresponding external columns of air which terminate at the mouth of the cave. The consequence is, that the heavy cool air descends from the fissures, and streams out into the cave, appearing as a cold current, and the hotter the day—that is, the lighter the columns of external air—the more violent will be the disturbance of equilibrium, and therefore the more palpable the current.

The evaporation which takes place as the air-currents descend through the moist rock-fissures likewise tends to lower their temperature. Several naturalists have attempted to explain the phenomena of ice-caves in a similar manner, as being produced by cold currents still further refrigerated by the evaporation caused in the moist and porous rocks through which they pass. But to this theory there are weighty objections, as in many ice-caves there is no current whatever, and in all the cases of cold-air streams investigated or mentioned by De Saussure, the lowest temperatures observed were still considerably above the freezing-point, and consequently incapable of converting water into ice.

Mr. Browne believes that, in many cases, the phenomenon may be satisfactorily accounted for by the position and surroundings of the caves in which it occurs; though, no doubt, cold currents and evaporation may often have an influence in maintaining the low temperature of ice-caves.

In every one of the fourteen natural glacières which he visited, the level at which the ice was found was considerably below the level at the entrance of the cave; so that, on ordinary principles of gravitation, the heavy cold air within could not be dislodged by the lighter warm summer air without. Heat naturally spreads very slowly in a cave like this; and even when some amount of heat does reach the ice, the latter melts but slowly, for ice absorbs 60° C. in melting; and thus when ice is once formed, it becomes a material guarantee for the permanence of cold in the cave.

Another means for preventing the encroachments of the hotter seasons is the dense covering of trees and shrubs, which, in the case of many of the glacières, shields their entrance or their roof from the rays of the sun, and thus keeps off the effects of direct radiation. Mr. Browne found all the glacières that came under his observation thus protected, with the single exception of that of St. Georges, where, in consequence of an incautious felling of wood immediately near the mouth, trunks of trees had been laid horizontally over it, to prevent the rays of the sun from striking down on to the ice. He moreover invariably found that the entrances to the caves were more or less sheltered against all winds—a very important condition, as air-currents from without would infallibly bring in heated air, in spite of the specific weight of the cold air stored within. There can be no doubt, too, that the large surfaces which are available for evaporation have much to do with maintaining a somewhat lower temperature than the mean temperature of the place where the cave occurs. Another great advantage which some glacières possess must be borne in mind, namely, the collection of snow at the bottom of the pit in which the

entrance lies. This snow absorbs, in the course of melting,
all the heat which strikes down by radiation, or is driven
down by accidental turns of the wind; and the snow-water
thus forced into the cave will, at any rate, not seriously
injure the ice.

It is easy to understand how, in caves thus protected
against the influence of summer heat, a great part of the
ice accumulated during the winter may be preserved, and

ENTRANCE TO THE GLACIÈRE OF ST. GEORGES.

that, for an explanation of the phenomenon, it is by no
means necessary to have recourse to cold blasts descending
from the interior of the rock in which they are situated. It
is indeed a common belief that the ice-caves are colder in
summer than in winter, and consequently contain a greater
abundance of ice during the former season; but this belief
may well be considered as one of those popular fallacies,
which—though, by dint of repetition, they come to be common articles of faith—have in fact no substantial proofs to
support them.

CHAPTER XVIII.

ROCK-TOMBS AND CATACOMBS.

Biban-el-Moluk, the Royal Tombs of Thebes—The Roman Catacombs—Their
Extent—their Mode of Excavation—Touching Sepulchral Inscriptions—Antony
Bosio, the Columbus of the Catacombs—The Cavaliere di Rossi—The Catacombs
of Naples and Syracuse—The Catacombs of Paris.

THE remoteness of caves and grottoes from the busy haunts
of life, their eternal silence and their nightly gloom,
have ever pointed them out as fit resting-places for the dead.
From the earliest times they have been used as sepulchral
vaults, and where nature neglected to hollow out the rock, it
has often been excavated for this purpose by the hand of
man.

Thus the Pharaohs of Egypt rested not in temples and
mausoleums reared in the heart of cities, but they chose the
desert-ravine for their sepulchre, and hid their tombs in deep
excavations in the earth.

A more impressive scene can hardly be imagined than that
which is afforded by these splendid memorials. Of all such
monuments which still mark the site of ancient Thebes,
perhaps none are more striking to the traveller than the
royal tombs—Biban-el-Moluk—which the pride of monarchs,
whose very name is now a mystery, excavated four thousand
years ago in the bosom of the Libyan mountains.

'The next morning at daybreak,' says Warburton,[*] 'we
started for the Tombs of the Kings. I was mounted on a fine
horse, belonging to the shiekh of the village, and the cool air
of the morning, the rich prospect before us, and the cloudless
sky, all conspired to impart life and pleasure to my relaxed
and languid frame. I had been for a month almost confined
to my pallet by illness, and now, mounted on a gallant barb,
sweeping across the desert, with the mountain breezes

---

[*] 'The Crescent and the Cross.'

breathing round me, I felt a glow of spirits and exhilaration
of mind and body to which I had been long a stranger. For
a couple of hours we continued along the plain, which was
partially covered with wavy corn, but flecked widely here
and there with desert tracts. Then we entered the gloomy
mountain gorges, through which the Theban monarchs
passed to their tombs. Our path lay through a narrow
defile, between precipitous cliffs of rubble and calcareous
strata, and some large boulders of coarse conglomerate lay
strewn along this desolate valley, in which no living thing of
earth or air ever met our view. The plains below may have
been, perhaps, once swarming with life and covered with
palaces ; but the gloomy defile we were now traversing must
have ever been as they now are, lonely, lifeless, desolate—
a fit avenue to the tombs for which we are bound.

'After five or six miles' travel, our guide stopped at the
base of one of the precipices, and, laying his long spear
against the rock, proceeded to light his torches. There was
no entrance apparent at the distance of a few yards, nor was
this great tomb betrayed to the outer world by any visible
aperture, until discovered by Belzoni. This extraordinary
man seems to have been one of the few who have hit off in
life the lot for which Nature destined them. His sepulchral
instincts might have been matter of envy to the ghouls ;
with such unerring certainty did he guess at the place containing the embalmed corpses most worthy of his body-
snatching energies.

'We descended by a steep path into this tomb, through a
doorway covered with hieroglyphics, and entered a corridor
that ran some hundred yards into the mountain. It was
about twenty feet square, and painted throughout most
elaborately in the manner of Raphael's Loggia at the Vati-
can, with little inferiority of skill or colouring. The door-
ways were richly ornamented with figures of a larger size,
and over each was the winged globe or a huge scarabæus.
In allusion, probably, to the wanderings of the freed spirit,
almost all the larger emblems on these walls wore wings,
however incompatible with their usual vocations; boats,
globes, fishes, and suns, all were winged. On one of the
corridors there is an allegory of the progress of the sun

through the hours, painted with great detail; the god of day sits in a boat (in compliment to the Nile he lays aside his chariot here), and steers through the hours of the day and night, each of the latter being distinguished by a star. The Nile in this, as in all other circumstances of Egyptian life, figures as the most important element; even the blessed souls for its sake assume the form of fishes, and swim about with angelic fins in this River of Life. One gorgeous passage makes way into another more gorgeous still, until you arrive at a steep descent. At the base of this, perhaps 400 feet from daylight, a doorway opens into a vaulted hall of noble proportions, whose gloom considerably increases its apparent size. Here the body of Osirei, father of Rameses the Second, was laid about 3,200 years ago, in the beautiful alabaster sarcophagus which Belzoni drew from hence, the reward of his enterprise. Its poor occupant, who had taken such pains to hide himself, was "undone" for the amusement of a London conversazione.

'There are numerous other tombs, all full of interest; but as the reader who is interested in such things will consult higher authorities than mine, I shall only add that the whole circumstance of ancient Egyptian life, with all its vicissitudes, may be read in pictures out of these extraordinary tombs, from the birth, through all the joys and sorrows of life, to the death, the lamentations over the corpse, the embalmer's operations, and finally the judgment and the immortality of the soul. In one instance the Judge is measuring all men's good actions in a balance against a feather from an angel's wing; in another, a great serpent is being bound head and foot, and cast into a pit; and there are many other proofs, equally convincing, of the knowledge that this mysterious people possessed of a future life and judgment.'

But not the kings alone; the illustrious, the wealthy, the whole nation, reposed in rock-tombs magnificently sculptured or rudely excavated, according to the means of the defunct. Behind the ruins of the stately temples of ancient Thebes, which extend from Gourna to Medinet Abou, and fill the narrow strip of desert between the inundated fields and the foot of the mountains, lies an interminable necropolis, whose graves, like the cells of a bee-hive, one close to the other, are hewn in the rocky ground of the plain, or in the slopes of the neighbouring hills.

These grottoes, originally destined for sarcophagi and mummies, are now occupied by fellahs and their herds, as they were in the fifth and sixth century by pious anchorites ; and, being roomy and situated at a considerable height above the plain, may be considered as the most healthy dwellings of the country.

The oldest graves are hewn in the mountains ; and at a later period, when the rocky plain at their foot alone gave room for these excavations, it gradually became invaded by the dead. In the more splendid of these mausoleums, high gates and walls inclosed deep courtyards, scooped out of the rock, and from these long corridors led to subterranean halls, profusely decorated with sculptures and paintings.

Similar cities of the dead existed in Upper Egypt, near the cities of the living, wherever the adjoining rocks allowed them to be excavated. Those of Syout—the ancient Lyco-polis, where along an extent of several miles the whole declivity of the Libyan mountains is perforated with graves rising in terraces to their very summit—of El Kab (Eilei-thyia), of Assuan (Syene), of Madfunch (Abydos), of Kau (Antæopolis), and of a hundred other places, would in any other country excite the wonder of the traveller; here, where along the Nile one gigantic necropolis follows upon another, they hardly attract any attention.

But the ancient Egyptians not only embalmed human bodies and preserved them in rock-tombs ; they also converted into mummies the various animals to which they paid divine homage, and deposited them in subterranean cavities. This honour was paid to Apis, the ox-god, to the sacred Ibis, to dogs, cats, and even to the repulsive crocodile.

Besides the Rome of the Cæsars and the Rome of the Popes, there is a third Rome, scarcely less remarkable than the other two. The two former, gilded by the warm sun-beam, proudly rise above the banks of the Tiber with their ruins, palaces, and churches, while the latter lies hidden beneath the earth.

From the cupola of St. Peter's—the most favourable

site for a panoramic view of the chief ruins and edifices
of the Eternal City—the eye is also best able to embrace
at one glance the general topography of the catacombs,
or of subterranean Rome. Fifteen great consular roads,
over which the victorious legions once marched out to subju-
gate the world, radiate from the centre of the town, and
furrow the arid Campagna, until they are finally lost in the

GALLERY WITH TOMBS.

hazy distance. To the right and left of these causeways
the catacombs have been hollowed out in the depths of the
earth, and, though separated by the Tiber into two distinct
regions, yet their various subdivisions trace a vast circle, the
size of which may be measured by the circumference of the
town itself. To form some idea of their extent, we must
fancy an intricate wilderness of galleries and arched alcoves
with their layers of sarcophagi, one above another; their
lucernaria, for light or ventilation; their stairs, straight or
winding; and all this, not on one level only, but floor
beneath floor—one, two, three, four, five—hewn out on a
labyrinthine, yet harmonious and economic plan. Network

is perhaps a feeble description of these vast and intricate mazes; a spider's web seen through the glass of a naturalist, or rather four or five spiders' webs, one within the other, would seem a more fitting illustration.

Such is the immensity of this city of the dead, that, according to the opinion of those who have made themselves best acquainted with all its subdivisions, their galleries, supposing them ranged in a line, would form a street 900 miles long, and lined by no less than six millions of tombs!

So vast a necropolis would command attention in any country, or as the memorial of any age, or of any part of the human race; but to us the catacombs of Rome are doubly interesting, as the mysterious crypts which served the first confessors of our faith for the purposes of sepulture and sometimes of concealment. Their extent at once precludes the idea of their having been excavated in a clandestine manner, as was at one time erroneously believed; and, moreover, history tells us that, apart from some passing storms of persecution, the Christians had as little reason as the Jews, their religious ancestors, for making a secret of their faith, or of their places of interment. From the times of the Apostles their community constantly grew and multiplied throughout the Roman world—in Rome especially, the centre of that world—and there can be little doubt that from Nerva to the middle of the reign of Marcus Aurelius (from 96 to about 166), and so onward to the great persecution under Decius (A.D. 249-256), the Christians, if exposed here and there and at times to local persecutions, were growing in unchecked and still expanding numbers. But as the living community increased, so also must the number of its dead, a reverence for whom was, among the survivors, not only a solemn duty but a deep-rooted passion. The Christians not only inherited from the Jews the ancient usage of interment, but their reverence for the dead was strengthened by the belief that Christ had risen bodily from the grave, and that a bodily resurrection was to be their own glorious privilege. Hence the burning of the dead, customary among the wealthier pagans, was to them a profanation; and as the body of the slave was as holy as that of his master, it claimed the same right of decent burial.

But where was room for the spacious burial-places required for so vast a community?

Within the walls of the city interment was very properly forbidden by the law, and at a convenient distance beyond its crowded precincts large plots of ground could hardly be obtained; but the construction of subterranean cemeteries on a vast scale was greatly facilitated by the geological formation of the land. Three different kinds of stone compose the groundwork of the Roman campagna: the *tufa litoide*, as hard and durable as granite, which furnished the materials for the palaces and temples of the Cæsarian city; the *tufa granolare*, which, though consistent enough to retain the form given it by the excavators, cannot be hewn or extracted in blocks, and the loose *tufa friabile*, or *pozzuolana*, which has been extensively used from the earliest ages for mortar or Roman cement. It is evident that neither the hard lithoid nor the loose tuff were suitable for the excavation of the catacombs, while this purpose could be admirably attained in the courses of the granular tuff, which, though not too hard to be worked, is yet solid enough to make walls for long and intricate passages, to be hewn into arches vaulting over deep recesses for the reception of coffins, and to support floor below floor down to the utmost depth to which the formation reaches.

Neither in the stone quarries nor in the sand-pits of ancient Rome is there the slightest sign that they were ever used for the purposes of sepulture, while in the granular tuff not a yard seems to have been excavated except for the making of tombs, which line the walls throughout their prodigious length, as close to one another as the berths in the sides of a ship. Though the most ancient catacombs were excavated by the Jews, yet these excavations are of a very limited extent when compared with those of Christian origin, where, instead of the seven-branched candlestick and other sacred emblems of the Jewish persuasion, every ornament or inscription marked or painted upon the walls bears witness to the faith of those who were deposited—to use the peculiar and appropriate expression—within these narrow cells. Everywhere we see Christian symbols only—the horse emblematic of strength in the faith; the hunted hare, of persecution; the dove and the cock, of the Christian virtues of vigilance and meekness; the peacock and the phœnix, of resurrection; the anchor, of hope in immortality; the palm-leaf, of the martyr's triumph over death, and many others.

The sepulchral inscriptions are short and simple, but often extremely tender and touching; recalling to memory, in a few brief words, the innocence and purity of life, the beauty and the wisdom, or the amiable, peace-loving character, of the deceased. The pompous or desponding tone of the heathen mortuary inscriptions disappears; the Christian 'sleeps and sleeps in peace.' There is no sign of affectation or hypocrisy in these simple epitaphs, and the ennobling influence of the new creed upon the spirit of man is, perhaps, nowhere more conspicuous than in the unostentatious inscriptions traced upon the tombs of its first confessors.

The use of the catacombs for the various purposes of interment, assembly, or concealment began, no doubt, with the first persecution under Nero. Most of the inscriptions, however, bear the date of the fourth or fifth century, and some are evidently of a much later period. When this custom may have ceased is uncertain, but in the Middle Ages the catacombs were entirely forgotten, and remained blocked up, until towards the end of the sixteenth century, when they were re-opened and explored by the indefatigable and courageous Antony Bosio, who devoted thirty-three years of his life to this labour. During his frequent wanderings through the Roman Campagna, this zealous archæologist once found, to the left of the Appian Way, near to the church of Sancta Maria in Palmis, a brick vault in a field covered with rubbish. He immediately presumed it to be the entrance of a catacomb, and descended through the narrow opening. Fired with scientific ardour, he penetrated further and further into recesses untrodden for centuries by the foot of man. The passage soon became so narrow and low as to oblige him to creep, but neither the difficulty of the exploration, nor the fear of being crushed to death by the crumbling stones, could restrain him; and thus, day after day, he continued his perilous search, until, finally, a

P

complete subterranean city revealed itself, although he could not ascertain its limits; for, however far he might probe its intricacies, new passages were still branching out on every side, and the maze descended in several successive stories deeper and deeper into the earth. After Bosio, other archæologists have continued his researches, and in our days the Cavaliere de Rossi—to whose indefatigable zeal is due, amongst others, the discovery of the Catacomb of Callistus, where many of the early popes have been entombed—has all but completed the topography of subterranean Rome.

Besides the ancient metropolis of the world, several other old Italian towns possess remarkable catacombs or subterranean burial-places. Those of Naples, historically far less interesting than those of Rome, are executed on a far more spacious plan. They are situated not beneath the town itself, but in a neighbouring mountain, where they have been excavated to a distance of more than two miles. Large galleries, eighteen feet broad, and fourteen or fifteen high, branch out into a number of smaller passages, while the walls on both sides are pierced, like those in Rome, with horizontal sepulchral cavities, six, or even seven, one above the other.

The Catacombs of Syracuse, under the site of the ancient Achradina, are the largest and best preserved known. They are all excavated in the solid rock, and form lofty vaults, very different from the narrow and dangerous burrows of subterranean Rome. A broad gallery runs through the whole of the labyrinth, and from this many other passages of an inferior width branch out, leading to large circular vaults with openings at the top for the admission of light. Many of these have been closed, as they were equally dangerous for the people in their neighbourhood and their cattle, so that torches are necessary for visiting them. Along the walls are a number of niches, which served as sepulchres, so that these excavations, which originally were quarries, and during the flourishing times of the city were used by its vast population for various household purposes, were ultimately converted into a city of the dead.

The Catacombs of Paris, though ancient as quarries, are of a very modern date as places of sepulture. Until the end of the reign of Louis XVI. the principal burying-ground of Paris had been the Cemetery of the Innocents, near the church of the same name. Originally situated beyond the walls of the town, it had in course of time been so surrounded by the growing metropolis as to occupy its centre. Here, during nearly ten centuries, numberless bodies had been deposited, so that, from the malaria it engendered, it became a constant source of danger to the living. At length the cemetery became so intolerable a nuisance that its suppression and conversion into a public market-place was decreed in 1785.

The question now arose where the bones to be displaced should be deposited, and, from their proximity to the town and their extent, the ancient quarries were chosen as the most favourable spot for a vast subterranean necropolis. But these immense excavations, which, having been abandoned for several centuries, had in many places fallen in, needed a full year for repairs before they could be safely used for their new purpose. At length, on April 7, 1787, they were solemnly consecrated, and the same day the workmen began to remove the bones from the Cemetery of the Innocents—an operation which needed more than fifteen months for its completion. Gradually many others of the ancient cemeteries of Paris were in like manner removed to the catacombs, so that they are said to contain the bones of more than three millions of bodies. All these bones are symmetrically piled up along the sides of the galleries ; the apophyses of the large thigh and arm bones are disposed in front, so as to make a nearly uniform surface, interrupted from space to space by a row of skulls. Some of the crypts or sepulchral chambers are rather lugubriously decorated with festoons or pyramids of skulls and cross-bones, and many of the stone pillars which support the vaults have likewise received ornaments of the same sexton taste. Sixty-three staircases lead from different parts of the town into the catacombs, and are used by the workmen and agents appointed to take care of the subterranean necropolis ; but visitors are admitted only every three months by the entrance at the Barrière du Maine. A descent of ninety steps brings them to a narrow gallery, which conducts them, after several windings, to the more roomy vaults where the bones are deposited; and after wandering for some time among these gloomy memorials which are piled up on either side, they finally emerge into daylight through another gallery, similar to the first. No doubt Paris affords many a more pleasant ramble, but hardly one more interesting, or capable of making a deeper impression on the mind.

CHAPTER XIX.

CAVES CONTAINING REMAINS OF EXTINCT ANIMALS.

The Cave Hyena and the Cave Bear—The Cavern of Kirkdale—The Moa Caves in New Zealand—Various Species of Moas—Their enormous size.

BESIDES their picturesque beauty or their solemn grandeur, some caves are extremely interesting as containing the bones of *extinct* quadrupeds or birds. Unrivalled in point of antiquity by the oldest tombs erected by man, they carry us back to times which, though of comparatively recent date, are still so far removed from the present day as almost to terrify the imagination—for how many ages must have elapsed, and what changes of climate must have taken place, since the hyena or the tiger inhabited the grottoes of northern Europe?

In many cases the bones found in ossiferous caves must have been washed into them by currents of water, or else they may be the remains of animals that accidentally dropped in through holes in the roof; but there can be no doubt that frequently the caves in which the bones of extinct carniverous animals have been found, served as their dens while they were living. The abundance of bony dung associated with the remains of the hyena, as well as the great number of bones all belonging to one species which are frequently found congregated in the same cave, is strongly confirmatory of this opinion, no less than the circumstance that the walls of several ossiferous caves, when deprived of their stalactital covering, have been found smoothed or rounded off from the frequent ingress and egress of their former inhabitants as they squeezed themselves or dragged their prey through a narrow passage. Besides, the hyenas and bears of the present day frequently live in caves, thus justifying the inference that their extinct predecessors had the same habit.

The antiquity of many of the animal remains found in caves is proved not only by their being dissimilar to existing species, but by the manner in which they are entombed. For ages they accumulated on the floor of the cave, often to a considerable depth, mixed with mud or sand or fragments of rock—then, when, from a change of level or some other cause, the cavern became uninhabited, a thick crust of stalagmite was slowly formed, and burying them all, as under solid stone, preserved them undisturbed, until some accident revealed their existence after a time the length of which escapes all calculation. The dim vista into the past appears still more shadowy in the case of the cavern discovered by Dr. Schmerling at Choquier, about two leagues from Liège, where three distinct beds of stalagmite were found, and between each of them a mass of breccia and mud, mixed with quartz, pebbles, and in the three deposits the bones of extinct quadrupeds!

Ossiferous caves have been found and examined in many parts of the world: in France, in Belgium, chiefly in the valleys of the Meuse and its tributaries; in Austria and Hungary, in Germany and England.

It is remarkable that while the remains of a large and extinct species of bear, although found in our caves, are much more common on the Continent, the bones of the hyena form by far the largest proportion of those obtained from the English caverns. Thus under the incrustated floors of our rock-crevices and hollows, we find the proofs that our island was once inhabited by brutes which are now confined to Africa and the adjacent parts of Asia. But the extinct hyena of England was a much larger and more formidable animal than either the striped hyena of Abyssinia or the spotted hyena of the Cape, the latter of whom it most resembled. The abundance of these animals, and the length of the period during which they inhabited England, may be inferred from Dr. Buckland's account of the opening of the celebrated cavern of Kirkdale.

'The bottom of the cave, on first removing the mud, was found to be strewed all over, like a dog-kennel, from one end to the other, with hundreds of teeth and bones, and on some of the bones marks could be traced, which, on applying one

to the other, appeared exactly to fit the form of the canine
teeth of the hyena that occurred in the cave. Mr. Gibson
alone collected more than three hundred canine teeth of the
hyena, which must have belonged to at least seventy-five
individuals, and adding to these the similar teeth I have
seen in other collections, I cannot calculate the total number
of hyenas, of which there is evidence, at less than two or
three hundred.'

The grisly bear of the Rocky Mountains (*Ursus ferox*) is
the most ferocious of its race at the present day. It is about
nine feet long, and is said to attain the weight of 800 pounds.
Its strength is so prodigious that even the bison contends
with it in vain. But this huge and formidable animal was
surpassed in size and strength by the extinct bear (*Ursus
spelœus*) which once inhabited the caverns of Europe, at a
time when vast and interminable forests covered the land,
and the Rhine and the Danube flowed through wastes like
those through which the Mackenzie or the Yenissei now find
their way to the ocean.

From the proportions of the molar teeth, and from some
peculiarities of appearance and wearing, it has been inferred
that this extinct species lived chiefly on vegetable food; but
its prodigious strength, and the huge canines with which its
jaws were armed, enabled it to cope with its contemporaries,
the large Auerox and the teichorhine rhinoceros, and to defend itself successfully against the large lion or tiger whose
remains have also been found in the caverns of western
Europe, and which, if we may judge by the size of their
canine teeth, must have been more than a match for the
largest felides of the present day.

Besides the hyena and the bear, more than a hundred
species of extinct animals have been discovered in our
ossiferous caves. In those of Paviland, Glamorganshire,
bones of a primeval elephant have been found; and in that
of Wirksworth, Derbyshire, the almost entire skeleton of
a rhinoceros lay buried in a considerable mass of gravel
and osseous fragments. How the large creature came there
is a question that may well exercise the ingenuity of a
geologist.

Though in these and similar cases, the bone-caves have

been found to contain the remains of animals very different from those now existing in the same region, yet in general they show a remarkable relationship, in the same land or continent, between the dead and the living. Thus in the caves of Brazil there are extinct species of all the thirty-two genera, excepting four, of the terrestrial quadrupeds now inhabiting the provinces in which the caves occur, such as fossil ant-eaters, armadilloes, tapirs, peccaries, guanacoes, opossums, and numerous South American gnawers, monkeys, and other animals.

CAVE IN DREAM LEAD MINE, NEAR WIRKSWORTH, DERBYSHIRE.

The kangaroo, as is well-known, is peculiar to Australia, and caverns in that country have been described by Sir T. Mitchell, containing fossil bones of a large extinct kangaroo.

A singular wingless bird, the Apteryx australis, is found only in the wilds of the interior of New Zealand, where it takes refuge in the clefts of rocks, hollow trees, or in deep holes which it excavates in the ground. The caves of that country show us that it was preceded by other wingless birds of a gigantic structure—the Moas, by the side of which even the ostriches of the present day would shrink into comparative insignificance.

These wonderful creatures would probably have remained unknown to the present day, if, in 1839, the thigh-bone of a Moa had not fallen by chance into the hands of Professor Owen, who from this single fragment drew up a surprisingly correct notice of the bird. This memoir, sent out to New Zealand, gave a stimulus to further researches, and from the larger quantity of the Moa's bones now sent to England, Professor Owen built up the beautiful skeleton which, along with those of the Mylodon robustus, of the Mammoth, and of the primeval stag, forms one of the most conspicuous ornaments of the splendid Museum in Lincoln's Inn Fields.

But the reconstructive genius of our great palæontologist, not satisfied with this triumph, has detected a whole group of ostrich-like birds among the remnants of the past, which the New Zealanders, who, as we may suppose, are no adepts in comparative anatomy, all confound under the common name of the Moa ; and thus five species of Dinornis, the Palæopteryx, the Aptornis, and the comparatively small Notornis, have been, as it were, resuscitated by a miracle of science. A specimen of the last-named species of these birds was caught alive in a remote, unfrequented part of the south island of New Zealand in 1850 by some sealers, who, ignorant of the value of their prize, killed and devoured it as if it had been a common turkey. Fortunately, however, the skin of this unique bird, the link between the living and the dead, the last perhaps of a race coeval with the gigantic Moas, was preserved from destruction.

The largest species of Moa (*Dinornis robustus*) must have stood, when alive, about thirteen or fourteen feet high, since Dr. Thomson saw a complete leg, which stood six feet from the ground. Like the ostrich, this feathered giant was incapable of flight, its rudimentary wings being unable to raise it from the ground. It had three toes on each foot, and tradition says that its feathers were beautiful and gaudy. Portions of the eggs of the bird have been found among their bones, of a sufficient size to estimate the probable size of a whole egg, and the conclusion is that the hat of a full-grown man would have been a proper-sized egg-cup for it. From the structure of the toes, which were well adapted for digging up roots, and from the traditional report that the Moas were

in the habit of swallowing stones to promote digestion, it
may be concluded that they were herbivorous.

This is about as much as we know of the aspect and
habits of this colossal Struthionide, whose apparent con-
finement to the small New Zealandic group is one of the
most curious enigmas of its history. As it is extremely improbable that so gigantic a race was originally formed for so
narrow a sphere, we are led to believe that New Zealand is
but the remnant of a vast continent now whelmed under the
waves of the Pacific.

When the group was first peopled by the Maories, about
five hundred years ago, the Moas had already become excessively rare, and the last of them seem to have perished
during the seventeenth century, though the New Zealanders
believe that some of them still live in the remote and unfrequented wilds of their native land.

The bones of the various species of Moa have been
partly found in morasses, partly in the beds of mountain
torrents, but chiefly in various caves, which no doubt served
the living animals as dwellings or places of refuge.

One of these caverns, called by the New Zealanders ‘Te
Anaoteatua,’ or the Cave of the Spirit, was visited by Dr.
Thomson,* who describes it as very remarkable. It is
situated in the tertiary, extremely cavernous, and undermined
limestone mountain chain which extends along the west
coast of the northern island, and whose picturesque beauties
and natural curiosities are destined to occupy a conspicuous
place in some future guide-book. Its entrance, resembling
the gateway of an old castle, is concealed by a thick foliage of
shrubs, and a dark green creeper adheres to the limestone
rock and covers the opening. The cave extends in a tortuous
direction under the hill for upwards of a mile, and branches
out into several passages. From its roof and sides numerous
stalactites are hanging, some of them six feet long, and
composed of transparent calcareous spar, while others
have a red tint. In that part of the cave which Dr. Thomson explored, there were three openings in the roof at
different places, each from ten to fifteen feet in circum-

* ‘On the Moa Caves of New Zealand.’ Edinburgh New Philosophical Journal,
vol. lvi. 1854.

ference, through which light was seen streaming in, one hundred and fifty feet above the head. Immediately below these openings there were heaps of wood and *débris* washed down from the surface; but these openings did not throw much light into the cave, so that even during the day it was perfectly dark. A subterranean stream of water runs through part of it, and then disappears under the rock.

Before the introduction of Christianity, this cave was held in the greatest terror by all New Zealanders, and no one would have ventured to enter it; but as Moa bones fetch a high price, the love of money at length conquered the superstitious fears of some Christian natives, and their boldness was rewarded by a tolerably rich harvest of bones, although their search was made in a very hasty and imperfect manner.

Another cave situated in the same neighbourhood, and bearing the indigenous name of ' Te Anaotemoa,' or the Cave of the Moa, did not enjoy the privilege of spiritual protection, and the Maories were in the habit of resorting to it to procure the skulls of the Moas, to hold the powder which they used for tattooing, and their long bones for the manufacture of fish-hooks, before they became acquainted with the use of iron.

The Cave of the Moa is in a limestone hill, with two openings, one towards the north-east, and the other towards the south-west. The north-east opening has evidently been caused by the falling in of the roof, and is apparently of no great age; the south-west entrance is 14 feet high and 10 feet broad, and covered over with trees and bushes. The cave is 165 feet long, the greatest breadth 28 feet, and the height 60 feet. The roof is oval, and numerous stalactites drop gracefully from it, giving a cathedral-like effect to the whole. One part of the cave is floored with calcareous spar, another with a large deposit of soft stalagmites; and that which is furthest from the south-west opening is covered with earth and stones, which appear to have fallen down when the roof of the cave gave way. It is under this earth, and the soft deposit of carbonate of lime, that the Moas' bones are found. Dr. Thomson collected only four skulls in this cave, and the scarcity of them was accounted for by

their use in former days as powder-holders.   There was
nothing to lead him to think that these bones had been deposited in the cave by water, for he found a remnant of almost
every bone in the body, from the spine and the rings of the
trachea down to the last bone of the toes.   The bones belonged
to the largest and also to the smaller species of Moas.

It would require several days' labour of many men to clear
out the bottom of this cave properly, in order to see what
bones it contains ; but, as far as Dr. Thomson saw, there were
no osseous remains of man or of any animals, except Moas,
in it, or any marks of fire, sculpture, nor figures of any kind
on the walls of the cave.

There are, no doubt, many other grottoes and cavities containing Moas' bones in the islands, but they are carefully
kept secret by the natives, as, on account of the high price
paid for the bones, they may almost be regarded as small
gold-mines.

CHAPTER XX.

SUBTERRANEAN RELICS OF PREHISTORIC MAN.

The Peat Mosses of Denmark—Shell Mounds—Swiss Lacustrine Dwellings—
Ancient Mounds in the Valley of the Mississippi—The Caves in the Valley of
the Meuse—Dr. Schmerling—Human Skulls in the Cave of Engis—Explorations
of Sir Charles Lyell in the Cave of Engihoul—Caverns of Brixham—Caves of
Gower—the Sepulchral Grotto of Aurignac—Flint Implements discovered in the
Valley of the Somme—Gray's Inn Lane an ancient Hunting-ground for
Mammoths.

AMONG the various researches of geologists, there are
perhaps none of a more general interest than those
which relate to the antiquity of the human race. We would
gladly know whether those strata and caves which contain
so many relics of the past afford us also some insight into
the primitive condition of man—some indication of the times
when he first appeared upon the stage of life. Within the
last years many discoveries have been made which have
thrown light upon the subject, and proved that, though man,
geologically speaking, is of recent origin, and probably the
youngest born of creation, his ancestry still stretches back
at least as far as that remote period when the huge Mammoth ranged over the north, when the cavern bear and the
cavern hyena tenanted the excavations of our limestone
hills, and the uncouth form of the rhinoceros brushed
through the dense primeval forests of our land. In the following pages I intend briefly to point out some of the most
interesting of these discoveries.

The peat-mosses of Denmark, varying in depth from ten
to thirty feet, show their enormous antiquity by the changes
which have taken place in the vegetation of the land since
the first periods of their growth. At their lowest levels lie
thick prostrate trunks of the Scotch fir, which must once

have grown on their margin, but which is not now, nor has ever been in historical times, a native of the Danish isles. At higher levels the pines disappear, and are supplanted by the sessile variety of the common oak, while still higher occurs the pedunculated variety of the same oak, now in its turn almost superseded by the common beech. It is impossible to calculate the number of ages needed to bring about these vast changes in the forest scenery of Denmark—a very moderate estimate carries them back seven thousand years—but at every depth in the peat, and under all these various trees, implements and other articles of human workmanship have been found, which show that, since the appearance of man, the fir and the oak have successively flourished and disappeared.

In addition to the peat-mosses, the Danish 'shell-mounds' throw some light on the prehistoric ages of that northern land. These mounds, or 'refuse-heaps,' consisting chiefly of thousands of cast-away shells of the oyster, cockle, and other molluscs of existing species, may be seen at certain points along the shores of nearly all the Danish islands. The shells are plentifully mixed up with the bones of various quadrupeds, birds, and fishes, which served as the food of the savages by whom the mounds were accumulated; while scattered all through them are rude implements of stone, horn, wood, and bone, with fragments of coarse pottery, mixed with charcoal and cinders, but never any implements of bronze or of iron. Similar refuse-heaps are found near the huts of many wild nations of the present day; as, for instance, near the miserable wigwams of the Fuegians, who live chiefly upon limpets.

The most striking proof that the Danish refuse-heaps are very old is derived from the character of their imbedded shells. These, indeed, belong entirely to living species; but the common eatable oyster is now unknown in the brackish waters of the Baltic, and the eatable mussels, cockles, and periwinkles, which in the refuse-heaps are as large as those which grow in the open sea, now only attain a third of their natural size. Hence we may confidently infer that in the days when the savage inhabitants of the Danish coasts accumulated these heaps, the ocean must have had freer access than now to the Baltic, and mixed its waters through broader channels with those of that inland sea.

As in the peat-bogs the implements and weapons found buried with the Scotch fir are all made of stone, whereas those coinciding with the oak epoch are made of bronze, we may further conclude that the stone hatchets and knives of the refuse-heaps likewise belong to the same distant period when evergreen forests flourished in the Danish isles.

The ancient Swiss lacustrine dwellings, so frequently mentioned of late years, afford us another highly interesting glimpse into remote pre-historic ages. They seem first to have attracted attention during the dry season of 1853-4, when the lakes and rivers were unusually low, and when the inhabitants of Meilen on the Lake of Zurich resolved to raise the level of some ground, by throwing upon it the mud obtained by dredging in the adjoining shallow water. During these dredging operations they discovered a number of wooden piles deeply driven into the bed of the lake, and, among them, many stone instruments, fragments of rude pottery, fishing gear, and the bones of various animals; those which contained marrow being split open, in the same way as those found in the Danish shell-mounds.*

The ruins thus unexpectedly brought to light were evidently those of a village of unknown date, and since then many other hill-dwellings (more than a hundred and fifty in all) have been detected near the borders of the Swiss lakes, at points where the depth of water does not exceed fifteen feet. Such aquatic sites were probably selected as places of safety, since they could be approached only by a narrow bridge or by boats, and the water would serve for protection alike against wild animals and human foes. The relative age of the pile-dwellings is clearly illustrated by the nature of the relics that lie scattered among their ruins. In some, only bronze utensils or ornaments are found; in others all the articles are of stone. The former, indicating an advance in civilisation, are evidently of a more recent age; but, even among the villages of the stone period, some are of later date than others, as they exhibit signs of an improved state of the

* See also the article on Lacustrine Abodes in the Edin. Review, July 1863.

arts. Thus we have here a long perspective into an unknown past, for even the bronze villages are probably of an age long antecedent to the Roman period. The oldest stone settlements are perhaps as old as the times of the Danish refuse-heaps; but even among their ruins some domesticated animals occur, namely the ox, sheep, goat, and dog ; and the appearance of three cereals indicates that the population had already made some progress in the agricultural arts. Amber ornaments, which could only have found their way from the Baltic, as also hatchets and wedges of jade—of a kind not occurring in Switzerland or in the adjoining parts of Europe, and which some mineralogists would derive from the East— prove the existence of an active commercial intercourse, even at that early age, which the Swiss archæologists and geologists carry back as far as 7,000 years.

In the basin of the Mississippi, and especially in the valley of the Ohio and its tributaries, many large mounds have been found, which have served in some cases for temples, in others for outlook or defence, and in others for sepulture. Some of these earthworks are on so grand a scale as to embrace areas of fifty or a hundred acres within a single inclosure ; and the solid contents of one mound are estimated at twenty millions of cubic feet, equal to about one-fourth of the bulk of the Great Pyramid of Egypt. From several of these repositories pottery and ornamental sculpture have been taken, as also weapons made of unpolished hornstone and various articles in silver and copper. An active commercial intercourse must have existed between the Ohio mound-builders and the natives of distant regions, as mica from the Alleghanies, sea-shells from the Gulf of Mexico, obsidian from the Mexican mountains, and implements made of native copper from Lake Superior, have been found among the buried articles.

The extraordinary number of the mounds implies a long period during which a settled agricultural population, considerably advanced in the industrial arts, occupied the fertile valleys or the alluvial plains in the basin of the Mississippi, covered since then with vast forests, and tenanted by wild hunters without any traditionary connexion with their more civilised predecessors. The epoch when this people flourished, or the adverse circumstances which swept them away, are all equally unknown; but the age and nature of the trees found growing on some of their earthworks afford at least some data for estimating the minimum of time which must have passed since the mounds were abandoned.

Trunks, displaying eight hundred rings of annual growth, have been cut down from them, and several generations of trees must have lived and died before the mounds could have been overspread with that variety of species which they supported when the white man first set foot in the valley of the Ohio. In a memoir on this subject, General Harrison, who was skilled in woodcraft, observes that 'beyond all doubt no trees were allowed to grow so long as the earthworks were in use; and when they were forsaken, the ground, like all newly cleaned ground in Ohio, would for a time be monopolised by one or two species of tree, such as the yellow locust and the black or white walnut. When the individuals which were the first to get possession of the ground had died out one after the other, they would in many cases, instead of being replaced by the same species, be succeeded (by virtue of the law which makes a rotation of crops profitable in agriculture) by other kinds, till at last, after a great number of centuries (several thousand years, perhaps), that remarkable diversity of species characteristic of North America, and far exceeding what is seen in European forests, would be established.'

In all the cases hitherto mentioned, the remains or relics of prehistoric man, however remote a date we may assign to them, have been found associated with fossil shells and mammalia of living species. In the instances I am now about to relate, we advance a step farther back, and find man the contemporary either of extinct mammalia or such as could now no longer exist in the lands where once they throve. As the mere mention of the numerous caves in Belgium, England, France, and Germany, in which human bones or articles of human workmanship have been found embedded along with the fossil remains of extinct animals, would tire the reader's patience, I select from the number a few which have afforded the most convincing proofs of the antiquity of man. The credit of having given the first impulse to these fruitful investigations is due to the late Dr. Schmerling, of Liège, who, with untiring zeal, devoted several years of his life to the exploring of the ossiferous caverns which border the valley of the Meuse and its tributaries.   To gain access to many of the caves was in itself no easy task, as their openings could be reached only by a rope tied to a tree; and when we consider that, after these arduous preliminaries, Dr. Schmerling had frequently to creep on all fours through contracted passages leading to larger chambers, there to superintend by torchlight, week after week, and year after year, the workmen who were breaking through a stalagmitic crust as hard as marble, in order to remove, piece by piece, the underlying bone breccia nearly as hard, and that while thus directing their labours he stood for hours with his feet in the mud, and with water dripping from the roof on his head, in order to mark the position and guard against the loss of each single bone which they brought to light, we can scarcely praise too highly his rare devotion to the cause of science.

Among these caverns thus laboriously explored, that of Engis was found to contain the remains of at least three human beings, deeply buried under a thick floor of stalagmite. The skull of one of these, who may have been a beauty when the Meuse flowed at least fifty feet above its present channel, was embedded by the side of a mammoth's tooth.   Another skull was buried five feet deep in a breccia in which the tooth of a rhinoceros and several bones of other quadrupeds occurred.

On the right bank of the Meuse, on the opposite side of the river to Engis, is the cavern of Engihoul, where likewise bones of extinct animals, mingled with those of man, were observed to abound; but with this difference, that whereas in the Engis cave there were several human crania and but very few other bones, in Engihoul there occurred numerous bones of the extremities belonging to at least three human beings, and only two small fragments of a cranium.   None of the caves examined by Schmerling contained an example of an entire skeleton, and the bones were invariably so rolled and scattered as to preclude all idea of their having been intentionally buried on the spot.   As no gnawed bones or any coprolites were found, he inferred that the caverns of the province of Liège had not been the den of wild beasts, but that their organic and inorganic contents had been swept into them by streams communicating with the surface of the country. In 1860 Sir Charles Lyell, on a visit to Liège, determined still further to examine the Cave of Engihoul, into which Schmerling had delved in 1831, and engaged some workmen to break through the crust of stalagmite, with the intention of searching for bones in the undisturbed earth beneath. Bones and teeth of the cave-bear were soon found ; and, at the depth of two feet below the crust of stalagmite, three fragments of a human skull, and two perfect lower jaws with teeth, all associated with the bones of bears, large pachyderms and ruminants, and so precisely resembling these in colour and state of preservation as to leave no doubt that man was contemporary with the extinct animals.

Our English bone-caves have likewise afforded abundant proofs of the antiquity of our race. In 1858 a new series of caverns having been accidentally discovered near the sea at Brixham, by the roof of one of them being broken through in quarrying, the Royal Society resolved to have it scientifically and thoroughly examined. The united length of the galleries, which were cleared out under the superintendence of experienced geologists, amounted to several hundred feet. Their width never exceeded eight feet. They were sometimes filled up to the roof with mud, but occasionally there was a considerable space between the roof and floor. The numerous fossils discovered during the progress of the excavations were all numbered and labelled with reference to a journal in which the geological position of each specimen was recorded with scrupulous care.

As in many other bone-caverns, the underground passages and channels were generally found floored with a layer of stalagmite, varying in thickness from one to fifteen inches, and next below occurred loam or bone-earth from two to fifteen feet in thickness. In the latter were found remains of the mammoth, of the cave-bear, of the cave-lion, and other extinct mammalia. No human bones were obtained, but many flint knives, chiefly from the lowest part of the bone-earth, and one of the most perfect lay at the depth of thirteen feet from the surface, and was covered with bone-earth or that thickness.

At one point in the overlying stalagmite a perfect reindeer's horn was found sticking, and in another an entire humerus of the cave-bear—a convincing proof that both these animals must have lived after the flint tools were manufactured, or in other words that man in this district preceded the cave-bear. 'A glance at the position of Windmill Hill, in which the caverns are situated, and a brief survey of the valleys which bound it on three sides, are enough to satisfy a geologist that the drainage and geographical features of this region have undergone great changes since the gravel and bone-earth were carried by streams into the subterranean cavities. Some worn pebbles of hematite, in particular, can only have come from their nearest parent rock at a period when the valleys immediately adjoining the caves were much shallower than they now are. The reddish loam in which the bones are imbedded is such as may be seen on the surface of limestone in the neighbourhood; but the currents which were formerly charged with such mud must have run at a level seventy-eight feet above that of the stream now flowing in the same valley.' *

In 1861 Colonel Wood found, in a newly discovered cave of the peninsula of Gower, in Glamorganshire, the remains of two species of rhinoceros—R. teichorhinus and R. hemitœchus—in an undisturbed deposit, in the lower part of which were some well-shaped flint knives, evidently of human workmanship. This is the first well-authenticated example of the occurrence of human implements in connexion with R. hemitœchus—the first proof that this extinct brute, elsewhere the usual companion of the mammoth, has been coeval with man.

In 1852, on the side of a hill near the small town of Aurignac in the South of France, a sepulchral grotto was discovered which, though unadorned and rude, is of the highest interest, as in point of antiquity it surpasses all other burial-places known, and leads us back to times long anterior to the oldest traditions of our race. In that year a labourer observed that rabbits, when hotly pursued by the sportsman, ran into a hole

* Lyell, 'Antiquity of Man,' p. 101.

which they had burrowed in a talus of rubbish washed down from the hill above. Expecting no doubt to 'drag some struggling savage into day,' he reached as far into the opening as the length of his arm, and drew out, to his surprise, one of the long bones of a human skeleton. His curiosity being excited, he then began to dig a trench through the middle of the talus, and in a few hours found himself opposite a large heavy slab of rock, placed vertically against the entrance. Having removed this, he discovered on the other side of it an arched cavity almost filled with bones, among which were two entire skulls, which he recognised at once as human.

The good people of Aurignac, highly interested in the discovery, flocked to the cave; and as they probably expressed a wish to see the bones of what they supposed to be their forefathers enjoy a Christian burial, the mayor ordered these human relics to be removed from the lonely spot in which they had so long reposed in peace, and to be re-interred in the parish cemetery. But before this was done—having, as a medical man, a knowledge of anatomy, though, as it seems, a very imperfect idea of the ethnological value of the discovery—he ascertained that the bones must have formed parts of no less than seventeen skeletons of both sexes. Unfortunately the skulls were injured in the transfer, and the interment was so negligently conducted that when, after the lapse of eight years, M. Lartet visited Aurignac, even the place could not be pointed out into which the skeletons had been thrown.

The eminent antiquary, however, resolved systematically to investigate the ground inside and outside the vault; and, having obtained the assistance of some intelligent workmen, made the following interesting discoveries, which amply rewarded him for his trouble. Outside the grotto he found a layer of ashes and charcoal, about six inches thick, extending over an area of six or seven square yards, and going no further than the entrance of the cave, there being no cinders or charcoal in the interior. Among the ashes were fragments of sandstone, reddened by heat, which had once formed a hearth, not fewer than a hundred flint articles, and knives, projectiles, sling-stones, and chips. Here also lay scattered the bones of various animals, some belonging to species extinct for thousands of years in France, such as the mammoth, the Siberian rhinoceros, the reindeer, and the gigantic Irish deer; and among these bones those which had contained marrow were invariably split open, as if for its extraction, many of them being also burnt. The spongy parts, moreover, were wanting, having been eaten off and gnawed after they were broken—the work, according to M. Lartet, of hyenas—whose bones and coprolites were mixed with the cinders and dispersed through the overlying soil. These beasts of prey are supposed to have prowled about the spot, and fed on such relics of the funeral feasts as remained after the retreat of the human visitors.

In the cave itself, along with some detached human bones which had escaped removal to the churchyard, were also found the bones of animals and some rude works of art: flat pieces of shell pierced through the middle as if for being strung into a bracelet, and the carved tusk of a young boar, perforated lengthwise as if for suspension as an ornament. The bones of animals inside the vault differed in a remarkable manner from those of the exterior, as none of them were broken, gnawed, half-eaten, or burnt, like those which were found lying among the ashes on the other side of the great slab which formed the portal. They seemed also to have been clothed with their flesh when buried in the layer of loose soil strewed over the floor, as they were often observed to be in juxtaposition, and in one spot all the bones of the leg of an *Ursus spelœus* were lying together uninjured. When we consider that it is still the custom of many savage tribes to bury pieces of meat, such as the bear's fat haunch, with the bodies of the dead (so that they may not lack food on their long journey to the land of spirits), we gain a most interesting glimpse into the life of prehistoric man—the contemporary of the reindeer, the cave-bear, and the mammoth in the South of France. Armed with flint weapons, he had established his supremacy over the wild beasts of the forest. Rude and ferocious, as no doubt he was, he carefully preserved the remains of his friends and kinsfolk, and performed funeral feasts before the vaults in which their bodies were interred.

The lower parts of the Valley of the Somme—far above Amiens and below Abbeville, as far as the sea—are covered with a bed of peat, in some places more than thirty feet thick. Near the surface of these moor-grounds Gallo-Roman remains have been found, and, still deeper, Celtic weapons of the stone period; but the thickness of the superincumbent vegetable matter is far inferior to that of the underlying peat, so that many thousands of years must necessarily have passed since the first growth of that swampy vegetation. But, however remote its age, it is still of later date than the adjoining or underlying alluvial deposits of clay, gravel, and sand, which cannot originally have ended abruptly as they do now, but must have once been continuous further towards the centre of the valley. A long time must necessarily have elapsed between their deposition and subsequent denudation, and the first growth of the peat. In the lowest, and consequently oldest, of these beds, there has been found also a mixture of freshwater and marine shells, bones of the primitive elephant and teichorhine rhinoceros, and a number of flint implements shaped by the hand of man. Thus we have here convincing proofs that a race of savages inhabited the Valley of the Somme long before it was scooped out to its present depth, and when as yet not a trace existed of the thick bed of peat which now covers its lower grounds, and required so many thousand years for its formation.

Similar flint implements, in connexion with the bones of extinct animals, have been disinterred from ancient drift formations in many parts of England—in Surrey, Middlesex, Kent, Bedfordshire, and Suffolk. In the British Museum there is, among others, a flint spear-headed weapon which was found with an elephant's tooth near Gray's Inn Lane in 1716.

When, according to the old myth, Evander led Æneas over the site of future Rome, they saw the cattle grazing in what was to be the Forum, and heard them bellowing among the future dwellings of the rich senators and knights,

> —— 'passimque armenta videbant
> Romanoque Foro et lautis mugire Carinis.'

The change thus pointed out by Virgil was far less than that exemplified by these few relics of a time when possibly no intervening sea separated Britain from the rest of the world.

CHAPTER XXI.

TROGLODYTES OR CAVE-DWELLERS—CANNIBAL CAVES.

Cave-Dwellings in the Val d'Ispica—The Sicanians—Cannibal Caves in South
Africa—The Rock City of the Themud—Legendary Tale of its Destruction.

CAVES were probably the earliest habitations of primitive
man—the first rude shelter for which he had to do battle
with the hyena or the bear. As we have seen in the preceding chapter, some of the oldest relics of his prehistoric
existence have been discovered in caves; and at a far later
period, when he had already made some progress in the arts,
we find many races still adhering to the subterranean dwellings of their forefathers. Such a state of things may possibly be indicated in the mythical stories of the Greek and
Trojan heroes, who, near the base of Mount Etna, are said
to have found caverns tenanted by the troglodytic tribe of
the Cyclops.

In another part of Sicily the old cave-dwellings of the
Val d'Ispica deservedly attract the antiquarian's attention.
The vale is a narrow gorge, situated between Modica and
Spaccafurno; and throughout its whole length of about eight
Italian miles, the rock-walls on both sides are pierced with
innumerable grottoes, which at first sight might be taken
for the productions of nature; but on a closer inspection evidently show that they are the work of man, and have been
originally excavated to serve as dwellings. The luxuriant
vegetation of the Val forms a pleasing contrast with the
naked sterility of the plain in which it is imbedded; a small
rivulet flows along the bottom, and irrigates wild fig-trees
and stately oleanders; on a higher level grow broad-leaved
acanthuses and wild artichokes, while thick festoons of

cactus hang down from the top of the rocks and shade the
entrance of the grottoes. These are excavated at various
heights above the bottom of the vale, and often consist of two
or three stories, one above the other. A great part of the
rock-wall on the right bank of the brook has fallen in, and
exposes to sight the internal arrangement of the dwellings.
On the mounds of rubbish the visitor is able to ascend to
the entrances of the caves, which originally can have been
accessible only by ladders. The chambers are seldom more
than twenty feet deep and six feet high and broad. The
above-mentioned rock-slip has laid open, among others, a
dwelling of three stories, with flights of steps in a good state
of preservation, to which, on account of its superior dimensions, the neighbouring peasants have given the name of
Castello d'Ispica, and which may have been the abode of a
chieftain.

Several mortar-like basins or troughs, hollowed in the rock,
evidently served for the pounding of corn, so that the inhabitants, whoever they were, must have known at least some
of the arts of civilised life. A few of the rooms or burrows
(for their size hardly warrants a better name) are still tenanted
by shepherds, whose whole furniture consists of a kettle, a few
pots, and some skins to lie upon. Parthey, a learned German
traveller,* who visited many of the caves (the whole number
probably exceeding 1,500), nowhere found the least traces of
ornament about them; the doors and windows were mere
rough holes broken through the rock. Thus there can be no
doubt that the fragments of painted vases and sculptured
marble that have been found here and there in the caves
belong to a much later period, for a people sufficiently advanced in the arts to paint vases and to chisel marble could
not possibly have been content with a mole-like existence.
If we seek to ascertain the probable age of this cavern city,
all circumstances combine to throw back its origin to a very
ancient date. The total want of artistic decoration at once
forbids us to assign these extensive but rude excavations to
one of the Greek tribes which successively founded colonies in
the island after expelling the aborigines from their ancestral
seats. In the Roman, the Saracenic, or some later period, a

* 'Wanderungen durch Sicilien.'

band of fugitives might indeed have found a temporary retreat in this remote vale, but could hardly have remained concealed long enough to execute a work which evidently required long years of undisturbed toil. Thus it has been conjectured that these caves were the dwellings of the ancient Sicanians, a people who became known to history only when about to disappear for ever from the stage of the world, having left no memorial of their existence save perhaps these rude vestiges of an infant state of society.

We may justly presume that the aborigines of Malta, the Balearic Islands, and Sardinia, were likewise troglodytes when the first foreign colonists landed on the shores of these islands. And such, after the lapse of more than thirty centuries, are the Sarde shepherds of the present day, who frequently have no better dwellings than the caves of the rocks.

In Southern Italy the traveller meets with traces of the same primitive mode of life. Near a place called Iscalonga, in the province of Basilicata, Mr. Mallet found many inhabited caves, excavated in dry tufa of extreme antiquity. Some of the troglodytes came out to see him pass, and looked savage and queer enough in their rough brown blanket-cloaks, with peaked hoods and sheepskins.

Irrespective of climate, we at the present day find cave-dwellers among the tribes of the frozen regions of the North and the Arabs of the stony wastes bordering on the Red Sea; in the Libyan deserts, and in the sandstone rocks of Southern Africa—wherever nature has formed grottoes, or the soft material admits of an easy excavation.

Amongst the many interesting objects of the Transgariep country are the celebrated cannibal caverns which extend from the Moluta to the Caledon river. Thirty years ago they were inhabited by a race of men who, like the lion or the panther, were the scourge of the whole neighbouring country. Their mode of living was to send out hunting parties, who would conceal themselves among the rocks and bushes, and lie in ambush near roads, drifts, gardens, or watering-places, for the purpose of surprising and capturing women, children, or travellers. The influence of the chieftain Moshesh induced them to change their evil practices for a better mode of life, and they are now said to be an inoffensive people of agriculturists and traders. When Mr. James Henry Bowker visited the caves near the sources of the Caledon river, in 1868, he met at one of them an old savage, one of the most ill-looking ruffians he had ever beheld, who had formerly assisted at the cooking and bone-picking of many a human victim, and, like the Last Minstrel, seemed greatly to regret that

> 'Old times were changed,
> Old manners gone.'

and that

> ' The bigots of the iron time
> Had called his *harmless life* a crime.'

The largest cavern is situated amongst the mountains beyond Thaba Bosigo, the residence of the old chief Moshesh, about ten miles distant from the deserted missionary station Cana. The entrance to this cavern is formed by the overhanging cliff, and its arched and lofty roof is blackened with the smoke and soot of the fires which served for the preparation of many a horrible repast. At the further end of the cave some rough irregular steps led to a gloomy-looking natural gallery, where the victims not required for immediate consumption were confined. The cannibals, consisting of Betshuana and Kafir tribes, had the less excuse for their barbarity as they inhabit a fine agricultural country, which likewise abounds in game; but it is said that, having in a time of famine been reduced to the horrible extremity of eating human flesh, they acquired a taste for it, and continued to relish it as a delicacy even in times of abundance. Though they are now reported to be no longer cannibals, there is reason to fear that they have not yet entirely abandoned their diabolical way of living, for among the numerous bones which strewed the floor of the cavern and had chiefly belonged to children and young persons, Mr. Bowker found some that could hardly have been there many months. The skull of a child which lay before the mouth of the cavern afforded a touching example of 'life in death,' for a little bulb had sprung up from its cavity and covered it with a graceful tuft of drooping leaves.

At the distance of a ten days' journey from Medina on the road of the pilgrim caravans from Damascus, lies the deserted rock-city of the Themud, where a whole people had hewn their dwellings in the black rock, decorating the entrances with small columns on both sides, and tracing numerous inscriptions on the walls. No European traveller has ever visited this curious spot, for the Bedouins render the neighbourhood insecure, and the pilgrims to the holy cities allow no infidel to accompany them on their journey. How the subterranean city came to be deserted is still a secret of the past; but probably its ruin must be traced back as far as the first or second century of the Christian era, for Mahomet cites it more than once in his Koran as a warning to all true believers. According to the legend the prophet Saleh once came to Themud to convert the idolatrous inhabitants to the belief of the one true God. As a proof of his divine mission they required a miracle to be performed, upon which the prophet, raising his staff, struck the rock, which immediately opened and gave passage to a she-camel with its young one. But the obdurate pagans, still persisting in their incredulity, killed the camel, and the young one would no doubt have shared its mother's fate, if it had not speedily retired into the rock whence it had come forth. The sacrilegious crime of the Themud did not long remain unpunished. A dreadful earthquake destroyed them to the last man, and ever since the place is cursed. When the caravan passes by, the pilgrims raise loud shouts and hurry along as fast as they can, fearful of their dromedaries becoming shy from the wailings of the young camel, which is still supposed to exist in the rock.

CHAPTER XXII.

TUNNELS.

Subterranean London—The Mont Cenis Tunnel—Its Length—Ingenious Boring Apparatus—The Grotto of the Pausilippo—The Tomb of Virgil.

THE most renowned subterranean works of previous ages are generally of a religious character, as they have been executed to serve either as resting-places for the dead, or as temples in which gods or saints were worshipped. Thus the rock-tombs of the Egyptian Pharoahs and the sacred grottoes of India still bear witness to the feelings which conceived and realised the idea of these stupendous excavations.

Our own times furnish no similar examples of underground temples or mausoleums on a scale so grand as to command the admiration of posterity. We neither scoop out whole mountains nor deeply plunge into the entrails of the earth to reverence the dead or to testify our devotion; our subterranean labours all bear the stamp of practical utility. But never yet has the genius of man executed such wondrous excavations as those of the present day; and though the purposes for which they are performed may seem commonplace and prosaic when compared with the ideal aims of the unknown artists who planned the Pharaonic rock-tombs, or the temples of Ellora, yet the boldness of their conception entitles them to rank among the grandest architectural works, while the difficulty of their execution would have appalled the most enterprising engineers of any age.

Modern London alone has more subterranean wonders to boast of than all the capitals of the ancient world. As in the human frame numberless vessels and nerves provide for the circulation of the blood, and convey telegraphic signals through every part, so in the vast body of our metropolis an amazing system of subterranean communication carries off the sewerage of its millions of inhabitants, provides them with light and water, conveys intelligence from one end to the other of its enormous circuit almost with the rapidity of thought, transports thousands of travellers below the crowded streets, and, not satisfied with all these achievements, opens passages under the broad river to which it owes its boundless wealth.

But even the wonders of subterranean London, or of Paris —where troops moving underground can march from one fortified caserne to the other, so that, in a more literal sense

BORING MACHINE IN THE TUNNEL, MONT CENIS.

than Pompey, the rulers of France can say that they have but to stamp upon the ground to make legions start up from the soil—are surpassed by those which the railroad calls forth in its triumphal progress through the world.

The Alps themselves, with their eternal snows, no longer oppose a barrier to the locomotive; and we have lived to witness the completion of the most gigantic tunnel ever yet devised by man. The borer has slowly but indefatigably done its work in the entrails of Mont Cenis, and the subterranean junction of France and Italy has been at length achieved. The tunnel, which pierces 12,201 metres, or nearly seven miles, of solid rock, opens on the Italian side at Bardonneche, 1,291 metres (about 4,000 feet) above the level of the sea, and on the French side at Modane, at a height of 1,163 metres. From each opening the tunnel gently ascends towards its culminating point in the centre of the mountain, so as to allow the waters to drain off on either side. The direction of the excavations was determined by trigonometrical measurement, which of course required the nicest accuracy, as a deviation of but half a

BORING MACHINE IN THE SECOND WORKING GALLERY, MONT CENIS TUNNEL.

centimetre on both sides would amount to no less than 120 metres in the centre.

For this purpose a signal was erected on the mountain above the culminating point, or the centre of the tunnel; and opposite to each entrance, in the same longitudinal axis, a large theodolite pointed to the above-mentioned signal and towards a light in the interior of the tunnel, so that the slightest deviation from the requisite direction was rendered mathematically impossible.

The machines used in the drivage of the tunnel were worked by means of compressed air. The comminuted rock produced during the boring was continuously removed by a jet of water squirted into the hole by the machine—a precaution which was found to be of great economy in the wear of the tools. Under ordinary conditions the working speed was about 200 strokes per minute, at which rate from eight to ten holes, of 1·6 inches diameter and 35 inches deep, could be bored in the shift of six hours.

The tunnel is of an ⌓ section, 13 feet broad and 9¾ feet high; eight machines were employed at one time; they were mounted on a wrought-iron frame, which travelled on a railway, and could be removed to a safe distance when the blasting took place. From 65 to 70 holes were bored in the face of the rock, and then loaded with uniform charges, made up into cartriges, and primed with fuses of different lengths, so as to explode in groups at definite intervals. A horizontal series at about mid height was first fired, then followed two vertical side rows, then a group near the crown of the arch, and finally a horizontal series near the floor. In this way the rock was blown down in nearly uniform fragments of from six to eight inches in size. At first, owing to the hardness of the quartz-slate which had to be pierced, the work went on very slowly; but at a greater depth the mountain was found to consist of limestone strata, so that at length it advanced about thirteen feet per day.

The compressed air which set the borers in motion served also for the ventilation of the tunnel, as every stroke of the piston liberated sufficient pure air to drive back to a considerable distance from the workmen the suffocating gases produced by the blasting, which otherwise would render breathing impossible. Air-pumps placed before the entrance finally conducted the fumes of the gunpowder or dynamite into the open air.

When Louis Quatorze, that model of regal pomposity, succeeded in placing the crown of Spain on the head of his grandson Philip of Anjou, he bid him grandiloquently farewell with the words: 'Go, my son! the Pyrenees no longer exist.' With more truth is our age now able to declare that the Alps have ceased to be a barrier between nations.

There can be no doubt that the Mont Cenis tunnel will soon be followed by similar undertakings, some probably on a still grander scale. The governments of Italy, Switzerland, and Germany, have already decided upon piercing a tunnel through the Mount St. Gothard. In winter the traveller will no longer be obliged to wind slowly through the solitudes of the Schöllenen, to cross the Devil's Bridge, and then, after traversing the valley of Urseren, ·to pursue his perilous way up to the summit of the pass, before a zigzag descent, along yawning precipices, leads him into the sunny vale of the Ticino; but at Göschenen on the north he will plunge at once into the bowels of the mountain, and in an hour will suddenly emerge at Airilo into the genial south.

In early spring the change of climate will be almost miraculous; on one side winter still reigning supreme, the earth covered with deep snow and the cold wind moaning through the leafless trees—and then, after a short interval of darkness, the rich vegetation of an Italian vale bursting forth in all its first luxuriance, birds singing in the chestnut groves, and the green meads enamelled with myriads of flowers.

It is curious to compare these stupendous works with the most celebrated tunnel of ancient times—the Grotto of the Pausilippo, near Naples, which serves as a passage for the travellers who intend visiting Puzzuoli, Baiæ, or Cumæ, without being obliged to ascend the mountain or to cross the bay. It is about a mile long, from thirty to eighty feet high, and twenty-eight broad, so as to allow three carriages to pass abreast. Twice a year, in February and October, the last rays of the sun dart for a few minutes through the whole length of the grotto; at all other times a speck of gray uncertain light terminates the long and solemn perspective. The origin of this celebrated grotto is unknown. Some old chroniclers have imagined it as ancient as the Trojan war; others attribute it to the mysterious race of the Cimmerians.

The Neapolitans attribute a more modern, though not more certain, origin to their famous cavern, and most piously believe it to have been formed by the enchantments of Virgil, who, as Addison very justly observes, is better known at Naples in his magical character than as the author of the

'Æneid.' This strange infatuation most probably arose from the fact that the tomb, in which his ashes are supposed to have been deposited, is situated above the grotto, and that, according to popular tradition, it was guarded by those very spirits who assisted in constructing the cave. King Robert—a wise, though far from poetical monarch—conducted his friend Petrarch with great solemnity to the spot, and, pointing to the entrance of the grotto, very gravely asked him whether he did not adopt the general belief, and conclude that this stupendous passage derived its origin from Virgil's powerful incantations.

'When I had sat for some time,' says Beckford,* 'on a loose stone immediately beneath the first gloomy arch of the grotto, contemplating the dusky avenue, and trying to persuade myself that it was hewn by the Cimmerians, I retreated, without proceeding any further, and followed a narrow path which led me, after some windings and turnings, along the brink of the precipice, across a vineyard, to that retired nook of the rocks which shelters Virgil's tomb, most venerably mossed over and more than half concealed by bushes and vegetation. The clown who conducted me remained aloof at an awful distance whilst I sat communing with the manes of my beloved poet, or straggled about the shrubbery which hangs directly above the mouth of the grot.

'Advancing to the edge of the rock, I saw crowds of people and carriages, diminished by distance, issuing from the bosom of the mountain, and disappearing almost as soon as discovered in the windings of the road. Clambering high above the cavern, I hazarded my neck on the top of one of the pines, and looked contemptuously down on the race of pigmies that were so busily moving to and fro. The sun was fiercer than I could have wished, but the sea breezes fanned me in my aërial situation, which commanded the grand sweep of the bay, varied by convents, palaces, and gardens, mixed with huge masses of rock, and crowned by the stately buildings of the Carthusians and fortress of St. Elmo. Add a glittering blue sea to this perspective, with Caprea rising from its bosom, and Vesuvius breathing forth a white column

---

* Italy.

of smoke into the æther, and you will then have a scene upon which I gazed with delight for more than an hour, forgetting that I was perched upon the head of a pine with nothing but a frail branch to uphold me. However, I descended alive, as Virgil's genii, I am resolved to believe, were my protectors.'

CHAPTER XXIII.

ON MINES IN GENERAL.

Perils of the Miner's Life—Number of Casualties in British and Foreign Coal Mines—Life in a Mine—Occurrence of Ores—Extent and depth of Metallic Veins—Mines frequently discovered by Chance—The Divining Rod—Experimental Borings—Stirring Emotions during their Progress—Sinking of Shafts—Precautions against Influx of Water—Expense—Shaft Accidents—Various Methods of working Mineral Substances—Working in Direct and Reverse Steps—Working by Transverse Attacks—Open Quarry Workings—Pillar and Stall System—Long Wall System—Dangerous Extraction of Pillars—Mining Implements—Blasting—Heroes in Humble Life—Firing in the Mine of Rammelsberg—Transport of Minerals Underground—Modern Improvements—Various Modes of Descent—Corfs—Wonderful Preservation of a Girl at Fahlun—The Loop—Safety Cage—Man Machines—Timbering and Walling of Galleries—Drainage by Adit Levels—Remarkable Adits—The Great Cornish Adit—The Georg Stollen in the Hartz—The Ernst August Stollen—Steam Pumps—Drowning of Mines—Irruption of the Sea into Workington Colliery—Hubert Goffin—Irruption of the River Garnock into a Mine—Ventilation of Mines—Upcast Shafts—Fire Damp—Dreadful Explosions—The Safety Lamp—The Choke Damp—Conflagrations of Mines—The Burning Hill in Staffordshire.

FEW metals are found in a native state, nor are they commonly scattered in loose masses, nuggets, grains, or scales, over the surface of the earth. Hence the seeker's trouble is by no means confined to the task of gathering these masses, or of separating them by washing from the alluvial sand or gravel with which they are mixed; much more frequently they are chemically combined with other substances, from which a far advanced state of science is alone able to separate them, or deeply imbedded in subterranean mines, often so difficult of access as to tax for their working all the energies of man and all the resources of his metallurgic skill. The labours of the miner require indeed no less courage and presence of mind than those of the mariner. He no more knows whether he shall ever return from the pit into which he descends in the morning for his hard day's work than the sailor knows whether he shall ever revisit the port which he is leaving. He is perpetually at war with fire and earth, with air and water; and this eternal strife levies a no less heavy tribute of death and suffering than the storms of the raging seas.

In the year 1867, 1,190 persons perished in our 3,192 collieries, which employ a total of 333,116 workmen. Of these, 286 were killed by explosions; 449 by falls of rock; 211 by other subterranean causes; 156 in the shafts; and 88 above ground. In the same year, 293 lives were lost in the Prussian collieries, where 102,773 workmen find employment. Of these, 39 perished by fire-damp; 106 by fall of roof; 65 in shafts; 74 by casualties under ground; and 9 by casualties above ground. These melancholy lists may give us some idea of the number of serious but non-fatal accidents which are not mentioned in the official accounts. Every visitor to a coal-mine will meet with many pit lads who have been 'lamed' (or injured) several times in a few years, and who reckon events by the mournful chronology of their various 'lamings.' Collieries, as will be seen in the sequel, are indeed peculiarly liable to frightful accidents; yet the number of lives lost in the inspected ironstone mines of Great Britain, in 1866, amounted to 81. Extending our view to the mines of Austria and Russia, of France and Belgium, of Mexico and Peru, &c., &c., where the same dangers cause, no doubt, an equal amount of death or suffering, we may justly conclude that not a day, probably not an hour, passes that does not doom more than one miner to an untimely grave or to permanent mutilation.

But if mining is attended with a lamentable amount of individual suffering, the benefits derived from it by mankind in general are so important that the whole fabric of modern civilisation may be said to rest upon its basis.

Coal and the useful metals rule the world. Wherever they occur in large masses they establish the prosperity of a people on the surest foundations; and England is in a great measure indebted for her high station among the nations of the earth to the treasures hidden beneath her soil.

A large mine displays unquestionably some of the most interesting scenes of human activity. The restless industry pervading its subterranean caves and galleries impress s the visitor with feelings of wonder akin to those which he experiences when he first sets foot on a man-of-war; and if he feels giddy on seeing the sailor climb the loftiest masts, the sight of the yawning abyss into which the miner undauntedly descends seems terrible to his unaccustomed eye; and as he penetrates further and further into the recesses of this unknown world, his sensations are not rendered more agreeable.

The intricate passages branching out into a mysterious distance; the vaults and high halls faintly illumined here and there by a glimmering lamp; the dark forms emerging every now and then from some obscure recess, and then again plunging into night, like demon shades; the clanking of hammers, the rushing of waters, the creaking of wheels, the monotonous sound of machinery, or the loud explosion which, repeated by subterranean echoes, rolls like muttering thunder from vault to vault; the oppressive air in the low galleries, through which he can only move in a stooping position; the fear of being crushed any moment by a falling rock, or shivered to atoms by fire-damp combustion—all combine to produce an impression which can seldom be made by any scenes above ground.

The admiration which this imposing spectacle necessarily calls forth in his mind increases when he reflects how much skill and experience was required, and how many improvements and inventions had to be made, before mining could be brought to its present state of perfection.

Ores sometimes occur, like coal, in layers or beds, running parallel with the strata of the inclosing rocks, or in prodigious irregular masses. Most commonly, however, they are found in veins traversing the rocks in every conceivable direction, and filling the crevices and chasms which former terrestrial revolutions have rent in the hard stone. From the wild and titanic powers that have here been at work, it may easily be imagined how irregular the direction of these veins must be, and under how many various forms they must appear. Here they are horizontal, there vertical; here they form thin layers, there they fill chasms several hundred feet thick. Sometimes they split into several minor branches, or make abrupt bends; and frequently they have been rent, torn, or displaced in every possible manner by subsequent revolutions.

Their length is as various as their thickness or their direction. Some are short, while others traverse the rocks to a distance of many miles. Thus the argentiferous veins in the neighbourhood of Clausthal, in the Hartz, are three leagues long; and the famous Veta Madre, or chief lode of Guanaxuato in Mexico, has been traced for a length of eight miles.

With regard to depth, the lower extremity of hardly a single mineral bed or vein of any note has as yet been pointed out, though many have been worked to a considerable depth. Thus, one of the pits of St. Andreasberg, in the Hartz, is 2,485 feet deep, though, on account of its high situation in the mountains, not much more than 280 feet below the surface of the sea; while, in the coal mines of Valenciennes and Liège, the deepest shafts are sunk from 1,300 to 1,600 feet below the level of the ocean. The shaft of the colliery of Sacrée Madame, near Charleroi, is 800 yards deep, and that of Dukinfield Colliery 2,050 feet.

The great difficulty of carrying on mining operations at great depths will, probably, for ever prevent most metalliferous veins from being followed to their origin in the bowels of the earth; for while on high mountains the rarefaction of the atmosphere prevents respiration, the increasing pressure or impurity of the air at a depth of four or five thousand feet below the level of the sea must necessarily hinder the free expansion of the lungs. The weakness of our organisation soon sets limits to our progress, whether we wish to rise into the air or to penetrate into the interior of the earth; and it is only on spiritual wings that we soar aloft to the stars, or wander through the mysterious depths of our planet.

How have the ores been collected or precipitated in the veins or strata where they are often found in such enormous quantities? Partly they ascended as vapour from unknown depths, and then were condensed in the crevices, mixing with the gangue or the stones which filled the volcanic chasms; and partly their solutions, of which the mineral springs of the present day afford us so many examples, permeated the porous rocks, and saturated them on cooling, or were forced to relinquish their valuable contents by some more powerful chemical

affinity. Thus, numberless years before man appeared upon
the stage of his future empire, the means were provided
without which it would never have been possible for him to
establish his dominion over the earth, and to make himself
master of the treasures it bears on its surface.

When we consider the frequent upheavings and subsidences
of the earth-rind and the denuding power of water, which,
in the long series of ages, cuts deep ravines into the mountains and washes away whole strata, we cannot wonder that
many metalliferous veins emerge or crop out on the surface
of the earth, so that a fortunate chance sufficed for their
discovery. Thus, a poor Indian looking for wood first found
the rich deposits of silver that so long had been buried in
obscurity under the sterile soil of Copiapo in Chili (1632).
Partridges, in whose stomachs grains of gold were found, are
said to have led to the discovery of the rich mines of Kremnitz and Schemnitz in Hungary; and Ramm, a huntsman
of the emperor Otho the Great, having bound his horse to a
tree in a forest near Goslar in the Hartz Mountains, the fretful
steed, stamping with impatience and tearing up the soil,
pointed out the celebrated lode of the Rammelsberg, which is
worked to the present day.

But if in these and similar instances the treasures of the
subterranean world have revealed themselves spontaneously
to man, in many other cases laborious and costly investigations have been found necessary for their discovery.

As civilisation advanced, and the value of the metals, of
coal and salt, came to be more and more appreciated, nothing
could be more natural than the desire of no longer relying
upon the discoveries of chance or upon the mines bequeathed
by former ages but of sounding the mysterious recesses of
the earth, and forcing her, by dint of patient exploration, to
reveal the riches she conceals under her surface.

Thus, as far back as the eleventh century, the *divining* rod
came into practice, and found full credence in a superstitious
age. A forked branch of the hazel-tree, cut during a peculiar
phase of the moon, was the means employed in Germany for
the discovery of buried treasures, of veins of metal, of deposits
of salts, or of subterranean sources. But the miraculous rod
did not indiscriminately show its powers in every hand; it

was necessary to have been born in certain months, and soft and warm—or, according to the modern expression, *magnetic*—fingers were indispensable for handling it with effect. The diviner possessing these necessary qualifications took hold of the rod by its branches so that the stem into which they united was directed upwards. On approaching the spot where the sought-for treasure lay concealed, the magical rod slowly turned towards it, until finally the stem had fully changed its position and pointed vertically downwards. To increase the solemnity of the scene, the wily conjurers generally traced magical circles that were not to be passed, burnt strong-smelling herbs and spices, and uttered powerful charms to disarm the enmity of the evil spirits that were supposed to guard the hidden treasures.

At present, the divining-rod has lost its old reputation, and more rational means are employed for the discovery of mineral wealth. Relying on experience, tact, and geological knowledge, the investigator carefully examines the country where ores or coal are supposed to be concealed; and having fixed upon an appropriate spot, has recourse either to experimental digging or to boring for testing the truth of his opinion. These expensive explorations, though often unsuccessful, frequently prove highly remunerative; and many a saline spring, or coal seam, or metallic vein would, without their assistance, have remained unknown and unproductive.

The Prussian Government annually devotes a sum of 40,000 dollars to boring experiments, and has no reason to regret the outlay. To these it is indebted among others for the discovery of the saline springs of Rehme in Westphalia, now the resort of thousands of patients, and of the rock-salt beds of Speerenberg and Segeberg,* which will amply repay the loss of capital and labour incurred by other unsuccessful borings. It would be well if all governments paid the same attention to the development of subterranean wealth.

Boring through hard stone is necessarily a very tedious work, but as it proceeds it awakens not only in those who are directly interested in its success, but in every intelligent witness, all the stirring emotions of a game of chance. 'Of all branches of business,' says Williams—a thoroughly devoted

* See p. 439.

miner—'of all the experiments that a man of sensibility can be engaged in or attend to, there is, perhaps, none so amusing, so engaging and delightful, as a successful trial upon the vestigia, or appearances, of a seam of coal, or other mineral discoveries. When you are attending the people who are digging down or forward upon the vestige of the coal, and the indications are increasing and still growing better under your eye, the spirit of curiosity and attention is awakened, and all the powers of expectation are elevated in pleasing hopes of success. And when your wishes are at length actually fulfilled—when you have discovered a good coal of sufficient thickness, and all circumstances are favourable, the heart then triumphs with solid and satisfactory joy.' *

In a country abounding in mineral wealth like Great Britain, where vast treasures, whose discovery holds out the most brilliant prospects to avarice, are still hidden under the ground, and the restless spirit of enterprise is continually seeking for new sources of profit, it may naturally be supposed that searches for coal seams or veins of metallic ore must be frequently undertaken. Thus in all our mining districts there are professional master-borers, who engage for a fixed price to drill the hardest rock to any depth that may be required.† Their chief implements are boring-rods, made of the best and most tenacious Swedish iron; chisels fitted with screws, tipped with good steel, with a face of two and a half to three and a quarter inches, and generally eighteen inches in length; and a wimble, which is a hollow iron instrument like an auger, whose cavity is from six to ten inches in length, with an opening up at one side, with partial overlap, the better to receive and hold the chopped strata.

When the bore is intended to penetrate to a considerable depth, a lofty triangle of wood is set above the bore-hole. In boring, the lowermost rod is the chisel, which continually

---

* 'Our Coal and our Coal Pits, by a Traveller Underground,' p. 94.

† A great improvement has of late years been effected in rock-boring by the introduction of a borer, armed with black diamonds, which acts by rotation and constant pressure. A two-inch borer of this description perforates granite and sandstone at the rate of from two to four inches per minute. The hardness of the black diamond is double that of quartz, and it will perforate about a mile of sandstone without being worn out.

operates on the rock or stratum. The uppermost rod terminates in a stout ring, through which passes a cross-piece held
by two men in working, and which is also suspended to a
springing pole by a chain. One or more rods being pushed
into the ground, two men on a wooden stage take hold of a
cross stave at the end of the springing pole and work it
steadily up and down, while two men below, by means of
the cross-piece, at the same time heave the suspended rods a
few inches, and then allow them to fall by their own weight,
walking slowly round the hole. By these combined operations of chopping and scooping the workmen make slow but

PROCESS OF BORING.

sure way through whatever substance may be in contact with
the chisel. When the hole is too deep, and the added rods
become too heavy to be conveniently lifted by manual labour,
a brake or lever, or a horse-gin or steam-engine, is employed.
As the boring proceeds, it is also frequently necessary to
lower pipes into the hole made, to prevent the falling of
fragments from the sides of the cylinder.

When the position of a mineral vein is ascertained, its
direction known, and some reasonable conjecture made con-
cerning its extent, thickness, and value, measures must be taken
to obtain, by subterraneous excavation, the buried mineral,
and for this purpose vertical pits or shafts must be sunk,
and horizontal galleries—or, as they are sometimes called,

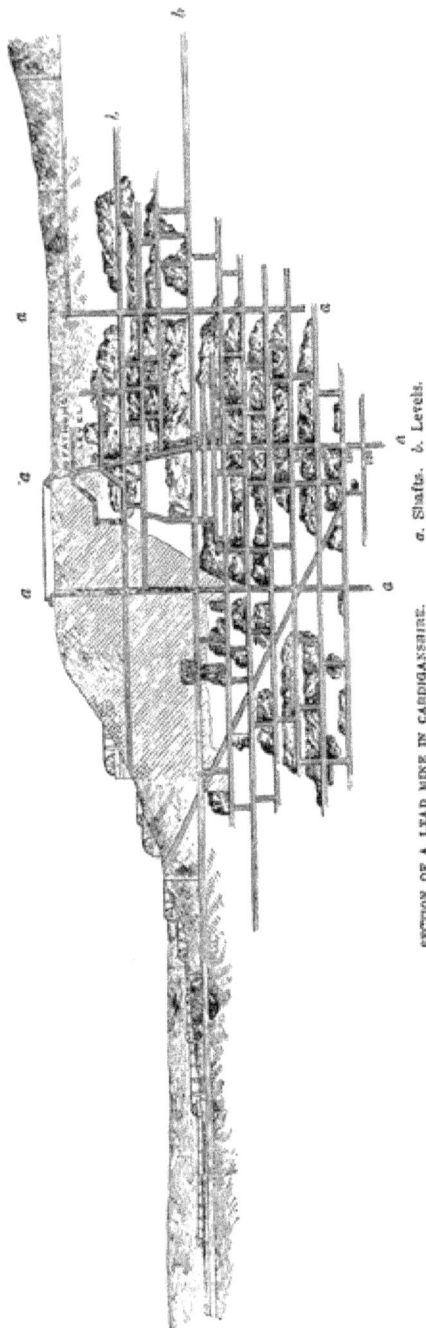

SECTION OF A LEAD MINE IN CARDIGANSHIRE. *a.* Shafts. *b.* Levels.

levels—must be driven, to prepare the way for its convenient extraction, and at the same time to carry off, so far as may be, the water which either rises into the mine from springs, or drains into it from the surrounding strata. Some idea may be formed of the extent of these works in many of the more considerable mines, when we learn that the total amount of sinking in the Consolidated Mines in Cornwall is stated to amount to more than twelve miles of perpendicular depth (including, of course, the winzes or underground shafts), and that the horizontal galleries extend to as much as forty miles in length. A mine like this is, in fact, a large subterranean town, with numerous lanes and avenues, passages and thoroughfares.

In sinking a shaft, danger is to be apprehended both from the falling in of loose and incoherent strata, and from the lateral springs, which sometimes empty themselves into the workings to an extent which it would at first appear hopeless to contend against. In such cases there is no safety to be obtained without walling the shaft with brick or stone, or securing it by timber or metal tubbing. For this purpose many of our coal-pits were formerly lined throughout with three-inch boards, nailed to a circular wooden framework called a crib, which was firmly attached to the sides of the pit at convenient distances. But this method, although it has been known to keep out a pressure of water equal to 100 pounds on the square inch, is not considered so safe as the metal tubbing now adopted in all difficult works. In comparatively solid ground the cast-iron tubs are forced down the shaft, but in soft ground they sink by their own weight. As they descend, fresh tubs are added, until the work is finished. When we consider that, in the coal-pits in the north of England, many shafts have a diameter of as much as fifteen feet, that in some cases they are sunk to a depth of nearly 300 fathoms, and that, under the most favourable circumstances, even where tubbing is not required to guard against the influx of springs, it is necessary to line almost the whole of the interior with bricks, to prevent the loose strata from falling or being washed in, we cannot wonder that as much as 100,000l. and ten years of labour have, in some cases, been expended before the seam of coal has been reached that was to repay all this vast outlay of capital and time. Unsightly and filthy as such a shaft may be, it is in reality a great triumph of architectural skill.

In most of our collieries one shaft serves for winding up the coal, for the passage of the men up and down, for ventilation, and for drainage by means of the engine-pumps. To answer these various purposes it is subdivided or bratticed by brick or wooden partitions into two, three, or four compartments; but this arrangement is very defective, for the safety of the workmen requires that in every large coal-pit, and indeed in every mine, there should be at least two separate shafts. When the partitions of the single shaft become injured or burnt, which has not unfrequently been the case, the ventilation of the pit may suddenly be deranged, and many lives have thus been endangered. The obstruction of a shaft by the breakage of an engine has caused some of the most appalling tragedies in mining history. On January 10, 1862, the beam of the pumping-engine broke at the Hartley Colliery in the Newcastle coalfield, and striking, like a huge catapult, with its enormous weight of forty tons, against the sides of the shaft as it descended, accumulated an enormous mass of rubbish and broken timbers at the depth of 138 yards from the surface. Two hundred and four colliers were thus shut out from all hope of rescue, for no exertion could possibly remove, in time enough to save them, the vast pile of ruins that obstructed their escape to the upper world.

The methods of working, winning, or excavating mineral substances naturally vary, according to the magnitude and direction of the beds, lodes, or seams in which they are contained. With a vein of moderate width, as soon as the preparatory labours have brought the miners to the point of the vein from which the ulterior workings are to ramify, the first object is to divide the mass of ore into solid rectangular compartments by means of oblong galleries, generally pierced ten fathoms below one another, with pits of communication opened up, thirty, forty, or fifty yards asunder, which follow the slope of the vein. These galleries and shafts are usually of the same breadth as the vein, unless when it is very narrow, in which case it is necessary to cut out a portion of the roof or the floor. Such workings serve at once the purposes of mining, by affording a portion of ore, and for the complete investigation of the nature and riches of the vein, a certain extent of which is thus prepared before removing the cubical masses. It is proper to advance first of all in this manner to the greatest distance from the central point which can be mined with economy, and afterwards to remove the rectangular blocks in working back to that point. This latter operation may be carried on in two different ways, by attacking the ore from above or from below. In either case the excavations are disposed in steps, similar to a stair, upon their upper or under side. The first is styled a working in *direct* or descending steps; the second a working in reverse, or ascending steps. By this method a number of miners are able to proceed simultaneously without interfering with each other.

In rich lodes and in the case of thick masses, the system of working in large chambers or extensive excavations, or by transverse attacks, is employed. Superficial deposits are worked like open quarries.

In British practice, coal is generally worked on the post (or pillar and stall) system, or on the long-wall system.

The pillar and stall principle is carried out in the working

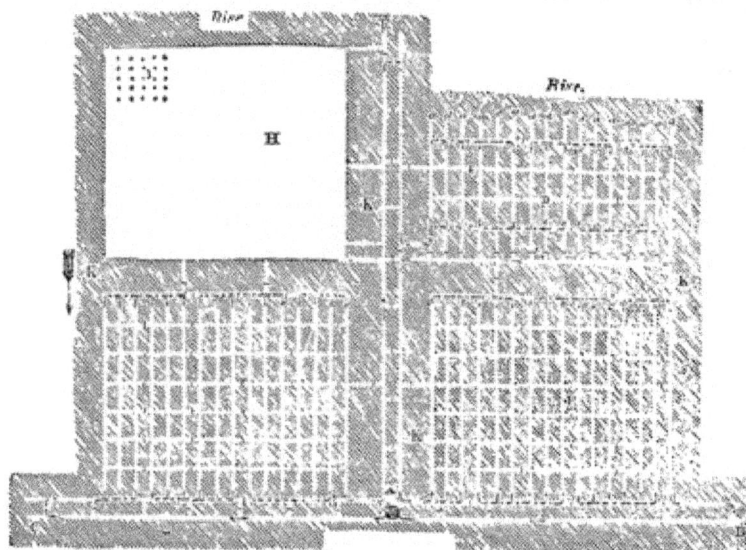

PART OF A COLLIERY LAID OUT IN FOUR PANELS.

A. Engine shaft: divided into three compartments, an engine pit and two coal pits
B, C. Dip head level
A, E. The rise or crop gallery.
K, K. Panel walls.

F, G. Two panels completed as to the first work.
D. Panel, with the rooms a a in regular progress to the rise.
H. Panel fully worked out.

away of a certain portion of the coal as a first measure, and in leaving the remainder in pillars, which are either to serve permanently for the support of the roof, or to be at some future time partially or totally removed. In the North of England, where this system has been brought to perfection, it is also named panel-work, because the whole area is divided into quadrangular panels, each panel containing an area of from eight to twelve acres, and round each panel is at first left a solid wall of coal of from forty to fifty yards thick. Through the panel walls roads and air-courses are driven in order to work the coal contained within these walls. Thus all the panels are connected together with the shaft as to roads and ventilation, and each district or panel has a particular name, so that any circumstance relating to the details of the colliery can be readily referred to a specified place..

By this plan, of which the above illustration gives a distinct idea, the pillars of a panel may be worked out at any time most suitable for the economy of the mine, and the loss of coal amounts to no more than about a tenth, instead of a third or a half by the old methods.

When the pillars of a panel are to be worked out, the most distant range is first attacked, and as the workmen cut away the furthest pillars, props of wood are placed between the pavement and the roof, within a few feet of each other, as shown by the dots at I. This is continued till an area of one hundred square yards is cleared of pillars. This operation is termed 'working the *goaf*.' The only use of the prop-wood is to prevent the stratum which forms the ceiling over the workmen's heads from falling down and killing them by its splintery fragments. Experience has proved that before proceeding to take away another set of pillars, it is necessary to allow the last made goaf to fall. The workmen then begin to draw out the props. This is a most hazardous employment. Knocking down the more remote props one after another, they quickly retreat under the protection of the remaining props. Meanwhile the roof stratum begins to break by the sides of the pillars and falls down in immense pieces; while the workmen still persevere, boldly drawing and retreating till every prop is removed. Nay, should any props be so firmly fixed by the top pressure that they will not give way to the blows of heavy mauls, they are cut through with axes, the workmen making it a point of honour to leave not a single prop in the goaf. If any props are left, it causes an irregular subsidence of the strata, and throws more pressure on the adjacent pillars. The miners next proceed to cut away the pillars nearest to the sides of the goaf, setting prop-wood, then drawing it, and retreating as before, until every panel is removed excepting small portions of pillars which require to be left under dangerous stones, to protect the retreat of the workmen. While this operation is going forward and the goaf extending, the superincumbent strata, being exposed without support over a larger area, break progressively higher up; and when strong beds of sandstone are thus giving way, the noise of the rending rocks is very peculiar and terrific; at one time loud and sharp, at another hollow and deep. As the pillars of the panels are removed, the panel walls are also worked progressively backwards to the pit bottom, so that only a small portion of the coal is eventually lost.

When mines are fully worked, the main shaft is fre-

GENERAL VIEW OF MINING OPERATIONS.

quently continued down to other seams of coal, which are excavated in the same way as the first seam. In such cases the workings communicate with each other by shafts called 'staples,' which are sunk at intervals between the seams of coal.

The principle of long-wall working (Shropshire and Derbyshire method) is the extraction of all the available coal by the single process of first working, maintaining the roads by means of stone-walls or wooden props. This system is applicable with advantage only to thin seams which lie near

s

to the surface, and in which the workings may be of a very limited extent.

According to the various hardness of the minerals or of the rocks in which they are embedded, different means and implements for dividing the masses are employed. In loose earth, sand, and clay, shovels and scrapers suffice; in gypsum, coal, and rock-salt, the pick becomes necessary; in many hard slates gads must be driven into the small openings made with the point of the pick. In still harder stones, forming the great majority of those which occur in veins and strata, recourse must be had to the explosive power of gunpowder, which is also largely employed in our coal-

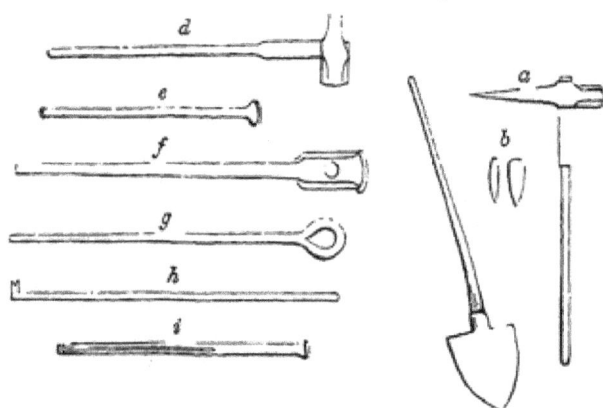

TOOLS USED BY MINERS IN CORNWALL.

a. Pick. b. Gad. c. Shovel. d. Mallet. e. Borer. f. Claying-bar. g. Needle or nail.
h. Scraper. i. Tamping-bar.

mines. The tools used for this purpose are the borer, an iron bar tipped with steel, formed like a thick chisel, which is held by one man straight in the hole with constant rotation on its axis, while another strikes the head of it with the iron sledge or mallet; the scraper, for clearing out the hole from time to time; the claying-bar, a tapering iron rod, which, after the introduction of some tenacious clay into the cavity, is forced into it with great violence, and, condensing the clay into all the crevices of the rock, secures the dryness of the hole; and the nail, a small taper rod of copper, which, after the charge has been introduced, is inserted to reach the bottom of the hole, which is now ready for tamping. For this purpose, any soft species of rock free from flinty particles, which might provoke a premature explosion, is introduced in small quantities at a time, and rammed very hard by the tamping-bar, which is held steadily by one man and struck with a sledge by another. The hole being thus filled, the nail is withdrawn by putting a bar through its eye and striking it upwards. Thus a small perforation or vent is left for the safety fuse, a woven cylinder containing gunpowder, and protected by a coating of tar. The fire being applied, the men retire to a safe distance.

Often, in order to lose as little time as possible, a number of shots are fired together, so that the explosions, pouring out their tongues of flame in rapid succession, and awakening the subterranean echoes far and wide, afford a highly interesting spectacle. But woe to the miner if he be too hasty to return to his post, for it often happens that a treacherous shot goes off several minutes after being lighted, and, exploding in the face of the imprudent workman, disfigures him for life, or kills him on the spot.

Accidents in blasting arise also from other causes, such as a negligent handling of the powder while preparing the charge, or some delay in retiring after the fuse has been kindled. It was an incident of the latter kind which some years back called forth the following instance of heroism.

'In a certain Cornish mine,' says Thomas Carlyle,[*] 'two miners, deep down in the shaft, were engaged in putting in a shot for blasting; they had completed their affair, and were about to give the signal for being hoisted up. One at a time was all their coadjutor at the top could manage, and the second was to kindle the match and then mount with all speed. Now it chanced, while they were still below, one of them thought the match too long, tried to break it shorter, took a couple of stones, a flat and a sharp, to cut it shorter, did cut it off the due length, but, horrible to relate, kindled it at the same time, and both were still below. Both shouted vehemently to the coadjutor at the windlass, both sprang at the basket; the windlass-man

---

* 'Life of Sterling,' p. 278.

s 2

could not move it with them both. Here was a moment for
poor miner Jack and miner Will! Instant horrible death
hangs over both, when Will generously resigns himself.
"Go aloft, Jack, and sit down. Away! In one minute I
shall be in heaven!" Jack bounds aloft, the explosion
instantly follows, bruises his face as he looks over; he is
safe above ground; and poor Will? Descending eagerly,
they find poor Will, too, as if by miracle, buried under rocks
which had arched themselves over him, and little injured;
he too is brought up safe; and all ends joyfully, say the
newspapers, which have duly specified the event.

'Such a piece of manly promptitude and salutary human
heroism was worth investigating. It was investigated and
found to be accurate to the letter, with this addition and
explanation, that Will Verran—an honest, ignorant, good
man, entirely given up to Methodism—had been perfect in
the "faith of assurance;" certain that *he* should get to
heaven if he died, certain that Jack Roberts would not,
which had been the ground of his descision in that great
moment; for the rest, that he much wished to learn reading
and writing, and find some way of life above ground instead
of below. By the aid of the Misses Fox and the rest of that
family, a subscription (modest *Anti*-Hudson Testimonial)
was raised for this Methodist hero; he emerged into daylight with fifty pounds in his pocket; did strenuously try, for
certain months, to learn reading and writing; found he
could not learn those arts, or either of them; took his
money and bought cows with it, wedding at the same time
some likely milk-maid.'

Several attempts have recently been made to diminish
the dangers of blasting, by substituting gun-cotton or
nitro-glycerine for gunpowder, or by firing charges by
means of the electric battery, but hitherto without much
success.

An enormous quantity of gunpowder is consumed for blasting in many of the larger mines. In 1836, 64,000 pounds
were used in the Consolidated Mines in Cornwall, and 90,100
pounds in the Fowey Consolidated Copper Mines. The total
amount of gunpowder consumed in the Cornish and Devonian
mines in the year 1837 reached 300 tons, which cost 13,200*l.*

This one item may serve to give some idea of the enormous working expenses of a large mining concern.

Sometimes the rock is so tenacious as to render boring too tedious and expensive an operation, and fire becomes necessary for subduing the solidity of the stone. In this manner the ancient mine of Rammelsberg, near Goslar, is forced to yield its treasures, and whole forests are anually consumed in order to loosen the hornstone and indestructible spar of its metalliferous veins. Every Saturday morning the fire is applied to the numerous piles of billets and faggots that have been distributed throughout the course of the week. Those in the upper floors of exploitation are first burned, in order that the inferior piles may not obstruct by their vitiated air the combustion of the former. Thus at four o'clock in the morning the fires are kindled in the upper ranges, and then from pile to pile the firemen descend towards the lower floor, which occupies them till three o'clock in the afternoon. The vaults of Rammelsberg now afford a truly magnificent spectacle. The rising flames, flickering against the walls of the vaults and ascending in broad sheets towards their roof; the dense clouds of smoke rolling towards the air-vents; the crackling of the wood; the loud detonations of the stones rent by the expansive force of heat from the primitive rock; the lurid glare of the conflagration; the naked workmen with their mighty stirring-poles, flitting like dark spirits before the blazing pile; the intense heat, and the air loaded to suffocation with sulphurous fumes—all combine to produce a picture worthy of Dante's ' Inferno.' During the Sunday the noxious vapours engendered by the conflagration have time to disperse; and on the Monday morning the workmen detach, with long forks of iron, the ores that have been loosened by the flames.

The ore being extracted from its bed, it becomes necessary to bring it to the light of day, an operation which is of course of the greatest importance, and not seldom requires the aid of complicated machinery, particularly in coal mines, where large masses have to be conveyed as economically and speedily as possible to the upper world. A great improvement has been effected of late years in the facility of transporting minerals underground, by the intro-

duction of small tramroads, and the saving of expense thus
effected sometimes amounts to one-half the former cost.
Many of our larger mines are provided with miles of this
subterranean railroad, and the advantage is greater because
for the most part there is a slight descent from the workings
to the bottom of the shaft, to allow of a more complete
system of drainage than could otherwise be attained. But
frequently where the galleries are low, narrow, and crooked,
the carriage is still effected by means of sledges, barrows, or
little waggons, which are with difficulty dragged or pushed

CONVEYANCE OF MINERALS UNDERGROUND.

along, over planks or uneven and muddy roads; and sometimes even the interior transport of the ore is executed on
the backs of men—a most disadvantageous practice, which is
gradually wearing out.

In great mines—such as the coal and salt mines of Great
Britain, the salt mines of Wielitzka, the copper mines of
Fahlun, the lead mines of Alston Moor—horses have long
been introduced into the workings, to drag heavier wag-
ons, or a train of waggons attached to one another. In
some cases these animals are brought to the surface at stated

intervals, but generally when once let down the pit they for ever bid adieu to the light of day. Strange to say, this *unnatural* mode of existence, which would soon undermine the health of man, agrees admirably well with that of the horses, of whom the greatest care is taken, as their useful services are duly appreciated by their owners. They are abundantly fed with hay and oats of the best quality; their stalls are large and well ventilated; and as they labour in a mild and equable temperature, they remain free from many complaints to which horses are liable. Their good condition and sleek shining coats prove that they have no sentimental longing for green fields or the bright sunshine.

In a few of the largest collieries it has been found advantageous to establish underground stationary engines, which bring the trains of waggons, by means of an endless rope, along the galleries to the bottom of the shaft. In other mines, such as those of Worsley in Lancashire, subterranean canals are cut, upon which the mineral is transported in boats.

In the European mines the ores are usually lifted from the bottom of the shaft to the surface by means of steam power, or horse-gins; but in Spanish America, men, and even women, are employed for this purpose. The steep ladders, which they ascend with heavy weights upon their backs, consist merely of the thick trunks of fir-trees, into which, at intervals of every ten or eleven inches, deep notches have been cut; but these rude steps are mounted with perfect security and ease by the sure-footed Indians.

In many European mines, where the workmen are let down or raised by means of ropes, wire cables, or chains, they sometimes sit on transverse round pieces of wood or in a kind of chair, consisting of two strong leather belts, one of which serves as a seat, and the other for supporting the back. In Wielitzka ten of these chairs are attached to the cable at distances of seven or eight feet one from another. The persons seated in the uppermost and lowest chairs direct the descent, and take great care to prevent the conveyance grazing against the sides of the shaft, for were it to be hooked fast by a nail or any other projection, a fall of several hundred feet would be the almost inevitable and fatal consequence. The old method of descending into a colliery was by a corf or strong basket, hooked on by a chain to the rope that hung down the shaft. Stepping into this, the men would swing down the dark hollow, gaily and readily, but not always safely.

Thus, in the year 1835, in a colliery near Liège, seven workmen were already seated in a corf that was about to descend into the shaft, when one of their comrades, anxious to seize the opportunity, came hurrying along, and, in spite of all remonstrances, jumped into it; but the rope, unable to bear the shock and the increased weight, suddenly snapped, and all eight were precipitated into the abyss, from which not one of them came forth alive.

In the Swedish mines small barrels or tuns are generally used as vehicles of descent, and the workmen are uncommonly dexterous in preventing their little aërial skiff from striking against the rugged rock-walls, when it would run the danger of being wrecked. Women, and even children, who find occupation in the mines, are often seen standing on the narrow edge of one of these swinging, turning, or oscillating tuns, with an arm slung round the rope; and such is the power of habit that they will quietly knit where even a stout-nerved man would be appalled by the frowning cliffs above or the black abyss below.

In the year 1785 a girl, descending alone into the pit of Fahlun, and unable to direct the tun, could not prevent it from striking against a rock. Jerked out of her conveyance by the violence of the shock, she fell upon a narrow ledge, about one hundred feet from the bottom of the pit, to which she clung with all the energy of despair. Her position was indeed terrific, for the least motion would have precipitated her into the dark grave which seemed to yawn for her reception, and to have given her but a momentary respite in order to make her feel more bitterly the pangs of approaching death. Already her strength was giving way, already a cloud swam before her eyes, when some bold miners, venturing their own lives in the hazardous undertaking, succeeded in rescuing her from her awful position, and snatching her as it were from the very jaws of death.

Another method much adopted, and preferred by the pitman in our collieries, was passing down and up in the loop. The pitman inserted one leg into a loop formed by curving the terminal chain and hooking it back upon an upper link, and then twined his arm tightly round the rope above. In this way he descended through any depth, and, as he alleged, with greater safety than in a bucket, out of which he might be ejected, while nothing except the breaking of the rope could harm him in loop.

At present the safety-cage is generally used. This is

MINERS DESCENDING SHAFT IN OWEN'S SAFETY CAGE.

simply a vertical railway carriage running up and down upon guides, and thereby introducing into the shaft the improvements of the iron road. Into one of its square, narrow, compartments two or three men crouch together, others get into an upper compartment, and down the cage moves easily and safely, the men needing only to take care that hands or fingers do not hang beyond the edge, while four or five minutes of easy motion carry them down a thousand or fifteen hundred feet below the surface of the earth.

In many mines the workmen climb up and down the shafts
on fixed ladders, with landing stages for resting; and it may
easily be imagined how severely their strength must be taxed
when, after their hard day's labour, they have still to ascend
step by step a thousand and even two thousand feet, before
they can return to their families. Some of the Cornish mines
require a full hour to rise from the lowest depths to ' grass;'
and besides this considerable loss of time, diseases of the
lungs and heart, which often terminate fatally, or pre
maturely make the miner an invalid, are the consequence of
these fatiguing journeys.

' O thou grumbling clerk in London city,' exclaims the
author of 'Cornwall and its Mines,'* 'whose daily fatigues only
extend to the ascent into and descent from the trim omnibus
that takes you to or from Peckham or Kennington! only
think for a moment of travelling some four or five times the
height of St. Paul's daily—before and after work! O thou
querulous socialist, demagogue, or artisan, who canst sit
in a comfortable coffee-house, under a flaming gaslight,
immediately before and after work—or in your own snug
parlour, or by your own fireside or murmuring kettle—do but
think for a moment of the Cornish miner, and what he must
do before he can reach home or house! I fully believe that
the best cure for discontent and gloom in fortunate workmen
would be to put them upon the treadmill of a deep Cornish
mine—for a temporary treadmill it is.'

It had long been deemed of the utmost importance to
devise some easier mode of locomotion; but it was not till
1833 that the circumstance of two water-wheels having been
thrown out of work by the opening of the deep Georg Adit in
the Hartz mines suggested the idea of employing the pump-
rods for aiding the ascent of the miners. The trial was first
made with a portion of one hundred fathoms. This was
divided into twenty-one portions; and on each portion iron
steps were fixed, at intervals of four feet, while hand-holes
were fixed at convenient distances. A reciprocating motion
of four feet was given to each rod, and the miners stepped
to and fro from a bracket or ledge on one rod to the
parallel one on the other. As one rod is always descending

* London: Longmans, 1857.

while the other is ascending, and *vice versâ*, it is easy to understand how this alternate stepping on to the little platforms must lead to the ascent or descent of the miner. At the division between each two of the twenty-two portions, there is a larger platform on which he may rest awhile; and nothing is lost by his rest, for the reciprocal motion goes on, and is ready again for his use when he is ready for it. This first machine surpassed expectation. Short as the length of ascent was, many invalids of the district were now able to resume their underground labours, as the fatigue of mounting or descending was reduced, by the alternate action of the machinery, to a mere easy lateral motion.

The advantages of this new method in saving both time and power were so obvious that it was soon imitated in the other deep metalliferous veins of the Hartz; and at present power-ladders or *man-engines* of an improved construction, such as the substitution of a single rod for the double apparatus above described, are in very general use over the Continent, whence they have passed in a modified form into Cornwall, where they are worked by steam. In Fowey Consols Mine the machine extends to a depth of 1,680 feet. The rod is eight inches square, with twelve-inch platforms at intervals of twelve feet; and there are stationary platforms equidistant at the side of the shaft. When a miner is about to descend the steps on a movable platform, the rod descends and carries him down twelve feet; he steps upon a fixed platform while the rod rises again; he then steps upon another movable platform, and descends another twelve feet, and so on to the bottom. In ascending, there is simply a reversed process. It is a very interesting sight to witness the ascent or descent of bodies of miners at certain hours of the day and night. You see them passing each other in the shafts, in a kind of zig-zag course, of as great regularity as any zig-zag will permit. As one miner steps off the rod platform to one fixed platform, another steps on to it from another fixed platform on the other side. Thus there are two streams of miners moving in opposite directions along the same rod at the same time, and this curious spectacle is rendered doubly pleasing when we consider how much distressing toil has been alleviated by the employment of the man-engine. Machinery constructed on the same principle has been latterly adopted in the mines of Anzin, for transporting the coal step by step to the surface; and it is evident that when coal mines are worked at greater depths than at present, ropes, however strong, will no longer be able to sustain even their own weight, and the whole transport up and down a shaft of perhaps 3,000 or 3,500 feet must be performed by means of similar machines.

Wherever the excavated rock is not hard or solid enough to bear the superincumbent weight, the galleries of a mine must necessarily be supported by timbering or walling. Timbering is most used, frequently in the form represented in the annexed woodcut; and when we consider how miles upon miles of galleries are thus supported, we can easily imagine that whole forests must be engulfed in our mines. It has been calculated that for the total quantity of timber in use for mining purposes in Cornwall, it would require no less than 140 square miles of forest of Norwegian pine, averaging a growth of 120 years. The expense thus incurred is enormous; the cost for timber, duty free, in Cornish and Devon mines, amounted in 1836 to 94,138*l.* and is probably still larger at the present time. For timbering, no tree is more esteemed than the larch or the Norwegian pine, on account of its great durability in the wet; but whatever wood may be employed, it is necessary to peel off the bark, experience having shown that unless this is done the wood rots much more easily, as the fibres of the rind attract a far greater quantity of moisture than the smooth surface of the splint. Like the potato and the grape, subterranean timbering is exposed to the attacks of a fungus, producing what is called dry-rot. The parasite germinates in the sap which remains in the wood, or at least derives its nourishment from it. Its vegetation is at first scarcely perceptible; but soon its white fibres multiply, and form at length small sponges on the surface. The decomposition of the wood now advances with rapid strides, and terminates at last in the

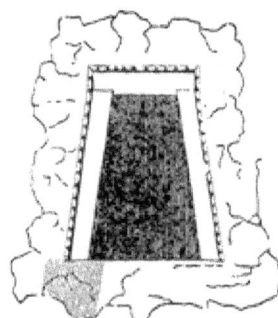

TIMBERING OF A MINE.

total destruction of the ligneous fibres. Not satisfied with depriving the roof of its support, the dry-rot likewise produces a vitiation of the air, so that wherever timbering is employed, it is reckoned among the great enemies of the mine and of the miner. Many remedies have been recommended, among which *kyanizing*, or saturating the wood with a solution of corrosive sublimate, is one of the most efficacious, though unfortunately too expensive to be of universal use. Mushrooms of various kinds likewise flourish upon the moist surface of the spars, and various insects collect near this parasitic vegetation. The timbering of a mine also affords very convenient lurking-places to the numerous rats which are met with underground, where they contrive to live upon the crumbs or offal of the miners' meals, or upon candle-ends and remains of wicks.

Not seldom the timbering of a gallery, weakened by rot, gives way under the pressure of the roof, which falls in with a tremendous crash, and sometimes buries the unfortunate miner under its ruins. Another disadvantage attending timbering is its liability to catch fire, and for this reason, wherever the cost is not found too great, the chief galleries of a mine are now usually constructed of stone.

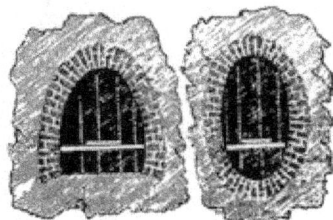

TRANSVERSE SECTIONS OF WALLED DRAIN GALLERIES.

Sometimes the two sides of a gallery are lined with vertical walls, and its roof is supported by an ogival vault or an arch. If the sides of the mine are solid, a simple arch is sufficient to sustain the roof; and at other times the whole surface of a gallery is formed of a single elliptic vault, the great axis of which is vertical, and the bottom is surmounted by a wooden plank, under which the water runs off.

The miner is generally in a state of perpetual warfare with the water, which threatens to inundate the scene of his labours; and as in a leaky ship the pumps must be kept continually working to prevent the vessel from sinking, so here also perpetual efforts are necessary to keep off the encroachments of this never-tiring foe. When a mine is situated above the level of a valley, or of the neighbouring sea, its drainage may be effected in a comparatively easy manner by means of sloping galleries, dry-levels or adits, which in many cases serve also for the transport of the ore or coal. In some mines these drainage levels are executed on a truly gigantic scale. Thus the Great Cornish Adit, which extends through the large mining district of Gwennap, begins in the valley above Carnon, and receives the branch adits of fifty mines in the parish of Gwennap, forming excavations and ramifications which have an aggregate extent of between thirty and forty miles, and which are in some places 400 feet below the surface of the ground. The longest branch is from Cardrew mine, and is five and a half miles in length. This great adit opens into the sea at Restronget Creek, and empties its waters into Falmouth Habour.

A similar great drain is Nent Force Level, in the north of

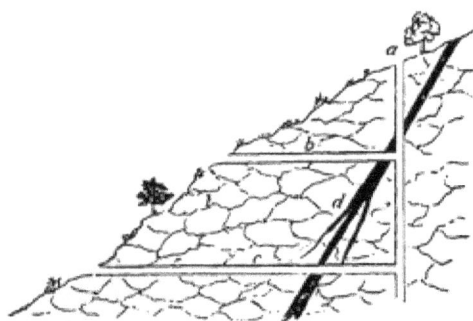

DRAINAGE OF A MINE BY ADIT LEVELS.
a. Shaft. b. Shallow adit. c. Deep adit. d. Mineral lode.

England, which drains the numerous mines in Alston Moor. It consists of a stupendous aqueduct nine feet broad, and in some places from sixteen to twenty feet high. For more than three miles it passes under the course of the river Nent, to Nentsbury engine-shaft, and is navigated underground by long narrow boats. At the distance of a mile in the interior, daylight is seen at its mouth like a star, and this star is continually enlarging upon you until you find yourself in open daylight. The ramifications of the Great Adit Levels of the mines of Freiberg in Saxony have a total length of seventy-two miles; but the most stupendous works of this description are those in the district of Clausthal and Zellerfeld, in the Hartz, where, as the mining operations have been carried on deeper and deeper, adits have been successively driven below adits. Four of these levels date from times previous to the seventeenth century; but, as they were found insufficient, the famous Georg Stollen was added to their number in 1777. This gigantic tunnel, which, piercing the hard rock, required twenty-three years for its completion, is above five miles long, and passes 900 feet below the church of Clausthal. It serves the double purpose of a draining gallery and of a navigable canal. The water is always kept at a height of from fifty to sixty inches; and the boats, which carry about five tons, are propelled by means of a chain attached to the vault, along which the boatmen drag or push them forward. In this economical manner about 20,000 tons of ore are anually brought to daylight. The boats are made and repaired in a subterranean wharf, which, though far from being one of the largest, may probably boast of being the deepest.

Until 1851 the Georg Stollen answered all the purposes for which it was constructed, but at the end of that period the increased depth and extention of the mines rendered necessary the addition of a new great adit level, which has been named the 'Ernst August Stollen,' in honour of the late king of Hanover. In spite of its vast dimensions, this magnificent work, which is six and three-quarter miles long, about ten feet high, and six and a half broad, required only thirteen years for its completion, and may justly be considered as one of the triumphs of modern engineering. The excavation was begun simultaneously at ten different points, and such was the admirable precision of the plans that all the junctions of the different sections of the gallery fitted accurately into each other.

Below the Ernst August Gallery (437 yards), the form of the country allows no deeper adit level to be driven; but to provide for the increasing vertical extention of the workings, a new underground gallery, without any opening to the surface, and at a depth of 262 yards below the former, is already in contemplation. The water is to be raised to the Ernst August Level by a special hydraulic machine placed in a vertical shaft, which will serve at the same time for raising the ore and for the passage of the miners. The expense is

calculated at about 60,000*l.* but will be amply repaid by the
new field of mineral wealth which it will open.   Thus in the
Hartz one magnificent work is but the precursor of another.

When a mine is so situated that drainage galleries cannot
be established, engines must be employed for pumping up the
waters.   Thus, in Cornwall, where most of the copper mines
open almost at the sea level, an enormous influx of water can
be kept in check only by an equally enormous steam-power.
In the United and Consolidated Mines between Truro and Redruth, seven steam-pumps, working with the united strength
of 2,000 horses, are kept constantly in motion, and raise
above 2,500,000 gallons of water in twenty-four hours.

In 1837 the whole quantity of water pumped out of the
earth by sixty Cornish engines attained the amazing aggregate of close upon thirty-seven millions of tons; but since
then mining has been carried on more extensively and deeper,
and consequently additional steam-power has become necessary to keep pace with the increasing waters.

'Even to the eye of an observer who is practised in
machinery* of great magnitude, the first sight and the subsequent examination of such engines is very gratifying.   To
watch the labour of a giant would be interesting; but to see
the giant not only labouring at ease amidst his enormous
work, but at the same time at the command of a child, who
should be able to stop him at any moment—this would be
doubly interesting.   Such is the case with the great Cornish
engines, for even the largest of them may be stopped by a
child of ten or twelve years of age.   Another peculiar feature,
too, of these engines is this, that they work with a quietness
—or absent of clash or clatter—which is in the inverse ratio
of their magnitude.   The water makes a great rush in the
pumps, but the engine itself is calm and comparatively noiseless—like a great mountain reposing in calm greatness while
a perpetual spring brawls at its feet.'

In the coal-fields of the North equally gigantic efforts
must be made to keep down the water.

In sinking to the coal at Dalton-le-Dale, eight or nine
miles from Durham, the borers penetrated the vast bed of

---

* 'Cornwall, its Mines and Miners.'   ²

saud beneath the magnesian limestone, which appears to contain the chief subterranean water-stores of the district. In this case their outburst was truly terrific, amounting, on June 1, 1840, to the enormous quantity of 3,285 gallons every minute. To oppose this formidable enemy the spirited proprietors of the mine at once proceeded to erect the necessary steam-power for pumping off 3,000 or 4,000 gallons a minute; but, the waters still increasing, it became necessary to meet them with a double and treble force, so that finally the floods had to be kept down by steam engines of an aggregate power equivalent to 1,584 horses and setting twenty-seven sets of pumps in motion.

Sometimes the influx of water into a coal mine is so enormous that no human contrivance can oppose it, and man is obliged to give up the conflict in despair. During the progress of the attempted winning of a pit at Haswell in the county of Durham, the engine power erected pumped out the water to the amount of 26,700 tons per day; but still the floods came in, and at last won the victory.

From the same cause many collieries have been closed, of late years, on the banks of the Tyne, and among these the famous Wallsend Colliery, which has given its name to the best kinds of coal.

The drowning of a coal mine not seldom occurs from the irruption of water accumulated in old wastes or ancient workings occupying a higher level in the vicinity. The growing pressure of such a body of water upon the beds or barriers below becomes enormous; and then the water, testing every weak point of the body opposed to its escape, at length unexpectedly rushes into the space which it finds open before it. All the works below are completely filled, and the mines are for a time rendered useless, or, it may be, for ever abandoned. This was the cause, in 1815, of the celebrated accident at Heaton, near Newcastle-on-Tyne, in which ninety lives were lost. The water flowed from two adjoining old collieries which had been abandoned seventy years before. A barrier of six feet withstood a pressure of thirty fathoms of water. But an irruption was aided at last by a natural fissure of the rock, and the catastrophe followed before any adequate protection could be interposed.

T

About thirty years later a tremendous calamity of the like kind, after an outlay of 100,000l., totally ruined the Baghil coal mines, in Wales. The water came from adjoining mines, which had been long abandoned.

If correct plans and descriptions of all the ancient workings had been preserved, these accidents, which happen frequently, might easily be prevented, as an exact knowledge of the localities would enable the owners to leave sufficiently strong barriers in parts where they are now often most inadequate. Such is the importance of accurate mining records, that thousands of pounds would be freely given at this moment by many owners for a knowledge of old works of which no plan exists and which no memory can now recall. Fortunately, the legislature has now taken steps to introduce a system of registration such as has already existed long ago in Prussia, Austria, and Belgium, and which, at least, will answer the purpose of obtaining greater security for the future.

Sometimes an enormous fall of rain, descending on the neighbouring country, finds its way into the mines through fissures in the ground or by breaking through galleries, and causes irreparable mischief; sometimes even a whole river bursts into the works and ruins them for ever. Thus, in 1856, the South Tamar Consols, in Devonshire, was flooded by the giving way of the bed of the Tamar, under which the workings were carried.

Even the sea has been known to take fatal vengeance for the undermining of her domain. Workington Colliery extended to the distance of 1,500 yards under the Irish Channel, and the workings, being driven considerably to the rise, were brought at length within fifteen fathoms of the bottom of the sea. The pillars of coal which supported the overlying strata were hardly strong enough to support the roof, but the imprudence of a manager eager to produce a larger quantity of coal weakened even this insufficient support by working it partly away. Heavy falls of the roof, accompanied by discharges of salt water, gave repeated warnings of the impending catastrophe, which took place on July 30, 1837. So violent was the irruption that many persons at a distance of hundreds of yards observed the whirlpool commotion of the sea as it rushed into the gulf beneath. Some few workmen near the shaft had time to escape, but thirty-six men and boys and as many horses were destroyed by the waters, which in a few hours entirely filled the excavations, the extension of which had tasked the labour of years.

The heroic devotion of a miner has invested with a more than ordinary interest the inundation of the mine of Beaujonc, near Liège, which took place on February 28, 1812. One hundred and twenty-seven persons were in the pit when the waters burst in from some old workings. Thirty-five had time to make their escape through the shaft; twenty-two were drowned in their eagerness to reach it; the majority, severed from the upper world by an impassable gulf, remained behind. Hubert Goffin, the overman, could have ascended in the tub; but, though the father of six children, his sense of duty would not allow him to desert his post, and he resolved to save all his men or to perish with them. As the rising waters forced the prisoners to seek a higher level, the boys burst out weeping and the boldest began to despair; but Goffin revived their courage by reminding them that their friends without would make every effort to save them. As one day after another passed, the prisoners suffered all the horrors of hunger, which some of them endeavoured to appease by devouring the candles they had brought with them. Others went to the water in the hopes of finding the body of some drowned comrade. Two of the pitmen quarrelled, and were on the point of coming to blows. 'Let them fight,' said the others; 'if one of them is killed, we will eat him'—a declaration which at once put an end to the dispute. To satisfy their thirst they had nothing but the foul water of the pit. Some made vows to all the saints, others complained in their delirium that they had to wait for their meals. In the midst of these scenes of horror, Goffin alone retained his courage; and exhorting, consoling, encouraging, and reproving on all sides, appeared as the guardian angel of his despairing comrades.

Meanwhile every effort was being made from without to bring them succour. Although as soon as the accident took place the pumps were incessantly at work, the water had risen to a height of 14 metres in the shaft on the following morning, and as it was still rising there was reason to fear that the captives, blocked up in a constantly narrowing space, would soon be suffocated. It was resolved to strike a gallery from the neighbouring pit of Mamonster to Beaujone, a distance of 175 metres. Unfortunately, only two men could hew at a time; but such was the ardour with which the work was prosecuted, that on the morning of March 4 a shout of triumph announced that the longed-for communication was effected, and that the prisoners were alive. Crawling through the narrow passage, they were wrapped up in woollen blankets and strengthened with soup and wine before being hoisted to the surface. Goffin and his son, a lad of twelve, were the last to leave the pit; as a brave sea captain, after some great catastrophe, never thinks of his own safety till he has satisfied himself of that of his men. With the exception of those who were drowned immediately after the accident took place, all were saved. The joy of some families, the despair of others, may be imagined when the final count was made. As a reward for Goffin's admirable conduct, he was decorated with the cross of the Legion of Honour and received a pension of 600 francs. Nine years later he was killed by an explosion of fire-damp, and thus the hostile elements with which he had so long waged a successful war triumphed over him at last.

The sudden irruption of an immense body of water into a mine naturally causes a compression of the air in those galleries which are cut off from all communication with the shaft. This pressure, which may rise to three or four atmospheres—or, in other words, may be three or four times greater than that of the external air—not only produces symptoms of suffocation and cerebral congestion in the unfortunate miners who are exposed to it, but, forcing its way through fissures in the roof or violently rupturing it, sometime produces the effect of an explosion of gunpowder, throwing the earth to a distance, and even overturning houses. One of the most interesting of the accidents of this kind on record occurred in 1833, in an extensive Scotch colliery, into which the waters of the river Garnock had broken through a cavity in its bed. As the stream poured into the mine the opening gradually enlarged, until at length the whole body of the river plunged into the excavations beneath. The river was affected by the tides, and this engulfment took place at low water; but as the tide rose the sea entered with prodigious force, until the whole workings, extending for many miles, were completely filled. No sooner, however, had this taken place, than the pressure of the water in the pits became so great that the confined air, which had been forced back into the high workings, burst through the surface of the earth in a thousand places, and many acres of ground were seen to bubble up like the boiling of a cauldron. Great quantities of sand and water were also thrown up, like showers of rain, during a period of five hours, and an extensive tract of land was laid waste.

Besides the danger of being crushed to death by a fall of rock, or immured in a living tomb by an obstruction of the shaft or an irruption of water, the miner has another, and often still more formidable enemy to encounter in the noxious gases frequently evolved in coal-pits. Thus in all well-regulated mines the greatest attention is paid to ventilation, so that no part of the workings may be left without a proper supply of air. In ordinary cases the natural currents, which set in different directions through the shafts and galleries, may sufficiently purify the atmosphere; but in the coal mines which are peculiarly subject to the evolution of foul gases, artificial or mechanical means must be resorted to for driving away the hurtful vapours as quickly as they form. To establish a proper air-current, the usual method is to keep a large fire continually burning at the bottom of one of the two shafts of the pit, or of one of the two compartments of the single shaft, and the difference of temperature thus caused between the column of air of the *upcast* shaft and the *downcast* becomes the motive power which impels or drags the air-current in obedience to it.* Yet this meets but half the difficulty; for

* Mr. Samuel Plimsoll ('Letters on the Iron Trade,' *Times*, February 10, 1868) informs us that in the Belgian coal mines the ventilation is carried on in a more economical and effective manner. Here no furnaces are lighted at the bottom of the upcast, because one-twentieth of the coal required for a furnace will make steam for an engine to work fans which act somewhat in the manner of huge paddle-wheels in steam-ships, and by rapid rotation over the shaft produce a draught which the incoming air rushes to meet, and thus powerfully promote ventilation. These fans they can work and control, and are therefore independent of those

the air-current, which naturally tends to the shortest passage, must be forced to do its duty in every corner of the pit, and not suffered to escape through the upcast shaft before it has performed the longest circuit. For this purpose a great number of mechanical contrivances are adopted, in the shape of 'stoppings,' of brick, or wood, or stone, all so placed as to divert and drive the air-current into the several galleries of the pit, and to make it perform every kind of complex movement, from turning back upon its own right or left to turning over in a somersault upon itself. The most curious and admirably simple contrivance is that of splitting the air by means of a wooden erection, which meets and cuts the current in two, and sends one part on the one hand and another on the other hand. In fact, what is commonly practised in minutely irrigating a meadow is also effected in thoroughly airing a mine. We may form some idea of the underground travels which the air is thus obliged to perform, by being forced along from split to split, when we hear that at Hetton Colliery the ventilating current in the total equals no less than 196,000 cubic feet of air per minute circulating through the mine at a velocity of 18 feet 3 inches per second.

The foul gases evolved in coal mines are either heavy or light. The most remarkable of the former is the *choke-damp*, or *black-damp*, the name given by the miners to carbonic acid gas. From its great specific gravity (1·527), this gas rests on the floor of the mine and gradually accumulates, having no tendency to escape beyond a slow mixture which takes place with atmospheric air; while the light *fire-damp*, or carburetted hydrogen, which, though not immediately fatal when breathed, explodes on the slightest contact with flame, tends to rise to the surface. The quantity of fire-damp which is poured out into the workings of some mines is very considerable, and constantly varying. Some seams of coal are much more full of it than others, and in working these, which are technically called *fiery seams*, it is not uncommon for a jet of inflammable air to issue out at every hole made for the reception of the gunpowder before blasting.

atmospheric influences to which some of our greatest calamities have been ascribed —the damp, heavy atmosphere of early winter. In the great colliery of Sacrée Madame, near Charleroi. one of these fans will draw 34,000 cubic metres (about 918,000 cubic feet) per minute.

In the celebrated Wallsend Colliery, in an attempt made to work the Bensham Seam (an attempt which ended in a fearful accident), Mr. Buddle said, in evidence before a committee of the House of Commons :—'I simply drilled a hole into the solid coal, stuck a tin pipe into the aperture, surrounded it with clay, and lighted it. I had immediately a gas-light. The quantity evolved from the coal was such that in every one of these places I had nothing to do but to apply a candle, and then could set a thousand pipes on fire. The whole face of the working was a gas-pipe from every pore of coal.'

The force with which the gas escapes on some occasions from clefts or joints is so great as to prove much previous compression. These sudden outbursts are locally termed *blowers*. Their issues and effects are surprising. In one minute they have been known to foul the air to a distance of 300 yards, and their noise is described as like that of rushing waters, or the roar of a blast furnace. They are not merely dangerous from their inflammable vapours, but also from the pieces of coal which their tension not seldom forces from the roof, and whose fall maims or kills the unfortunate workmen beneath.

The fire-damp is very liable to accumulate in old workings or *goaves*, which thus, unless completely isolated by stone and mortar, become a highly dangerous neighbourhood to the other parts of the mine. The immense quantity of gas evolved from a goaf of about five acres in Wallsend old pit affords a striking example of the danger of all such accumulations. A four-inch metallic pipe was conducted from the bottom of the pit to the surface of the ground and a few feet above it, when, a light being applied, a hissing streamer of flame flashed forth and burned night and day. The amount of gas thus drawn off from the mine was at first computed at about 15,000 hogsheads in twenty-four hours. Long did the little pipe continue to pour out in streaming flame thousands upon thousands of hogsheads of escaping fire-damp. The total issue might have illuminated a little town.

The explosion of inflammable gas is the most fearful enemy the collier has to encounter. Three or four cubic inches of carburetted hydrogen, when ignited, produce a detonation like that of a pistol-shot; half a cubic foot, enclosed in a bottle and set fire to, shivers the bottle into fragments; hence we may judge how terrific the effects must be when a blower pours forth its thousands of cubic feet into the galleries of a mine, and the careless approach of a light lets loose the demons of destruction. The explosion of a large subterranean powder-magazine would not be more terrific. Often without a moment's warning the unfortunate pitman is scorched and shrivelled to a blackened mass, or is literally shattered to pieces against the rugged sides of the mine.

It may easily be imagined that many efforts have been made, and many contrivances suggested, to disarm the fire-damp of its terrors; but Sir Humphry Davy's safety-lamp was the first invention which successfully coped with it. The power of the safety-lamp lies in the *non-communication of explosions through small apertures*, and the discovery of this natural law, as well as its practical application, is one of the greatest exploits of Sir Humphry Davy. A cylinder an inch and a half in diameter and seven inches long, formed of wire gauze, with 784 apertures to the square inch, surrounds the light of the lamp. When the miner, armed with this apparatus, enters an atmosphere, tainted with fire-damp, a light blue flame fills the cylinder; but, as if chained by some magic power, it is unable to transgress its bounds; and as in our Zoological Gardens we quietly view the beasts of the forest behind their iron grating, so the miner looks calmly upon his powerless foe.

Unfortunately, the negligence or the obstinate and blind perversity of the miner too often renders even this splendid invention ineffectual.

The safety-lamp requires to be kept in perfect order, and unless certain precautions are taken while using it, it loses its protecting power. Thus, although perfectly secure when at rest, it seems certain that the rapid motion communicated by the swing of the arm during a hurried transit through the mine has in many cases produced an explosion. Blowing out the lamp is

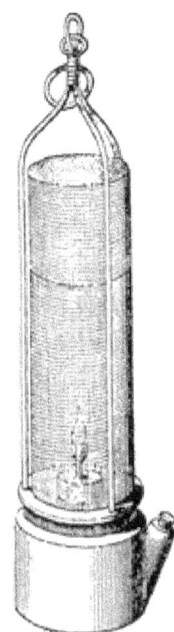

SAFETY LAMP.

likewise attended with danger, as the flame is then easily driven through the gauze, and, like a tiger escaping from his den, may spread terror and havoc around; but the chief cause of accidents consists in the small quantity of light diffused by the safety-lamp, and the consequent dislike acquired by the miners to its use.

Though in all well-regulated coal-pits one or several workmen are exclusively employed in keeping the lamps in the most perfect order; though they are handed to the miners burning and well closed, and fines are imposed upon any attempt to remove the gauze; yet the best regulations cannot possibly exclude the chance of accident resulting from the almost inconceivable carelessness of men whose daily tasks lead them into imminent danger. Thus it is a melancholy but unquestionable fact that the number of accidents from fire-damp since the introduction of the Davy lamp has been many more in a given number of years than before that invention. This has, no doubt, partly arisen from the larger number of persons employed on the whole; but it is to be feared that it has chiefly happened from dangerous portions of a mine being taken into work which without the *Davy* could not have been attempted, and partly also from the extreme carelessness of the workmen in removing the wire gauze.*

Unfortunately, the fatal effects of this rashness are not confined to the foolhardy miner who thus casts away the shield that preserves him from danger, but generally extend to many of his innocent comrades. In the year 1856 an explosion which took place in the Cymmer coal-pit killed 110 persons, and in the year 1857 170 workmen in the Lundshill colliery were swept from life to death with the rapidity of lightning. In the year 1858 the fire-damp levied a contribution of 215 victims in the coal-pits of England, while in 1859 it was satisfied with 95. But in 1860, as if to make up for this deficiency, it raged with double violence; for after

---

* Recent improvements have done much to render the Davy lamp a more perfect instrument of safety. These more or less insure increased illumination, prevention of bad usage by locking, and more perfect combustion. By an ingenious contrivance, one of these improved lamps cannot be opened without previous extinguishment.

having already claimed a tribute of 80 lives, the dreadful explosion which took place in the Risca Colliery on December 1 destroyed no less than 142 men and boys. The fire-damp explosion which occurred at the Oaks Colliery in 1866 swept away the unprecedented number of 361 victims, and in the same year 91 workmen perished from the same cause at Talk-o'-th'-Hill Colliery.

In these dreadful catastrophes the most terrible agent of destruction is not always the burning and concussion of the actual blast of fire-damp, but the choke-damp which succeeds the explosion. For the carbon of the inflammable gas, uniting with the oxygen of the air, produces that deadly poison, carbonic acid gas, which, from the disturbed ventilation of the pit, soon spreads far and wide through the galleries; so that the poor colliers who are caught in an exploded pit have two chances of death against them—one from burning, and the other from suffocation. The effect of death by the one gas or the other is very distinctly seen in the countenances of the dead. The men killed by the fire-damp are marked with burns and scorching, and their features are more or less distorted or disfigured. On the other hand, where men have been suffocated by choke-damp, their features are placid and simply inanimate. The fragile and faulty separations (whether doors or stoppings of any kind) having been broken down, there is an end to a hope of safe retreat even for men totally unharmed by the flames, for at once the air takes the shortest course between the entrance and exit, and leaves the shattered parts unventilated. Whatever after-damp is then and there generated exerts its effects in full, and those that cannot rush to the shaft are suffocated. In the explosion at Risca it was declared by the surgeon to the pit that of those who were killed no less than seventy persons died from the effects of after-damp who had not been near the fire. In the great Haswell explosion, several years since, seventy-one deaths out of ninety-five were occasioned by choke-damp; and at the explosion in the Middle Dyffryn Pit in 1852 no less than seven-eighths of the deaths proceeded from this cause. Persons of great experience attribute at least seventy per cent. of the deaths in fiery mines to after-damp, while some advance them even to ninety per cent.

After relating so many frightful disasters, too frequently caused by imprudence and rashness, it is a more pleasing task to mention a few instances of warnings taken in time. At Walker Colliery on the Tyne, in the year 1846, a hugh mass of coal weighing about eleven tons was forced from its bed, and a great discharge of gas succeeded. Two men who were furnished with Davy lamps were working where this discharge took place; one of them had his lamp covered with the falling coal, and the other had his extinguished. They groped their way to warn the other miners, and then all, extinguishing the lamps as they went, safely escaped to the bottom of the shaft, and were drawn up.

A few months after a second discharge from another part of the same colliery took place. A bore-hole having been made, a violent noise like the blowing off of steam was heard, and a heavy discharge of gas filled the air-courses for a distance of 641 yards and over an area of 86,306 cubic feet. At 400 yards from the point of efflux a mining officer met the foul air, felt it blowing against him, saw the safety-lamp in his hand enlarge its flame, and drew down the wick. Still the gas continued to burn in his lamp for ten minutes, making the wires red hot, and then the light went out—a hint not lost on the owner, who quickly followed its example. At a distance of 641 yards from the efflux of the gas he met four men and boys whose lamps were rapidly reddening. At once they had the self-possession to immerse them in water, and thus escaped all danger of explosion.

The disastrous effects of the fire-damp are not confined to the loss of human life; they are also extremely injurious to the workings, tearing up galleries, shattering machinery, or even setting fire to the mine—an accident which may also be caused by spontaneous combustion * or by the negligence of the workmen. These fires are often subdued by isolating the burning coal-seam, by means of dams or clay walls, or by

---

* Experience has proved that when sulphuret of iron undergoes a chemical change into vitriol it disengages a sufficient quantity of heat to set fire to the coal with which it is often found mixed.

filling the mine with water; but not seldom they last for years, and assume dimensions which mock all human efforts to extinguish them.

At Brûlé, near St. Etienne, a coal mine has been on fire for ages. The soil on the surface is barren and calcined, and the dense sulphurous fumes, escaping from innumerable crevices, give the country a complete volcanic aspect.

In the carboniferous basins of Staffordshire and of Saarbrück and Silesia there are likewise coal mines which have been on fire for a long period. At Zwickau in Saxony the first accounts of one of these subterranean conflagrations date as far back as the fifteenth century, and the fire still burns on. The hot vapours which rise from the surface have since 1837 been put to an ingenious use. Conducted through pipes into conservatories, they ripen the choicest fruits of the south, and produce a tropical climate under a northern sky.

In a Staffordshire colliery which had been on fire for many years, and which was called by the inhabitants Burning Hill, it was noticed that the snow melted on reaching the ground, and that the grass in the meadows was always green. Some speculators conceived the idea of establishing a school of horticulture on the spot, and imported colonial plants at a heavy expense. These flourished for a time, but one day the subterranean fire went out, and as the heat it had imparted to the soil gradually diminished and departed, the exotic vegetation likewise drooped and died.

CHAPTER XXIV.

GOLD.

The Golden Fleece—Golden Statues in ancient Temples—A Free-thinking Soldier
—Treasures of ancient Monarchs—First Gold Coins—Ophir —Spanish Gold
Mines—Bohemian Gold Mines—Discovery of America—Siberian Gold Mines—
California—Marshall—Rush to the Placers—Discovery of Gold in Australia—
The Chinaman's Hole—New Eldorados—Hydraulic Mining in California—
Quartz-crushing.

GOLD is probably the metal which has been longest known
to man. For as it is found only in the metallic state,
its weight and brilliancy most naturally have attracted
attention or awakened greed at a very early age. Thus
we read in the Bible that one of the rivers flowing from
Paradise 'compasseth the whole land of Havilah, where
there is gold, and the gold of that land is good.' Gold is also
mentioned among the riches of Abraham; and when the
patriarch's servant met Rebecca at the fountain of Nahor, he
presented the damsel with a ' golden earring of half a shekel
weight, and two bracelets for her hands of ten shekels
weight of gold,' undoubtedly the first trinkets on record.

The mythical history of Greece has likewise been thought
to point to a very ancient knowledge of gold, and the story
of the search for the 'Golden Fleece' has by some been
explained as an expedition undertaken in quest of the metal;
for the use of sheepskins or woollen coverings, to collect and
retain the minutest particles of gold during the operation of
washing, is common in many auriferous countries. From
the great value which the ancient nations attached to its
possession, gold was largely used for the decoration of their
temples, and many of their idols were made of gold. Such,
among others, was the image of Belus, seated on a golden
throne in the great temple of Babylon; that of Apollo at

Delphi, and the magnificent statue of the Olympian Zeus, composed, by the hand of Phidias, of ivory and gold, and still less remarkable for its costly materials than for the consummate beauty of its workmanship.

Pliny relates that a massive golden statue of the goddess Anaitis was taken by Marc Antony in his war against the Parthians. The Emperor Augustus, dining one day at Bononia with an old veteran who had taken part in the campaign, asked him whether it was true that the sacrilegious soldier who had first laid hands on the goddess had been suddenly deprived of the use of his eyes and limbs, and had thus miserably perished. 'I myself am the man,' answered the smiling host; 'you are dining from off her thigh, and to her am I indebted for all the plate in my possession.'

The wealth of monarchs was estimated less by the extent of their domains than by the gold which they possessed, and as each successive conqueror added to the spoils of vanquished nations, the treasures accumulated by single despots grew to an almost fabulous amount. Every schoolboy knows that the vast treasures of Crœsus fell into the hands of Cyrus, who, according to the rather questionable authority of Pliny, acquired in Asia Minor no less than 24,000 pounds weight of gold, without reckoning the vases and the wrought metal. To this treasure his son Cambyses added the gold of Egypt, and Darius Hystaspis the tribute of the frontier nations of India. Thus the gold of almost the whole known world was accumulated in one single hoard, which, after the taking of Persepolis, fell into the hands of Alexander the Great. Plutarch relates that 10,000 teams of mules and 500 camels were needed for the transport of this wealth to Susa, where Alexander was cheated out of a great part of it by his treasurer. Rome, the subsequent mistress of the world, naturally absorbed the greater part of the riches of Tyre and Carthage, of Asia and Egypt. Sixty-six years after the third Punic war the public treasury contained 1,620,831 pounds weight of gold, and still greater wealth was accumulated under the Cæsars. As the empire declined, the hoards amassed in the times of its increasing power were once more dispersed. A considerable part, however, found its way to Constantinople, and after many a loss, caused by the repeated disasters of a thousand years, the remnant fell at length into the hands of the victorious Turks.

The time when gold was first coined is unknown. The oldest specimen in the Imperial Cabinet at Vienna is from Cyzicus, a town of Mysia, and bears the date of the seventh century before Christ; the next coin in point of antiquity is Persian, and was probably struck under the reign of Cyrus. According to Pliny, gold was first coined by the Romans in the year 547 after the foundation of the city. During the empire of the Chalifes Abuschafar-al-Monsur established a mint at Bagdad, in which silver coins (dirhems) and gold coins (dinars)* were struck. The Visigoths in Spain likewise had golden coins; but in the other western mediæval States they first appear, after a long interval, under Lewis the Pious, son of Charlemagne; in Venice, in 1290; and in Bohemia, under John of Luxembourg. The gold of the Carolingian monarch probably proceeded from the spoils of the old west Roman Empire; that of the Venetians (zecchins or ducats) was, no doubt, obtained, like that of the Phœnicians of old, by trading with the gold countries of Africa and of the distant East. The Florentines, the rivals of Venice, likewise obtained wealth by trade, and struck gold coins, which, from their being stamped with a flower, the arms of Florence, were called fiorini, or florins. The coins of the kings of Bohemia were made from indigenous gold.

It is hardly necessary to remark that since those times the use of gold coins has been constantly increasing with the progress of trade and civilisation; but even now, in many African and Asiatic countries which possess large quantities of gold, no coins are struck, but the metal is weighed, and thus serves as a medium of exchange, in the same manner as in the times of Abraham or Jacob.

The countries from which the ancients obtained their chief supply of gold were the Indian Highlands, Colchis, and Africa. The seat of Ophir, which furnished this precious metal to the Phœnician and Jewish traders, is unknown. While some authorities place it on the east coast of Africa, others fix its situation somewhere on the west coast of the

---

* These names were borrowed from the Greek Drachma and the Latin Denarius.

Indian peninsula; and Humboldt is even of opinion that the name had only a general signification, and that a voyage to Ophir meant no more than a commercial expedition to any of the coasts or isles of the Indian Ocean, just as at present we speak of a voyage to the Levant or the West Indies. The golden sands over which the Pactolus, a small river of Lydia, rolled, gave rise, it has been said, to the wealth of Croesus.

The richest auriferous land in Europe was the Iberian peninsula, which for centuries yielded a golden harvest, first to the Carthaginians, then to the Romans, and at a still later period to the Visigoths and the Moors. During the middle ages Bohemia was renowned for its gold, and the accounts that have reached us of the times when her auriferous deposits first began to be extensively worked remind us of the scenes which our own age has witnessed in California or Australia. Bloody conflicts frequently arose between gold-diggers and peasants because the former devastated the fields and meadows and left them permanently sterile. Even now in many parts of the country long ranges of sand hillocks and rubbish mounds remain as memorials of the mediæval gold-diggers. Frequent famines arose in the land, as many of the inhabitants gave up agriculture for mining.

A new epoch in the history of gold began with the discovery of America. We all know by what prodigies of valour the Spaniards obtained possession of the treasures of Montezuma and of the Peruvian Incas, and how frequently acts of a fiendish cruelty, inspired by the love of gold, and aggravated by a bloodthirsty fanaticism, tarnished the lustre of their arms.

More recently, about the year 1836, rich deposits of auriferous sand were discovered in Siberia, and soon raised the frozen regions of the Jenisei to the rank of the first gold-producing country in the world.* But the fame of the Russian mines was soon eclipsed by the eventful discovery of the Californian placers.

It was in January 1848, a short time after the incorporation of the province with the United States, that one James Marshall, who had contracted to build a saw-mill on the

* 'The Polar World,' p. 231.

land of Captain Sutter, about sixty miles east of Sacramento, discovered the glittering particles in the mud of the brook on which he was at work. Trembling with excitement, he hurried to his employer, and told his story. Captain Sutter at first thought it was a fiction, or the wild dream of a maniac; but his doubts were soon at an end when Marshall laid on the table before him a few ounces of the shining dust. The two agreed to keep the matter secret, and quietly to share the golden harvest between them. But, as they afterwards searched more narrowly, their eager gestures and looks happened to be closely watched by a Mormon labourer employed about the neighbourhood. He followed their movements, and the secret was speedily divulged.

It appears that Marshall did not escape the ordinary lot of discoverers, for a few years later he was wandering, poor and homeless, over the land which was first indebted to him for its enormous development.

The intelligence of the Californian gold treasures soon spread over the world, and a wonderful flood of immigration began into the newly-proclaimed Eldorado. An innumerable crowd of adventurers from every part of the New World, from the Sandwich Islands, from Europe, from Australia, came pouring in over the Rocky Mountains, through Mexico, round Cape Horn, or across the Pacific, all eager to seize fortune in a bound or to perish in the attempt. Every week dispatched its thousands to the diggings, and saw its hundreds of successful adventurers return to dissipate their earnings in the gambling saloons of the infant metropolis. In less than ten years California numbered more than half a million of inhabitants, and San Francisco, from an obscure hamlet, had risen to the rank of one of the great commercial emporiums of the world.

Science had little to do with the discovery of gold in California; but the case was different in Australia. As early as 1844 Sir Roderick Murchison directed attention to the remarkable resemblance between the Australian cordillera and the auriferous Uralian chain. Two years later, his surmises about the hidden treasures of that distant colony were confirmed by some samples of auriferous quartz sent to him from Australia. Relying upon this fact, he advised some

U .

Cornish emigrants to choose Australia for their new home, and to seek for gold among the *débris* of the primitive rocks. His opinion having become known at Sydney, through the newspapers, researches were made, which proved so far successful that in 1848 gold was found in several places in South Australia.

The first important discovery was, however, not made before the year 1851, when Mr. Hargraves made known to Government that rich gold deposits were situated to the north-west of Bathurst, on the Summerhill and Lewis rivulets, which flow into the Macquarie. When the geological Government inspector arrived at Summerhill Creek, on May 19, he found that about four hundred persons had already assembled there, who, without any other mining apparatus than a shovel and a simple tin pot, gained, on an average, from one to two ounces of gold daily.

Soon after, still richer deposits were found near the Turon and the Meroo, two other branches of the Macquarie. Here a native shepherd, in the service of Dr. Kerr, found three quartz blocks, of which the largest contained sixty pounds weight of pure gold. It may easily be supposed that the whole neighbourhood became at once the scene of active researches, which at first proved fruitless, until at length a fourth quartz block was discovered, and publicly sold for a thousand pounds. Other discoveries were made within the bounds of New South Wales; but even the richest of them were soon to be obscured by the treasures of the neighbouring colony.

As late as 1836 Port Philip had remained an unknown land, for it was not until then that its first settlers, attracted by the richness of the pastures, arrived from Tasmania. Soon a small town arose on the Yarra-Yarra, and, though badly chosen as a port, Melbourne soon rose to importance. In 1850 the district was made an independent colony, which received the name of Victoria. Here the traders and sheep-drivers now mourned over the news from Sydney. The best workmen had already left for the gold-fields, and if the exodus went on increasing, nothing remained for them but to follow the example, or quietly to await the ruin of their hopes—a patience which agrees but little with the Anglo-Saxon character.

To prevent the impending evil, a reward of 200 guineas was immediately set upon the discovery of a gold-field within 120 miles from Melbourne, and soon after the world was astonished by the intelligence of the fabulous riches of Ballarat, at the source of the River Lea. The first consequence of this discovery was that the towns of Geelong and Melbourne, both not above sixty miles from Ballarat, were immediately deserted by their inhabitants, and that, within a few weeks, more than 3,000 gold-diggers had collected on the spot, who were gaining, on an average, their ten or twenty pounds a day. But here, also, there was no definite resting-place, for new prospects of dazzling wealth constantly allured the crowd to new and still more distant fields of enterprise. Twenty thousand people, meeting with fair success, would migrate in a day, abandon their claims, and rush upon the new tract. The passions of human nature were roused by one of the strongest of its instincts; and madness and suicide, arising from excess of joy and wild despair, were far from uncommon occurrences. The whole order of society was inverted, and the labourer became of more importance than the employer of labour. The scum of the adjoining colonies boiled over and deluged the land with vice and crime. Bush-ranging extended over every portion of the country, and even the streets of Melbourne became the scenes of robbery and murder. The diggers were of all nations: Germans, French, Italians, American, Irish, Californians, and Chinese—these last being the best conducted of this motley population, which as early as 1860 numbered 50,000. To this strange people one of the most remarkable of the Australian gold discoveries is due. The immigration-tax, which had been vainly devised to check their influx (for they are objects of the greatest antipathy to the white gold-diggers), drove them to a surreptitious mode of entering the colony; and, landing at Gurchen Bay in South Australia, and taking a course thence over the frontier across the Grampian ranges, they came upon a deposit of marvellous richness, in the neighbourhood of Mount Ararat. In one of their first encampments, while picking up the roots of grass and prying for gold, they found the celebrated 'Chinaman's Hole,' which yielded 3,000 ounces in a few hours. This led to the greatest rush which had ever been known in the gold-fields, for 60,000 people congregated there in a few weeks, and before a month had elapsed an immense town was systematically laid out. Shops, taverns and hotels, theatres and billiard-rooms, sprang up in the desert, like the mystic trees of Indian jugglers, and were quickly followed by a daily mail and a daily newspaper. Thus, within the space of two months, the magical power of gold converted a wild

GOLD WASHING IN AUSTRALIA.

mountain gorge into a teeming city, where frontages were nearly as valuable as in the heart of London.

The great social disorganisation which distinguished the first few years of the Australian gold discoveries has long since passed away, together with much of the excitement natural to a transition state. Order now universally pre-vails, and the occupations of life are pursued with as much regularity as in the oldest States. The growth of the colony, which scarcely thirty years since was a mere unknown waste, is not the least marvellous of the many marvels that have been worked by gold. In the year 1851 the population of the province was 77,345 persons, of whom 28,143 were located in Melbourne. In 1860, it had already increased to 462,000, and probably the next few years will find it augmented to a million, while Melbourne already rivals our larger cities in size and wealth.

The wonderful discoveries in California and Australia having made gold the all-absorbing topic of the day, it is not surprising that new Eldorados were now eagerly sought for wherever the geological formation of a country held' out the hope of similar treasures.

Fresh and highly productive gold-fields have, within the last few years, been opened in British Columbia, and, still further to the north, in the Arctic wilds of the infant colony of Stikeen. Numerous diggers are at work in New Zealand, and in the deserts to the north of the Cape. A system of auriferous veins has been discovered in North Wales; the county of Sutherland, the Ultima Thule of our isle, claims to be ranked among the gold-producing regions; and numerous adventurers are on their way to the frozen deserts of Lapland, where the glittering metal is said to abound in the basin of the Ivalo.

At no former period of the world's history has gold been so eagerly sought for over such extensive areas in all parts of the globe; never have larger quantities of the precious metal been added to the accumulated hoards of ages. No doubt this vast influx of wealth has in many cases been productive of evil consequences; but its beneficial influence upon the progress and happiness of mankind far outweighs the injury it may too frequently have caused by rousing the worst passions of our nature. An astonishing impetus has already been given to commerce and industry; competence and wealth have been diffused over many lands; deserts have been transformed into growing empires; and a vast continent, long despised as the convict's prison, has been raised in the social scale to a height almost commensurate with its geographical importance.

The mineral formations in which gold originally occurs are the crystalline primitive rocks, the compact transition rocks, and the trachytic and trap rocks, which, by their disintegration have, in the course of ages, enriched large alluvial tracts with particles of the precious metal. Torrents and rivers washed them down from the heights, along with the worthless rubbish of their original matrix, and finally deposited them in the gulleys and ravines of the lower grounds. Hence the alluvial territories have always been the chief sources of auriferous wealth, and this circumstance explains how countries which at one time abounded in gold have long since ceased to be of importance. For the comparative ease with which the metal could generally be obtained by digging and washing, and the greed which stimulated the researches of thousands, could not fail to exhaust even the richest placers. No one now dreams of searching for gold in the sands of the Pactolus or of the Golden Tagus; no modern Argonaut sails to Colchis in quest of the golden fleece; the fields of Bohemia are no longer ransacked by gold-seekers; and a like fate probably awaits many of our modern Eldorados.

The annual product of gold in California has fallen off nearly 60 per cent.* from the highest point at any time reached, and the population engaged in mining has been diminished in a still greater ratio. In addition to manual labour, capital and skill are at present required, and share the profits that formerly accrued to labour alone. But the Californian gold fields are far from being exhausted, for in no part of their vast extent, which spreads almost continuously over seven degrees of latitude, have the mines, other than the shallow placers, been exhausted or worked to any great extent. Even in the older and more popular districts few of them can be said to have been much more than thoroughly prospected, the majority not having been opened at all; while in the more remote sections of this gold field still less work has been done, large districts remaining but partially employed. Should even the placers be ultimately

---

* The yield in 1872 was about 18,143,314 dollars, but the neighbouring states and territories of Nevada, Oregon, Utah, Colorado, and Wyoming probably produced as much again. From their discovery in 1851 to the end of 1871, the gold mines of Victoria produced 40,750,000 ounces, worth 162,700,000*l.* The Russian gold mines in Siberia and the Oural yield about a million ounces annually. The total produce of the globe may be computed at five million ounces.

exhausted, the gold-bearing banks and rocks are doomed to give way to the crushing machine, to the resistless force of powder, and to the power of water, which in every age has been the gold digger's necessary auxiliary, but is now used in California as a giant strong enough to remove mountains from their foundations.

Tunnels long and deep, ditches from twenty-five to fifty miles in length, and huge iron pipes, sometimes crossing chasms of a thousand feet vertical depth, convey it to the gold-bearing 'bank,' where hydraulic machines, such as Craig's Globe Monitor and Dictator, or Fisher's Knuckle Joint and Nozzle, discharge a thousand inches of water, equal to 1,579 cubic feet, per minute, with a velocity of about 140 feet per second. The discovery of the most powerful explosives, such as nitro-glycerine, giant-power, &c., and furthermore the invention of a diamond drill, have, by preparing the ground and thus assisting the action of water, raised hydraulic mining to one of the foremost and most lucrative pursuits in California. In many cases, however, blasting is not required, for a quantity of water uninterruptedly striking the bank with one-tenth of the velocity of a cannon-ball suffices to produce its 'caving.' After an hydraulic mine has been opened for washing operations, and water under a high pressure connected through an iron pipe with one of the improved hydraulic nozzles, the real mining work can commence. A screw is turned, and presently a stream of water from five to seven inches in diameter plays with magnificent force against the opposite bank of gravel. The water issuing from the nozzle is to the touch as hard as a bar of steel, and retains, when thrown from a good nozzle, its cylindrical and condensed shape till it strikes the gravel bank, where its effect is soon visible. At the first shock a thousand rays of water fly in all directions; a little later the stream has buried itself deep in the bosom of the bank, and the water boils and hisses over the lips of the aperture, carrying with it gravel, sand, clay, and whatever matter may be at hand. The opening widens; flakes of gravel tumble in all directions; an arch, wide and deep, is made on the gravel-bank, the 'jambs' of the arch to the right and left are demolished by turning the jet of water upon them, and the first 'cave' in the hydraulic mine takes place. Care is taken to secure as soon as possible a large open front so as to occupy two, three, or more hydraulic nozzles, according to the supply of water and general capacity of the works. These different nozzles being supplied from the same distributor can open a 'cross fire' upon any point within two hundred feet from their mouths, and thus do excellent execution. Whole hills are in this manner successively levelled to the ground and forced to yield their treasures.

The blasting of rocks takes place on a stupendous scale. Six hundred kegs of powder, each containing twenty five pounds, are often fired simultaneously, and will dislodge at once from fifty to sixty thousand cubic yards of rock.

An ample supply of water during all seasons of the year is of course of vital importance to the success of an hydraulic mine, and calls forth all the resources of engineering skill.

Thus gold mining has made prodigious strides since the primitive times when the crevicing knife and spoon and the gold rocker first did their duty; or since the spring of 1852, when an unknown miner (whose ingenuity would have deserved a better remembrance) put up the first simple hydraulic machine on his mining claim at Yankee Tim in Placer County. From a small ditch on the hill side a flume was built towards the ravine, when the mine was opened; the flume gained height above the ground as the ravine was approached, until finally a head or vertical height of forty feet was reached. At this point the water was discharged into a barrel, from the bottom of which depended a hose about six inches in diameter, made of common cow-hide and ending in a tin tube about four feet long, the latter tapering down to a final opening or nozzle of one inch.*

The ores of auriferous quartz are of course treated in a different way from the alluvial *débris*. After having been crushed and pulverised by powerful machinery, the finely-powdered quartz is then treated with mercury, a method well known to the ancients. This metal dissolves and unites with the gold, producing an amalgam, which is easily separated from the dross, and, by straining and distillation, yields the gold.

* From 'Statistics of Mines and Mining West of the Rocky Mountains,' by Rossiter W. Raymond, United States' Commissioner of Mining Statistics, 1873.

CHAPTER XXV.

SILVER.

Its ancient Discovery—Its uses among the luxurious Romans—The Mines of Laurium—Silver Mines of Bohemia, Saxony, and Hungary—Colossal Nuggets—Silver Ores—Silver production of Europe—Mexican Silver Mines—The Veta Madre of Guanaxuato—The Conde de la Valenciana—Zacatecas and Catorce—Adventures of a Steam-Engine—La Bolsa de Dios Padre—The Conde de la Regla—Ill-fated English Companies—Indian Carriers—The Dressing of Silver Ores—Amalgamating Process—Enormous Production of Mexican Mines—Potosi—Cerro de Pasco—Gualgayoc—The Mine of Salcedo—Hostility of the Indians—The Monk's Rosary—Chilian Mines—The Comstock Lode.

LIKE that of gold, the first discovery of silver precedes the historic times, and must no doubt be sought for in the remotest antiquity ; for as it is not seldom found in a native state, its brilliancy could not fail to attract attention at a very early age. Veins of silver ore, moreover, not seldom crop out on the surface—a circumstance which likewise greatly facilitated the accidental discovery of the metal. Thus Diodorus relates that streams of melted silver flowed out of the calcined soil where some shepherds had set fire to a forest in the Pyrenees. The cunning Phœnician merchants, who were trading in the neighbourhood, bought the metal for a trifle from the natives, who, ignorant of its value, gladly exchanged it for some worthless trinkets.

Our earliest annals show us silver in common use among the more polished nations of antiquity, both for ornamental purposes and as a means of exchange. When we read in the Bible that Abraham weighed to Ephron the Hittite ' four hundred shekels of silver, *current money* with the merchant,' for the purchase of the field of Machpelah, where Sarah was buried, we cannot possibly doubt that, long before the patriarch's time, great quantities of silver must have circulated among the traders of the East, and that even then it belonged to the discoveries of ancient days.

Homer describes the shields and helmets of his heroes as inlaid with silver, and in Northern Asia silver ornaments have been found in the tumuli of the Tschudi, a mysterious people who have left no vestiges of their existence save their tombs.

The Babylonian, Assyrian, and Persian monarchs imposed large tributes of silver on the conquered nations of Asia, and at a later period the treasures thus amassed by a long line of despots came into the possession of Alexander the Great, and finally of the Romans, who absorbed all the riches of Carthage and the East. In the reign of Augustus, the tables of the wealthy senators groaned under silver dishes weighing from one hundred to five hundred pounds; silver statues of their ancestors decorated their apartments, and they not seldom performed their luxurious ablutions in baths of silver. Mirrors of this metal were in frequent use among a people to whom the art of applying a lustrous amalgam to the back of a plate of glass was unknown.

The skill of the artist not seldom added an inestimable value to the intrinsic worth of an embossed or chiselled silver vase, and Pliny mentions several works of this kind which enjoyed a world-wide celebrity.

The most ancient silver mines of which we have any historical account were situated in the mountainous parts of Chaldæa and in Spain. With the development of the Grecian States, Eastern Europe also began to furnish its contingent. Silver mines were discovered and worked in the Pangæan Mountains, between Macedonia and Thrace, in the islands of Siphnos and Cyprus. Athens derived a considerable part of its revenue from the mines of Laurium in Attica. At first the profits derived from this source were distributed among the citizens, until Themistocles persuaded the general assembly of the people to devote them to the construction of ships of war. The battle of Salamis was won by the galleys built with the money thus obtained, so that the silver mines of Laurium have been the means of adding some of its brightest pages to the history of Ancient Greece.

After the fall of the Roman Empire, the exhausted or neglected mines of the East, and of the Iberian peninsula, ceased to be the sources from which silver flowed over the world.

The riches amassed during ages of civilisation were now scattered among barbarous hordes, or buried in the ruins of cities destroyed by fire; and as no new influx replaced the losses caused by accident or the slow wear of time, the precious metals gradually became more rare and of increasing value, until Germany began to open a new era in mining history, and for a time to take the lead among the silver-producing countries of the globe. The first discovery of this metal in Bohemia dates back as far as the seventh century ; and in the tenth the Rammelsberg in the Hartz Mountains began to yield its still unexhausted treasures. In the twelfth century the silver mines of Saxony and Hungary were discovered, and at a much later period those of Kongsberg in Norway.

Many of these deposits are remarkable for the richness of their ores, and for the large masses of native or crystallised silver which they have sometimes yielded. From the mine of Himmelsfürst, near Freiberg in Saxony, lumps or nuggets weighing above a hundred pounds have more than once been extracted. The largest single block ever known was discovered at Schneeberg, in the Saxon Erzgebirge, in the year 1477. It consisted of silver-glance and native silver, and measured no less than seven yards in length and three and a half in width. On hearing of this magnificent prize, Duke Albrecht of Saxony visited the mine, where he eat his dinner from the block. Agricola Bermannus relates that, during the repast, he exclaimed, 'Frederick is a wealthy and powerful emperor, but he has never dined from a table such as this.' The subsequent smelting of this wonderful mass produced 40,000 pounds of solid silver.

Large nuggets of solid silver have likewise been found at Kongsberg in Norway. A block weighing 530 pounds, which was discovered in 1666, is still preserved as a curiosity in the Copenhagen Museum; and as a convincing proof that the ancient riches of these northern mines are not yet exhausted, a still larger block, weighing 750 pounds, was disinterred as late as the year 1834.

Though native silver is found in many localities, our chief supply of the precious metal is derived from the ores in which it is found combined with other substances, such as silver-glance (sulphuret of silver); antimonial silver; red-silver (sulphuret of silver and antimony); horn silver (chloride of silver), &c. A considerable quantity is likewise obtained as an accessory product from the lead mines, by separating it from the galena or lead-glance, which usually contains a small percentage of silver.

At present the mines of the Austro-Hungarian Empire, which in the year 1851 produced 122,950 marks of silver, are the richest in Europe. In Bohemia the mines of Birkenberg, near Przibram, yield on an average 40,000 marks a year, considerably more than the celebrated mine of Schemnitz in Hungary, which in the year 1854 produced 26,064 marks.

Prussia obtains from her mines in the Hartz Mountains in Nassau and the county of Mansfield about 100,000 marks; and the kingdom of Saxony, which, in proportion to its small extent, holds the first rank among the silver-producing countries in Europe, produces annually about 53,000 marks. In Great Britain, silver is accessorily obtained from the lead mines, to an amount of 524,307 ounces in 1873. France produces 26,800 marks, Sweden and Norway 6,000 marks, and Italy 4,500; while Spain, which in ancient times enriched Tyre and Carthage with the rich produce of her silver mines, yields at present but an insignificant quantity.

On adding together the various sums above-mentioned, we find that, exclusively of the Russian Empire, which obtains the greater part of its silver (about 65,000 marks a year) from the Altaï Mountains in Siberia, the whole annual production of Europe may be estimated at about 400,000 marks.

This quantity, large as it is, sinks into comparative insignificance when compared with the enormous masses of silver with which, ever since their discovery and conquest by Cortez and Pizarro, Mexico and Peru have enriched the world.

The Mexican silver mines, which deserve a particular notice from the immensity of their produce and the interesting details with which their history abounds, are mostly situated on the back or on the western slopes of the Cordillera of Anahuac, at elevations which sometimes approach the line of perpetual snow. But little is known of their first history. The lodes of Tasco and Pachuca appear to have been worked soon after the conquest, and in 1548, twenty-eight years after the death of Montezuma, the mines of Zacatecas were opened, though situated above 400 miles from the capital. Muleteers travelling from Mexico to Zacatecas are said to have discovered the silver mines of Guanaxuato.

Many of my readers will no doubt be surprised to hear that the Mexican silver ores are generally poor. On an average they contain only from three to four ounces of silver per cwt., much less than the ores of Annaberg, Marienberg, and other districts of Saxony. Their comparative poverty is, however, amply redeemed by their abundance and the facility with which they are worked.

After these preliminary remarks I will now briefly describe the chief Mexican silver mines. The Sierra de Santa Rosa, a group of porphyritic hills, rises in the centre of the province of Guanaxuato. Partly arid and partly covered with the evergreen oak, it attains an absolute height of from 8,000 to 9,000 feet; but, as the neighbouring plain is nearly 6,000 feet above the sea, it hardly attracts the notice of the traveller accustomed to the vast proportions of the Andes.

The southern slope of this porphyritic range is crossed by the famous Veta Madre of Gaunaxuato, the richest silver lode as yet discovered in Mexico. This enormous vein, which traverses the country for upwards of eight miles, with an average width of from 120 to 135 feet, is, however, not productive throughout its whole extent; but the ore occurs in branches and bunches, leaving intermediate spaces of dead and unproductive ground. Among the numerous mines that have been opened along its course, the Valenciana exhibited, at the beginning of the present century, the almost unparalleled example of a mine which, during a period of more than forty years, never yielded its proprietors less than an annual income of from 80,000l. to 120,000l. The part of the Veta in which it is situated had remained unexplored till 1760, when Obregon, a young Spaniard, began to work it, with the assistance of some friends who advanced him the necessary capital. In the year 1766 the diggings had reached a depth of 240 feet, and the expenses were far greater than the proceeds. But Obregon clung with the passionate ardour of a gambler to the hazardous enterprise on which he had staked all his hopes of fortune. In 1767 he entered into

partnership with a small shopkeeper, Otero, who was destined
soon to share the fabulous riches that were about to reward
his perseverance. Already in the following year the produce
of the mine considerably increased, and in 1771 it began to
yield enormous masses of sulphuret of silver. From that time
till 1804, when Humboldt left Mexico, it never produced
less than 560,000*l.* worth of silver annually, and the net
profits of the partners amounted in some years to 240,000*l.*

Under the title of Conde de la Valenciana, Obregon maintained the simple habits and the urbanity of character which
had distinguished him in poverty. When he began to work
his mine, the goats were feeding on the spot where ten years
later a thriving town of 8,000 souls had started into existence.
Guanaxuato, the capital of the State of the same name, is
indebted to the neighbourhood of the richest silver mines in
the world for its origin and prosperity. Inclosed in a narrow
valley, its houses rise in terraces one above the other; and
the contrast of the magnificent abodes of the rich mining
proprietors with the miserable huts of their dependants adds
to the singular appearance of the place. The Mexican miner
is, however, not so poor as his wretched dwelling might lead
us to suppose. In some measure he shares the fortunes of the
proprietors of the mines, as he is entitled to part of the ore;
so that when the vein is more than usually productive his
weekly profits may amount to as much as a hundred dollars.
Yet he never thinks of purchasing a piece of land, or of repairing his hut, when favoured by fortune; but foolishly
squanders his money in drinking and gambling, and seldom
returns to his work before his last farthing has been spent.

The population of Guanaxuato naturally fluctuates with
the prosperity of the mines. In 1806 and 1807, when they
were in the highest state of activity, it amounted to 90,000
souls; during the wars of independence it sank to 20,000,
but since then it has again risen to 60,000.

Next to the mines of Guanaxuato those of Zacatecas are
distinguished by their richness. They are likewise situated
on the great central plateau of the cordillera in a wild mountain region whose forbidding aspect forms a strange contrast
to the riches concealed under its surface. In 1826 the
United Mexican and the Bolaños Company undertook the

working of these mines ; and two years after the lattter had the good fortune to find an exceedingly rich vein, which up to the year 1834 produced no less than 1,680,316 marks of silver.

Before 1770 the populous district of Alamos de Catorce in the State of San Luis de Potosi, was still a complete desert. About this time a free negro, named Milagros, who made a scant livelihood as an itinerant musician, having lost his way, was obliged to pass the night in the forest. On the following morning he found a few drops of silver on the spot where he had made a fire, and, on a closer examination, discovered the rich cropping-out of a bed of argentiferous ores. He lost no time in establishing the right of property which he derived from his discovery, and opened the shaft Milagros, which, in a few years, made him a wealthy man.

Soon after Don Barnabé de Zepeda discovered the chief vein of Catorce— the Veta Madre—which continued to be worked with great success until the revolution. This event having proved as destructive to the draining machines of Catorce as to those of Guanaxuato and Zacatecas, a contract was made with an English house for the furnishing of a steam-engine, the first ever seen in Mexico. It was landed at Tampico in May 1822, but arrived at Alamos six months later, as the carts which dragged the heavy piece of machinery broke down every moment on the wretched roads which lead to the central plateau on which the mines are situated. This, however, was but the prelude of new difficulties ; for as the neighbouring forests could not furnish wood fit for the purpose, it was necessary to order iron pump-tubes in England, which did not arrive before 1826, so that the engine could not be set to work until four years had passed after its arrival in Mexico. The history of Catorce affords many remarkable examples of good fortune ; but as most of the rich *mineros* (mine-proprietors) were men of low birth and without education, they squandered their treasures as fast as they acquired them. Medellin, the proprietor of the mine Dolores, once spent 36,000 dollars on a christening party ; and at times, when the share of the hewers amounted to one-third of the extracted ores, a common miner would stake two or three thousand dollars on the issue of a cock-fight.

Among the first settlers at Catorce was an ecclesiastic named Flores, who bought for 700 dollars a newly-opened mine, which received the significant name of ‘La Bolsa de Dios Padre,’ or ‘God the Father’s Money-Bag.’

Never was a small capital more profitably invested, for, at a depth of about twenty yards, a deposit of such enormous richness was discovered that in less than three years the profits of Padre Flores amounted to three millions and a half of dollars.

The mines of Pachuca, Real del Monte, and Moran, began to be worked soon after the conquest. The Veta de la Biscayna, the chief vein of the district, yielded immense profits from the sixteenth to the beginning of the eighteenth century, when, in consequence of insufficient drainage, the mines were drowned. An enterprising hidalgo, Don José Bustamente, then began to drive a draining gallery, 7,000 feet long, which, however, was only finished after his death by his partner, Don Pedro Terreros, a merchant of Queretaro. Its immense cost was amply repaid, for Terreros extracted no less than fifteen millions of dollars from the mine, and was ennobled under the title of Conde de la Regla. His liberality was worthy of his wealth, for besides a gift of two ships of the line—one of them of 112 guns—to King Charles III. of Spain, he lent the court of Madrid a million of dollars, which, it is almost superfluous to remark, were never repaid. He built the enormous factories of La Regla at a cost of more than 400,000l., and left his children a property rivalling that of the Conde de la Valenciana. Since 1774, however, the profits of the Biscayna, which was now worked 300 feet below the adit, began to diminish; for though it still continued to yield enormous quantities of ore, the water flowed in so abundantly that twenty-eight baritels,* each requiring forty horses, and worked at an expense of 2,000l. per week, was incapable of mastering it. At the death of the old Conde the works were abandoned, until 1791, when his heirs once more set all the baritels in motion; but the proceeds not covering the expenses, the works were again abandoned.

In the year 1824, when the frenzy of mining speculations

---

* A very primitive contrivance for raising the water in skin bags.

was at its height, a company was formed in London, with a capital of 400,000l., for the purpose of working the mine of Tlalpujahua, in the State of Mejoacan, which had long since been abandoned. Without any accurate information, a number of supervisors and workmen were sent to the spot in 1825, when they found all the mines under water. Yet, in spite of this rather unpromising state of affairs, the company, not satisfied with its first acquisitions, entered into new engagements, so that at the end of 1825 it had contracted for no less than eighty mines, for which it bound itself to pay the proprietors annuities amounting to more than 50,000 dollars during the first three years. Operations were now begun in many places at once, but every one of them ended in disappointment; and in 1828 the company, after spending every farthing of its capital, vanished into 'airy nothing,' abandoning the mines to their original proprietors, and leaving the ill-fated shareholders to mourn over their credulity and folly.

Subsequently another English company undertook the working of the mines of Real del Monte, and after spending no less than 15,381,633 dollars, against a produce of 10,481,475 dollars, was dissolved in 1848. Mr. Buchan now undertook the management for a Mexican company, and almost immediately struck the great Rosario vein, which opened a long career of prosperity, and yielded a profit of a million of dollars in 1867. Every fortnight a conducta, or escort, of one hundred and fifty armed men * conveys the silver, in bars of seventy pounds, inclosed in an iron safe, to the capital.

It would be vain to seek in the Mexican mines for the scientific arrangement which is to be found in most of the English or German subterranean workings. One of their chief faults consists in a want of communication between their various parts, so that they resemble those ill-arranged houses where, to go from one room to another, one is obliged to traverse long and crooked passages. Hence the impossibility of introducing in most of the mines an economical transport by means of tramroads and waggons, and hence also the necessity of conveying the ore to the surface by human labour. The native Indians, however, are ad-

* _Illustrated London News_, No. 1477, Saturday, April 11, 1868.

mirable carriers, for they will climb steep ladders, with 240
to 380 pounds, and perform this hard work for six hours
consecutively.   Their muscular strength seemed truly astonishing to Humboldt, who, though of a vigorous constitu-
tion and having no weight to carry but his own, felt himself utterly exhausted after ascending from a deep mine.

Most of the Mexican silver is extracted from the ores by
the process of amalgamation. For this purpose the ore is
first crushed either by rollers, or, more generally, by stamps,
called in Mexico *molinos*, which in principle resemble those
used in the tin mines of Cornwall, but are not so powerful,
and are worked either by water-power or by mules.

STAMPING MILL.

The crushed ores are then conveyed to the *arrastres*, or
grinding-mills, which are usually arranged in rows in a
large gallery or shed. The ore, having been brought into a
finely divided state, is next allowed to run out of the arrastre
into shallow tanks or reservoirs, where it remains exposed
to the sun until it has the appearance of thick mud, and in
this state the process is proceeded with. The *lama*, as it
is called, or slime, is now laid out on the *patio*, or amalgamation floor (which is in some places boarded, and in others
paved with flat stones), in large masses called *tortas*, from
forty to fifty feet in diameter, and about a foot thick ; and
so extensive are the floors that a large number of these
tortas are seen in progress at the same time. When the
ore has been laid out in masses on the patio, the operations
necessary to produce the chemical changes commence. The

first ingredient introduced is salt, in the proportion of fifty
pounds to every ton of ore, and a number of mules are made
to tread it, so that it may be dissolved in the water and
intimately blended with the mass. On the following day
another ingredient is introduced, called in Mexico *magistral.*
It is common copper-pyrites, or sulphide of copper and iron,
pulverised and calcined, which converts it into a sulphate.

About twenty-five pounds of this magistral are added for
every ton of ore in the torta, and the mules are again put in,

GRINDING MILL.

and tread the mass for several hours. Chemical action now
commences, and new combinations between the decomposed
mineral substances are in progress. Quicksilver is then
introduced, being spread over the torta in very small
particles, which is effected by passing it through a coarse
cloth. The quantity required is six times the estimated
weight of the silver contained in the ore.

The quicksilver being spread over the surface, the mules
are once more put in, and tread the whole until it is well
mixed. Great skill is now required to watch the progress

x 2

of the amalgamation, and to decide whether any one of the ingredients that have been added to the ore is in deficiency or excess.

The amalgamating process being at length ended, the mass is washed in large vats, through which streams of water are made to pass, so as to drive away the lighter particles of the mud and to leave the heavy amalgam at the bottom of the tub. After being strained through the strong canvas bottom of a leathern bag, in order to separate the superfluous mercury, the amalgam is finally heated in large retorts, when the quicksilver is volatilised and the pure silver remains behind. As a considerable quantity of mercury is thus lost, this metal has always had a great influence on the working of the Mexican mines. When, in times of war, the importation of quicksilver from Europe was stopped, thus causing a considerable increase in its price, the ores accumulated in the magazines, as their poverty made it impossible to meet the additional expense; and then it not unfrequently happened that proprietors, possessing ores to the amount of several millions of dollars, were unable to pay their current expenses. In the last century, the Mexican mines annually required sixteen thousand hundredweight of mercury, which was furnished chiefly by the mines of Almaden, Huancavelica, and Idria. The sale of mercury to the various mining proprietors was a Government monopoly.

The quantity of silver produced by the Mexican mines in 110 years, ending with the first year of the present century, amounted to about ninety-eight millions of pounds troy, and the total value of the gold and silver produced from 1689 to 1803 to about 285 millions of pounds sterling.

The production was most abundant in the years 1805, 1806, and 1809, when it reached the sum of twenty-six and twenty-seven millions of dollars. It then fell enormously during the revolutionary wars; but the English mining mania of 1824 having furnished the necessary capital for the re-opening of a large number of mines, it again gradually rose, and even reached during some years the rate of its most flourishing period. But it never amounted to one-third of the value which is now annually extracted from our coal mines; and while the latter open new and unbounded sources of wealth by the activity they communicate to numberless branches of industry, Mexico—a prey to bigotry, ignorance, anarchy, and sloth—remains, in spite of her silver mines, as poor and as barbarous as ever.

In South America, the mines of Potosi, Cerro di Pasco, and Gualgaoc, are the most renowned for their richness. In 1545, an Indian, while pursuing some deer along the declivity of a steep mountain, took hold of a shrub, the roots of which, giving way, brought to view a mass of silver, the first discovery of the riches which have rendered Potosi the proverbial symbol of wealth. The Indian, wisely concealing his good fortune, repeatedly visited the mine, but his improved circumstances having been remarked by one of his countrymen, he was obliged to take him into the secret. Unfortunately a quarrel ensued, and the faithless confidant betrayed it to his master, Don José Villaroel, a Spaniard, whose extraordinary success in working the mines soon drew the attention of all America to the wild Cerro di Potosi. A town of 100,000 inhabitants soon rose in the desert, in spite of the wintry inclemency of the climate (12,842 feet above the level of the sea) and of the fabulous prices of provisions, as all the necessaries of life had to be conveyed from a vast distance over the pathless mountains.

But if living in Potosi was extravagantly dear, Mammon took care to provide his votaries with the necessary means; for the treasures which were here extracted from the bosom of the earth seem rather to belong to the world of Oriental romance than to that of sober reality. According to Humboldt, the mountain of Potosi, whose topmost mine is situated 15,384 feet above the level of the sea—considerably higher than the eternal snows of the summit of Mont Blanc —produced from 1545 to 1803 no less than 230,000,000l. besides the silver which was not registered, or had been carried off by fraud, and which may probably have amounted to as much again. During the period of their greatest prosperity—from 1585 to 1606—when they annually yielded 882,000 marks of silver, 15,000 Indians were occupied in the mines of Potosi. At present, however, their produce, though still considerable, has diminished to one-eighth, and the population of the town has shrunk in the same proportion.

The ores were at first reduced in a very imperfect manner, according to the old Indian method. On the mountains surrounding the town of Potosi, wherever the wind blew with sufficient force, portable smelting ovens of clay, in which numerous holes kept up a strong current of air, received alternate layers of ore and charcoal, and the lively blast soon separated the metal from the dross. The first travellers in the Cordillera describe with enthusiasm the magnificent aspect of more than 6,000 fires, which every evening blazed on the mountain crests of Potosi. Amalgamation was first introduced about the year 1571.

On the bleak Puna, or high table-land between the parallel chains of the Cordillera and the Andes, is situated the famous mining town of Pasco. Surrounded by a crescent of steep and naked rocks, its straggling buildings extend over an uneven ground, bordered by small marshes and lagunes. The shivering traveller, descending from the windy heights, is at first agreeably surprised by the sight of a large town in the midst of these dreary solitudes; but a nearer inspection of its narrow crooked streets, and of its miserable huts, with here and there a stately mansion, soon dissipates the fancies he may have formed at a distance.

The wild, forbidding aspect of the neighbourhood, and the rigorous climate, only a short day's journey from the loveliest valleys, prove the greatness of the subterranean treasures which could induce so large a population to settle in so inclement a region.

The mines of Pasco, like those of Potosi, were discovered, it is said, by an Indian shepherd, who, accidentally lighting a fire where the ores cropped out, found silver among the ashes. There are two chief lodes, with numerous branches, so that the whole neighbourhood may be considered as resting on a subterranean network of silver. The entrances to most of the mines are situated in the town itself; and, as every proprietor thinks only of his present profits, they are worked in so slovenly a manner that they frequently fall in. Tschudi, who visited some of the deepest of them, always thought himself extremely fortunate when, after descending on half-rotten steps or by mouldering ropes and rusty chains, he returned again to daylight without accident, and mentions an instance where three hundred workmen were buried in the ruins of a mine, in which the necessary props had been shamefully neglected.

When a mine is very productive, it is said to be in 'boya;' and when such periods of affluence take place in several of the mines at once, the population of Pasco sometimes increases to double or treble the usual number. The Peruvian miners are no less dishonest than their fellow-workmen in Mexico, and equally cunning in robbing their employers. On the other hand their patience and perseverance are unrivalled. Satisfied with the coarsest food, and with a still more miserable hut, they undergo an amount of bodily toil which no European could endure. Their only solace is the chewing of Coca, the mysterious plant to which they ascribe such wonderfully stimulating powers, and which has at length begun to awaken the curiosity of European chemists and naturalists.* When a 'boya' raises their earnings to a considerable sum, they spend it in drunkenness, and never think of returning to their work until their last farthing or their last credit is exhausted. The mineros, or proprietors of the mines, are almost equally uncivilised. Passionately devoted to gambling and mining speculations, they are commonly deeply in debt to the capitalists of Lima, who advance them money at the rate of 100 and 120 per cent.; and when a 'boya' favours them, this sudden increase of wealth is merely the prelude of new embarrassment.

According to law, all the produce of the silver mines should be sent to the Callana or Government smelting-house; but in order to avoid the heavy duties levied by the State, vast quantities are smuggled to the coast. The annual produce registered at the Callana amounts to about 300,000 marks, but perhaps as much again is exported in a clandestine manner. Besides the mines of Pasco, Peru possesses many others of considerable value, situated chiefly in the provinces of Pataz, Huamanchuco, Caxamarca, and Gualgayoc.

The famous Cerro de San Fernando de Gualgayoc, fourteen leagues from the town of Caxamarca, is an isolated mountain traversed by numberless veins of silver. Its summit is sharply

* The seventeenth chapter of 'The Tropical World' is devoted to the Erythroxylon Coca.

serrated by a multitude of tower-like or pyramidal pinnacles, and its steep sides are not only pierced by several hundred galleries for the extraction of the ores, but also by many natural openings or caverns, through which the dark blue sky is visible to the spectator standing at the foot of the mountain. The singularity of aspect is increased by the numberless huts, sticking like nests to the slopes of the fortress-like mountain wherever a ledge allowed them to be constructed. During the first thirty years after the discovery of the mines from 1771 to 1802, they yielded considerably more than thirty-two millions of dollars, and are still very productive.

One of the most celebrated silver mines of Peru was that of Salcedo, renowned alike for its richness and for the tragical end of its possessor. Don José Salcedo, a poor Spaniard who had settled in Puno, fell in love with an Indian girl, whose mother, on condition of his marrying her daugher, revealed to him the existence of a rich silver lode. The fame of Salcedo's wealth spread far and wide, and excited the envy of Count Lemos, the viceroy, who sought to obtain possession of the mine. As the good-natured and liberal Salcedo had become very popular among the Indians, this circumstance was made use of by the rapacious viceroy to accuse him of fomenting among the natives a spirit of rebellion against the Spanish yoke. He was cast into a dungeon, and the obsequious judges condemned him to death.

While in prison Salcedo begged the viceroy to submit the case to the high court of justice at Madrid, and to allow him to appeal to the mercy of the king, At the same time he offered, as an acknowledgment for this favour, to give the viceroy daily a bar of silver, from the day the ship left the port of Callao to its return from Europe. If we consider that in those times a journey from Peru to Spain and back required at least from twelve to sixteen months, we may form some idea of Salcedo's wealth. The viceroy, however, would not listen to the proposal, the very brilliancy of which probably inflamed still more his cupidity, and he ordered Salcedo to be hung. But his cruelty met with the disappointment it deserved; for when it became known that nothing could save Salcedo, his Indian friends destroyed the works of the mine, and so carefully concealed the entrance that it has remained undiscovered to the present day. After performing this work of retribution, the Indians dispersed, and neither promises nor tortures could wring the secret from those that were caught.

Though the mines of Peru have yielded and still yield vast quantities of silver, yet probably only a few of the richest lodes are worked; for the Indians, to whom other lodes are well known, will never reveal their existence to the white men. They know by experience how small a benefit they derive from the mines, which are to them but a source of severe labour. Thus they prefer leaving the treasures of the earth undisturbed, or use them only in cases of extreme necessity. In many provinces undoubted proofs exist that the richest silver mines are secretly worked by the Indians, but all efforts to discover them have proved fruitless.

A Franciscan monk at Huancayo, a desperate gambler and almost always in want of money, had by his friendly manners gained the goodwill of the Indians. One day, after a severe loss, he bitterly complained of his distress to one of his Indian friends. After some hesitation the man promised to assist him, and brought him on the following evening a bag full of rich silver ores. This gift he repeated several times. But the monk, greedy after more, begged the Indian to show him the mine—a request which, after repeated refusals, was at length reluctantly granted. On the appointed night, the Indian, with two of his comrades, came to the Franciscan's dwelling, took him on his shoulders, after first carefully blindfolding him, and carried him, alternately with his friends, a distance of several leagues into the mountains. Here they halted, and the Franciscan's bandage having been removed, he found himself in a subterranean gallery, where the richest silver ores sparkled from the walls. After feasting on the grateful sight and filling his pockets, he was carried back again in the same way. On his return he secretly loosened the string of his rosary, and let a bead drop from time to time, hoping by this means to be able to find the mine. But, on the following morning, as he was about to reconnoitre, his Indian friend knocked at the door, and saying, ' Father, thou hast lost thy rosary!' brought him a whole handful of the loose beads.

In 1850 the mines in the province of Copiapo in Chili yielded 335,000 marks of silver, nearly as much as the entire production of Europe, and the State of Nevada in North America bids fair to rival the riches of Peru. The ores are found on the eastern slopes of the Sierra Nevada, in the region of the Carson River, and have since 1859 attracted a stream of emigrants to the Washoe Mines. The Comstock Lode (which in 1872 produced 13,569,724 dollars) may be ranked among the richest mineral deposits ever encountered in the history of mining enterprise, and recently a new lode has been discovered in Nevada by Mr. Deidesheimer, a Freiberg miner; the value of which is estimated at no less than 1,000 million dollars!

Thus we find veins and deposits of silver ore scattered throughout almost the entire length of America; and no doubt many an unknown Potosi still lies concealed in the lonely ravines or on the bleak sides of the Andes and Rocky Mountains, awaiting but some fortunate discoverer to astonish the world with its treasures.

CHAPTER XXVI.

COPPER.

Its valuable Qualities—English Copper Mines—Their comparatively recent Importance—Dreary Aspect of the Cornwall Copper Country—Botallack—Submarine Copper Mines—A Blind Miner—Swansea—Smelting Process—The Mines of Fahlun—Their Ancient Records—Alten Fjord—Drontheim—The Mines of Rivaas—The Mines of Mansfeldt—Lake Superior—Mysterious Discoveries—Burra Burra—Remarkable Instances of Good Fortune in Copper Mining.

COPPER derives its name from the island of Cyprus, where it was extensively mined and smelted by the Greeks; but its first discovery is of much more ancient date, and loses itself in the darkness of the prehistoric ages. Weapons and tools of bronze—its alloy with tin—have been found both in the tumuli of extinct nations and in the lacrustine dwellings of the Swiss lakes, erected by an unknown people in unknown times. Among the Egyptians, the Greeks, and Etruscans, copper was in immemorial use; and the ancient Celtic nations fought their battles with copper or brazen swords, and felled the trees for the construction of their rude hovels with axes of the same metal.

As in many parts of the world native copper is found scattered over the surface of the earth in large lumps or masses, it naturally attracted the attention of barbarous tribes much sooner than iron, which very rarely occurs in the native state; and some fortunate chance or lucky experiment having shown that, when rendered malleable by heat, it could easily be hammered into any convenient shape, it soon became, and has ever since remained, one of the most valuable metals. Forming important compounds with tin (bronze) and zinc (brass), remarkably incorrodible as compared with iron, and nearly as tenacious in structure, but not so hard, it is recommended by its qualities for a variety of uses, and its consumption everywhere increases with the progress of civilisation and the extension of commerce. Fortunately copper is of such common occurrence that a mere enumeration of the localities where it is found would swell into a long and tedious list; it is enough to state that rich copper mines exist both in the Old World and the New, and promise an inexhaustible supply to the most distant generations.

In Europe England is the chief copper-producing country. Rich mines have been discovered and worked in Anglesey, Shropshire, Cheshire, and Staffordshire; in the counties of Wicklow, Cork, and Waterford; but by far the largest quantity is supplied by Cornwall and Devon.

'The history of Cornish copper,' says Mr. Warner, 'is as a mushroom of last night compared with that of tin. Lying deep below the surface of the earth, it would be concealed from the inquiries of human industry till such time as natural philosophy had made considerable progress, and the mechanical arts had reached their present state of perfection; for notwithstanding tin in Cornwall seldom runs deeper than fifty fathoms below the surface, good copper is seldom found at a less depth than that. Accordingly we do not find that any regular researches were made for copper ores in Cornwall till the latter end of the fifteenth century, when a few adventurers worked in an imperfect manner some insignificant mines. Half a century afterwards, in the reign of Elizabeth, though the product of the mines would naturally be greater than before, yet little advantage seems to have been derived to the country at large from the working of its copper. Writers hint at the mystery made of its uses by the merchants. In the next reign, however, all mystery was dispersed, the mines were inspected, their value determined, and a system was introduced of working them to greater advantage.'

Yet so wretched was the knowledge of mineralogy before 1712 that the yellow copper ore, at present so highly valued, was considered of no importance, and cast aside as worthless rubbish. Since the reign of George I. there has been so much improvement that, next to iron and coal, copper is now the most important of our mineral products.

The chief Cornish copper mines are situated in the districts

of Camborne, Redruth, and Gwennap, which are about the
dreariest of all British wildernesses. Few trees are to be
seen, few fields; furze and wild berries form the chief
vegetations of the niggard soil. Blocks upon blocks of stone
are scattered over these desolate moorlands, that have been
excavated, dug into hillocks, disturbed and turned over and
over again, sometimes by the primeval stream-works of the
old men or ancient miners, sometimes by more modern labour,
in search of metallic wealth. Off the roads these districts
are utterly impervious on wheels or on horseback, and the

THE BOTALLACK MINE, CORNWALL.

traveller can only walk, or rather flounder, over them by
jumping from patch to patch of firmer land. Yet this
scene of apparent poverty is in reality one of the very
richest portions of the kingdom, and conceals more wealth
beneath its sterile surface than has ever been produced by a
similar extent of the fairest fields and pastures.

The bluff promontory of Botallack, not far from Cape
Cornwall, conceals in its rocky entrails a copper mine, the
most singularly placed, probably, of any mine in the world,
for nowhere do the triumphs of industry appear in more
picturesque connection with the magnificent scenery of the

ocean. The metalliferous veins running along the cliffs into the sea vainly concealed themselves beneath the swelling surge; in vain a huge barrier of rocks seemed to render them inaccessible to the miner, as soon as it became known that here, buried under the ocean, lay treasures that would amply repay the cost and labour expended, and the danger encountered in seeking for them.

To those who stand below the cliff and look up from the sea, the view is fearfully grand, and remarkable for the combination of the wonders of art with the wonders of nature. The separate parts of an enormous steam-engine had to be lowered 200 feet down the almost perpendicular precipice, and a tram-road runs right up the face of the cliff. Lofty chimneys, pouring out dense volumes of black smoke, are seen perched on the verge and even on the ledges of a tremendous precipice, and the miner has built his hut over the sea-bird's roost. All these constructions seem at the mercy of every storm, and to the beholder from beneath they almost appear suspended in the air and tottering to their fall.

On one side of the cliff tall ladders scale the rock; but he must have strong nerves who can tread them fearlessly, the sea roaring under him and flinging its raging spray after him as he ascends, while in other parts mules and their riders may be seen trotting up and down the rocky tracks which the pedestrian visitor would scarcely dare to pass. A strange and restless life pervades a scene which nature seemed to have for ever removed far from the busy haunts of man.

A visit to this remarkable mine leaves an ineradicable impression on the mind. Descending ladder after ladder, and passing on from gallery to gallery, stepping over rough stones and awkward holes, now stooping down under masses of overhanging rock, and now climbing over stony projections beneath your feet, you are at length informed that you are vertically 120 feet below the sea-level, and horizontally 480 feet under the bottom of the ocean or beyond low-water mark, while still deeper down human beings are hewing the hard rock. The brine oozes through the metallic ceiling, and the sound of distant waters falls faintly upon the ear.

There are other submarine mines in the neighbourhood of

Botallack. In Little Bounds and Weal Cock the hardihood of the miners tempted them to follow the ore upwards, even to the sea ; but the openings made were very small, and, the rock being extremely hard, a covering of wood and cement in the former, and a small plug in the latter mine, sufficed to exclude the water, and protected the workmen from the fatal consequences of their rashness. 'In all these, and in Wheal Edward and Levant,' says Mr. Henwood,* 'I have heard the dashing of the billows and the grating of the shingle overhead even in calm weather. I was once, however, underground in Wheal Cock during a storm. At the extremity of the *level* seaward, some eighty or one hundred fathoms from the shore, little could be heard of its effects, except at intervals, when the reflux of some unusually large wave projected a pebble outward, bounding and rolling over the rocky bottom. But when standing beneath the base of the cliff, and in that part of the mine where but nine feet of the rock stood between us and the ocean, the heavy roll of the large boulders, the ceaseless grinding of the pebbles, the fierce thundering of the billows, with the crackling and boiling as they rebounded, placed a tempest in its most appalling form too vividly before me ever to be forgotten. More than once, doubting the protection of our working shield, we retreated in affright, and it was only after repeated trials that we had confidence to pursue our investigations.'

It seems that at times of great storms, even the miners, accustomed for years to these submarine caverns, have been terrified by the roaring of the sea. They have heard, as it were, mountain dashing against mountain, or as if all the artillery of England was booming over their heads. Yet their roof of rock, thin as it is in some parts, has hitherto shielded them against the sea, and will no doubt continue to defend them against it. On leaving these wonderful submarine excavations, the scenery of the upper world appears doubly beautiful.

The 'Traveller Underground'† tells us a remarkable story of a blind man who once worked in Botallack, and continued his perilous toils underground for a long period, from the

* 'Transactions of the Royal Geological Society of Cornwall,' vol. v. p. 11.
† 'Cornwall, its Mines and Miners.' London, 1860.

dread of being compelled to accept parish relief. By the fruits of his labour he supported a family of nine children; and such was his marvellous recollection of every turning and winding of the mine, that he became a guide to his fellow-labourers if by any accident their lights were extinguished. On being discharged from this employment (and they must truly have had rocky hearts who did discharge him), this poor blind man soon afterwards met his death in a melancholy manner. Being engaged as attendant on some brick-layers, who were building a house at St. Ives, he had to carry the hods of mortar up to the scaffolding. Stepping too far back from a platform, he fell, and died almost immediately.

The chief copper ores of Cornwall are the bisulphuret (containing nearly equal parts of copper, sulphur, and iron), the sulphuret, or grey ore of the miners (containing more than 79 per cent. of copper), and the black ore, an almost pure oxide; but when extracted from the mine, these ores are generally so mixed up with impurities that their average contents do not amount to more than $2\frac{1}{2}$ or $3\frac{1}{2}$ per cent. They have consequently to undergo various processes of picking, crushing, sorting, and washing, before they are rendered saleable and fit for export to the smelting works of Swansea, the grand emporium for copper. The reason why they are not smelted on the spot is that the fuel needed is more bulky than the ore, and it is cheaper to bring the copper of Cornwall to the coal of South Wales than to take the coal to the copper. In Swansea —but half a century ago a mere hamlet, and now a flourishing town of 36,000 inhabitants—we find all the conditions needed for the development of a vast industry. Coal and water-power in inexhaustible abundance, excellent roads and railroads, the nearness of the sea, canals which allow vessels of considerable burden to load and unload close to the smelting huts, so that their high masts rise alongside of the towering chimneys—these are the natural and artificial advantages to which Swansea owes its rank as the first copper manufacturing town in the world.

For, not satisfied with the abundant ores of Great Britain, its smelting works seek their materials in almost every copper-producing country of the globe. The rich ores of Chili and

Australia, of Cuba and North America, of Norway and Tuscany, of Spain and the Cape, all find their way to Swansea, which re-exports the metal to every part of the world.

Since the last ten years the production of copper in the United Kingdom has considerably decreased. In 1864, 201 mines yielded 214,604 tons of ore, from which 13,302 tons of copper were smelted; while in 1873 the 122 mines then working only produced 80,188 tons of ore, which yielded 5,240 tons of copper. The total quantities of British and foreign copper ores and regulus smelted, and copper produced, in England and Wales in 1873 amounted to 481,413 tons of ore and 31,996 tons of copper. In the same year the total export of unwrought British copper amounted to 12,896 tons, and of foreign unwrought and part wrought copper to 20,213 tons.

The whole smelting process is carried on in reverberatory furnaces.* In order to disengage the sulphur and other volatile impurities, the ore is first roasted in these powerful ovens, each of which holds forty hundredweight, and performs its office in six hours. The roasted ores are then mixed with a certain proportion of fluor spar, and smelted in smaller reverberatory furnaces. A ton is introduced at a time, and in each oven seven tons can be smelted in twenty-four hours. It would lead me too far were I to enter into more minute details. I will therefore briefly state that the copper is still obliged to pass four times through differently constructed furnaces before it is sufficiently pure to be rolled into sheets or to be granulated—a condition in which it is used for the fabrication of brass, as it then presents more surface to the action of the zinc, and combines with it more readily. To produce this granulation, the metal is poured into a large ladle pierced with holes, and placed above a cistern filled with water, which must be hot or cold according to the form of the grains required. When it is hot, round grains are obtained, analogous to lead shot, and the copper in this state is called *bean shot*. When the melted copper falls into cold water per-

---

* A reverberatory furnace is a furnace in which intense heat is produced by a flame which, while passing through a furnace, *reverberates* from the roof over the substance to be fused, the draught being created by means of a lofty chimney.

petually renewed, the granulations are irregular, thin, and ramified, constituting *feathered* shot.

The process of preparing the copper does not present the bustle and activity nor the glare and brilliancy of an ironwork. The smoke and vapour disengaged from the ore are of the most noxious and disagreeable kind, and impart to the whole neigh-. bourhood a singularly gloomy character. The stunted vegetation is so kept down by it that there are no trees ; and, instead of grass, a dry, yellow, sickly growth of chamomile barely covers the ground. When viewed from a neighbouring eminence at night, the livid glare from the chimneys, the rolling white clouds of smoke which fill up the valley beneath, the desolate-looking heaps of slag, and the pungent sulphurous vapours, remind the spectator of

> 'The dismal situation, waste and wild,
> The dungeon horrible on all sides round,'

where Satan lay weltering after his fall from heaven.

After England, Sweden, Germany, and Russia take the lead among the copper-producing countries of Europe.

The mines of Fahlun in Dalecarlia are no less remarkable for their picturesque appearance than the celebrated iron mines of Dannemora in the same province. A vast pit, 1,200 feet long, 600 feet broad, and above 180 feet deep, with precipitous, sometimes vertical, and occasionally even overhanging walls, opens before the spectator, who might fancy himself standing on the brink of an enormous crater. 'The aspect of this deep chasm,' says Professor Haussmann,* ' affords a desolate picture of ruin caused by improvidence and waste, as it owes its origin to the successive fallings in of subterranean excavations carelessly widened and left without sufficient supports. From the vast mounds of rubbish accumulated at the bottom of the pit, remnants of ancient shafts, formed of thick beams of wood, are seen protruding, but these show only a part of the devastation produced by the great falling in which took place in the year 1678. On the northern side of the pit is a broad and convenient wooden staircase, by which not only the miners, but also the horses used for working the subterranean machinery, descend to

* ' Geological Travels through Sweden.'

the bottom. Thence it gradually winds underground to a depth of 177 fathoms.

As is generally the case in Sweden, the ore of Fahlun forms considerable masses, the chief being a vast reniform lump 1,200 feet long and 600 broad at its upper surface, and gradually narrowing as it descends. Near this gigantic *stock* are situated similar deposits, which though of smaller dimensions are still very considerable. From the copper pyrites being deposited chiefly on the circumference or the outer shell of these reniform masses, which are themselves of extremely irregular outline, the mining operations are carried on with great difficulty, and exhibits a perfect labyrinth of crooked and winding galleries, situated at various depths, and supported by pillars or sometimes by walls—a peculiarity which explains the successive fallings-in that have formed the enormous pit of Fahlun. The mine has been worked from time immemorial, and is said to have been known even before the Christian era. The oldest document extant bears the date of the year 1347, and contains the privileges granted to the proprietors of the mine by King Olaus Smek; but still more ancient documents are mentioned, among others a purchase-deed of the year 1200. As the ores are poor, their abundance alone renders the working of the mine profitable, and though it appears to have seen its best days it still furnishes a considerable part of the entire production of Sweden, which in 1871 amounted to 33,426 cwt. During its greatest prosperity Fahlun is said to have yielded 5,000 tons of copper annually.

In 1719 a body, preserved from corruption by the vitriolic water with which it had been saturated, was found in an abandoned part of the Fahlun mines. When it had been brought up to the surface, the whole neighbourhood flocked together to see it; but nobody could recognise a lost friend or kinsman in its young and handsome features. At length an old woman, more than eighty years of age, approached with tottering steps, and casting a glance on the corpse, uttered a piercing shriek and fell senseless on the ground. She had instantly recognised her affianced lover, who had mysteriously disappeared more than sixty years previously, but whose image she still bore in her faithful memory. As he was not employed in the mines, no search had been made for him underground at the time. Most probably he had fallen, by some accident, into one of the numerous crevices by which the surface of the mines is traversed. Thus the tottering woman, weighed down with the double burden of infirmity and age, saw once more the face of her lover as she had looked upon it in the days of her youth.

In the sister kingdom of Norway, which in 1871 exported 676,000 cwt. copper ore, and produced 10,400 cwt. copper, the mines of the Alten Fjord are remarkable for their high northern situation (70° N. Lat.). A piece of copper ore found by a Lap woman in 1825 fell accidentally into the hands of Mr. Crowe, an English merchant in Hammerfest, who immediately took measures for obtaining a privilege from Government for the working of a mine. All preliminaries being arranged, he set off for London, where he founded a company with a capital of 75,000*l.* When Marmier visited the Alten Fjord in 1842, more than 1,100 workmen were employed in these most northerly mining works of the world; but probably the number has since decreased.

Although Drontheim or Tronyem is renowned in Norse history as the seat of many kings, yet the town seems as if built but yesterday. Repeated conflagrations have often reduced its wooden houses to ashes. The choir of the ancient cathedral, the finest edifice ever built in so high a latitude, is the only remaining memorial of old Tronyem; but the modern city is remarkably clean and well built, and gives evidence, by its outward appearance, of the prosperity of its citizens, which is partly owing to the fish-trade and partly to the neighbouring copper mines of Röraas. The tall chimneys of the smelting huts and other manufactories founded on the mineral riches of the country show that the spirit of trade is perfectly awake in the old capital of Saint Olaus, and that the abode of the ancient sea-kings is none the worse for having abandoned piracy for the more homely pursuits of modern commerce. The copper ores, which were first discovered in 1644, occur in the Röraas Mountains in extensive veins. The entrance, which resembles the mouth of a cave and leads into the mine by a gradual descent, is so broad that carts laden with ore and drawn by horses can freely pass in and out. When Professor Haussmann, of Göttingen.

visited the mine, long stalactites of ice hung from the roof of the entrance and covered its rugged walls with crystal drapery. The lights of the numerous workmen who opened the march made the ice glitter with all the colours of the rainbow, and then, as they went onward, illuminated the broad galleries, propped by mighty pillars, and branching into gloomy recesses. At length they halted, and all at once, on a given signal, the brilliant illumination was changed into the deepest darkness. A deathlike stillness now reigned in the vault, when suddenly a flash of lightning blazed through the gloom, a loud clap of thunder instantly followed, and, with crash on crash, the explosions of many charges of blasting-powder shook the walls of the neighbouring galleries. After the last shot was fired, the torches were relit, and joyously exchanging the usual salutation of German miners : ' Glück auf ! ' * the company moved on. ' I cannot find words,' says the Professor, in whose honour the impressive scene had been arranged, ' to express the pleasure I felt at this cordial reception given me in the high north by men unknown to me a few days since. It confirmed the experience I had already so often made before, that probably no profession so soon produces a friendly and intimate connexion between strangers as that of the miner.'

The copper production of the German Empire has risen considerably within the last year, and surpasses at present that of the United Kingdom. In 1870 it amounted to 175,000 cwt., the works in the county of Mansfeldt alone having increased from 40,000 cwt. in 1864 to 110,000 cwt. in 1872. These celebrated mines afford a striking example of the success obtainable in mining operations by perseverance and a wise economy. The whole thickness of the cupriferous bed of bituminous shale is no more than from eight to sixteen inches ; but as the ore, though poor, contains a small quantity of silver, this circumstance, assisted by good management and the application of science, has not only rendered it possible to work the mines for many centuries, but to render them so flourishing that in 1852 they produced 1,350 tons of copper and 31,800 marks of silver, leaving a net

* ' Good luck upwards ! ' or ' A happy return to daylight.'

profit of more than 20,000*l.* The hewers are obliged to perform their labour in galleries not more than twenty-two or
twenty-eight inches high, the narrowest limits within which
a man can possibly move and work. The boys who transport
the ore slide or creep with a truly wonderful rapidity along
the floor, dragging after them, by means of a sling attached·
to their foot, a waggon loaded with as much as five hundredweight of ore. The hewer's wages for seven hours of this
hard work are no more than two shillings; yet the miners
look very healthy and cheerful, a remarkable proof of the
wonderful effects of habit.

The vast empire of Russia produced 92,000 cwt. of copper
in 1871, chiefly from the mines of the Ural, belonging to
Prince Demidoff; but a large portion is furnished by the
Asiatic mines of the Altaï and of Nertschinsk in Transbaikalia. New deposits have lately been discovered in the land
of the Kirghise, near the Irtysch; and, as the ores are exceedingly rich, and coal is found near them, they will, no
doubt, become valuable in time.

During the last twenty years America has far exceeded
Europe in the production of copper. The inexhaustible mines
of Chili extend along the whole coast of the republic, and
are generally situated within a convenient distance from the
sea and near the best ports of the Pacific, such as Caldera,
Coquimbo, and Valparaiso. Originally the ores were all
sent to Europe to be smelted; but since 1865 the discovery
of coal near various parts of the coast has encouraged the
establishment of numerous smelting furnaces, so that Chili
now exports no less than from 40,000 to 45,000 tons of
metallic copper, besides furnishing large quantities of ore to
the smelting works of Swansea.

After Chili, no country has made such rapid strides in
copper-mining as the United States. The primeval forests
of Northern Michigan and Wisconsin would probably still be
the undisputed domain of the Indian hunter if the mineral
treasures of the soil had not been a prize too valuable to
escape the notice of our wealth-seeking age. Soon after
the first settlement of the French in Canada, some bold
adventurers had indeed penetrated as far as the distant
shores of Lake Superior, and given wonderful accounts of

the large masses of copper which they had seen scattered over the country; but the want of all means of communication hindered for a long time the advance of the miner.

The Chippeways, who for centuries had occupied the banks of the lake, where, like all other Indian tribes, they spent their time in hunting and fishing, never thought of availing themselves of the mineral riches of their territory. They indeed picked up now and then some pieces of copper, and sold them as curiosities to the fur-dealers with whom they traded; but they were still far too uncivilized to seek in the neighbouring hills for deposits of the valuable metal. Their traditions give no account of their first settlement in the country; they believed themselves to be aboriginals. Thus, when at length the land came to be geologically surveyed, the discovery of extensive prehistoric mining works created no small astonishment. These relics of an unknown people, whose existence and disappearance is one of the most interesting enigmas of ancient American history, are chiefly situated on the hill-crests of Isle Royale and in the Ontanagon district, where they may be traced for miles. Trees, many hundred years old, now grow in the hollows laboriously excavated by that extinct race in the hard rock with tools of stone or copper. Shafts, twenty or thirty feet deep, sunk in the hardest green-stone, have been discovered after felling the trees and removing the rubbish which, in the course of time, had been accumulated in the cavities. In many the old tools were found which served to excavate them—stone hammers of various sizes, or chisels of artificially hardened copper. On the hill behind the Minnesota Pit a mass of copper several tons in weight was found placed on wooden rollers, which proved that those unknown miners must have possessed a considerable mechanical knowledge, without which it would have been impossible to remove such heavy masses. In some galleries copper blocks were discovered from which pieces had been chiselled off, and the whole of the works gave proofs of a skill and persevering industry quite foreign to the unsettled habits of the wild and indolent race of hunters which, as far as memory reaches, has occupied these distant regions.

Although the expeditions of General Cass in 1819 and of Major Long in 1823 had drawn public attention to the copper of Lake Superior, still twenty more years passed before they became the object of mining speculations, which at once rose to a feverish height. Numerous companies were started in 1843, and mines were opened in many hundred places at once. The natural consequence of this copper mania was disappointment in most cases, and in 1847 the greater number of the mines, which had been opened with the most extravagant expectations, were abandoned. A few companies only withstood the crisis, and ultimately proved so remarkably successful as fully to retrieve the lost credit of the copper country, the annual yield of which at the present time is about 10,000 tons, and consequently surpasses that of Great Britain.

The copper occurs in the native state in veins intersecting the trap and sandstone, but also in scattered superficial masses along the chain of hills which extends from the western to the north-eastern extremity of Lake Superior. In no known locality have such large masses of copper been found. An enormous block was discovered in February 1857 in the Minnesota Mine. It was forty-five feet in length, twenty-two feet at the greatest width, and the thickest part was more than eight feet. It contained over ninety per cent. of copper, and weighed about 420 tons. A still more prodigious mass, sixty-five feet long, thirty-two feet broad, and four feet thick, was found in 1869. This king of copper nuggets weighed no less than 1,000 tons, and was worth 80,000l., or more than the greatest lump of gold that ever came to light in Australia or California.

Rich copper mines have likewise been discovered in the States of Maryland, Pennsylvania, Massachusetts, and New York, but chiefly in California, where since 1861 the small town of Copperopolis * has risen into importance. More than 30,000 tons of Californian copper ores (chiefly sulphurets) are now annually exported to the smelting-houses of the bay of Boston, which are likewise supplied by the ores of Chili and Canada, and form a new Swansea on the opposite

* This hybrid name, a vile compound of English and Greek, is enough to excite the wrath of a philologist.

shore of the Atlantic. The whole annual production of the
United States at present exceeds 20,000 tons of metallic
copper, mostly consumed in the country.

The mines of Cuba, which were very important, have
latterly fallen off; but in 1866 the exportation to Swansea
still amounted to 11,254 tons of ore.

Among the rich copper countries of the world I have
finally to mention South Australia, New South Wales, and
Victoria. The most extraordinary copper mine of modern
times for produce is that of Burra-Burra in South Australia.
It was started in September 1845, with a capital of 12,000l.,
subscribed by a few merchants and traders of Adelaide, and in
the following five years yielded no less than 56,428 tons of
ore, worth 738,108l. The gold discoveries momentarily put
a stop to its prosperity; but of late years the works have
been resumed, and other rich mines have been opened, so that
copper will long remain one of the staple productions of
Australia.

The history of some of our copper mines affords examples
of good fortune no less remarkable than those which
we find mentioned in the annals of the Mexican silver-
mining.

Tresavean Copper Mine, within a walk of Redruth, had
once or twice been abandoned as a failure. At length it was
taken up by parties who persevered in exploring it, and succeeded in discovering its wealth by an outlay of little more
than 1,000l. From 1838 to 1843 the profits averaged 30,693l.
per annum, and in 1833 630l. were divided per share, or in
all 60,480l. upon ninety-nine shares, each share having about
20l. paid up, so that in one year the profits surpassed more
than thirty times the capital invested.

Old Crinnis Copper Mine, near St. Austell, was in 1808
abandoned, after repeated failures, and declared by the best
miners of the day to be not worth 'a pipe of tobacco.' In
1809 Mr. Joshua Rowe, of Torpoint, and some co-adventurers,
notwithstanding the general contempt for the mine, began
working it again. As it still remained poor, the adventurers
dropped off one by one, leaving the entire cost of working
upon Mr. Rowe, who, after laying out a few additional
hundreds, was rewarded by the discovery of a rich mass of ore

at about ten fathoms from the surface. Upon this becoming known, the old adventurers again claimed their shares; but Mr. Rowe resisted their unjust pretensions, and won his lawsuit. In the short space of four years and a half, this mine made a clear profit of 168,000*l.*, besides paying 20,000*l.* for law expenses.

Another instance of remarkable success is afforded by the Devon Great Consols Mines, which were opened in the year 1844. The capital of the company which undertook their working was parted into 1,024 shares, with 1*l.* paid on each share. In the same year by November a rich copper lode was cut, and the profits paid working expenses without call. The lodes soon began to turn out so rich that in the six years between the dates of 1844 and of 1850 the company extracted and sold copper ores to the amount of 600,000*l.* After paying all expenses, the shareholders received about 207,000*l.*, or more than 200*l.* per share on 1*l.* paid. No more was called, and thus an average annual dividend of 35*l.*, equivalent to 3,500 per cent., fell to the lot of each share.*

Such instances, however, of good fortune are very rare; for mining in Cornwall, as elsewhere, is much more frequently attended with disappointment and loss. Sometimes an apparently rich produce is absorbed by still greater expenses, or veins very promising when first opened fall off below, and occasion immense loss to the adventurers. A sudden fall in the price of the metal is alone sufficient to render many of the poorer mines perfectly worthless for a long time. Hence nothing can be more hazardous than to invest capital in a mining concern; and if Shakspeare had foreseen the delusions of modern speculations in concerns of this kind, he could not more truly have characterised them than by saying—

> 'The earth hath bubbles as the water hath,
> And these are of them.'

---

* The Devon Great Consols still yield the largest quantity of ore of all our mines (8,716 tons in 1873), but the richer ores of South Caradon in Cornwall (5,293 tons in 1873) yield a greater quantity of copper. Next to these our most productive copper mines are West Huel Seton, West Huel Tolgus, Marke Valley, Glasgow Caradon in Cornwall, and Brook Wood and Gawton in Devonshire. The greatest smelting works in Swansea are those of Williams, Foster & Co.; Neville, Druce & Co.; Vivian & Sons and Mason & Elkington, where 20,938; 13,408; 11,334, and 11,071 tons of ore were smelted.

Yet the hope of suddenly getting rich, and the very risks and daring attending all mining undertakings, have an almost magical attraction ; and, in spite of numberless instances of loss or ruin, there will probably never bo a want of speculators willing to embark their fortunes on this unstable foundation

CHAPTER XXVII.

TIN.

Tin known from the most remote antiquity—Phœnician traders—The Cassi-
terides—Diodorus Siculus—His Account of the Cornish Tin Trade—The Age
of Bronze—Valuable qualities of Tin—Tin Countries—Cornish Tin Lodes—
Tin Streams—Wheal Vor—A Subterranean Blacksmith—Huel Whorry, a
Tin Mine under the Sea—Carclaze Tin Mine—Dressing of Tin Ores—Smelting
—The Cornish Miner.

TIN is one of the metals most anciently known to man.
Its first discovery is hidden—like that of silver, gold,
copper, and iron—in complete obscurity, for even the names
of the nations which first made use of it are not known.
Axes and lances, sickles and fishhooks of bronze—the well-
known alloy of copper and tin—occur among the ruins of
the ancient lacustrine habitations of Switzerland, and the tin
of these bronze utensils could only be obtained by commerce
from countries far remote, where it must, doubtless, have been
known for many ages, before it found its way into the heart
of Central Europe. Thus a few bronze implements picked up
among other rubbish in the muddy bed of an Helvetian lake
open a long vista into the obscure history of primitive man.

On turning from Europe to the East we find other proofs
that tin has been known from the most remote antiquity.
It is mentioned in the Book of Numbers (xxxi. 22) among
the spoil which the children of Israel gained by their victory
over the Midianites; and Ezekiel, in his prophetic warning
to the Tyrians, enumerates tin as forming part of their riches.

It is frequently noticed by Homer as a substance used for
architectural ornaments or for the embellishment of the
armour of his heroes, and its Greek name 'Kassiteros,' which
evidently represents the Sanscrit 'Kastira,' leaves no doubt
as to the part of the world from which it was first obtained.

The tin which, in times unrecorded by chronology, served
for the consumption of Western Asia, Egypt, and Greece, was
supplied by the mines of India to the Phœnician traders, who
conveyed it, either by land to Babylon, or by water to the
ports of the Red Sea. At a much later period that great
merchant people extended their maritime expeditions to the
West, and sailing along the Atlantic coast of Gaul, ultimately
discovered Cornwall, which afforded them a new and inexhaustible supply of tin. With the jealous spirit of trade, they
long made a profound secret of its position; but about 450
years before Christ, Herodotus speaks of the Cassiterides, or
tin islands, which some have supposed to be Britain.   Four
centuries later Diodorus Siculus, who lived in the times of
Julius Cæsar and Augustus, gives us an interesting account of
the ancient tin trade of Britain.   'The inhabitants of that
extremity of Britain which is called Bolerion' (probably

ST. MICHAEL'S MOUNT, CORNWALL.

Land's End), says the historian whose narrative is the more
deserving of attention as we are told that he visited all the
places he mentions: 'both excel in hospitality, and also, by
reason of their intercourse with foreign merchants, they are

civilized in their mode of life. These prepare the tin, working very skilfully the earth which produces it. The ground is rocky, but it has in it earthy veins, the produce of which is brought down and melted and purified. Then, when they have cast it into the form of cubes, they carry it to a certain island adjoining to Britain and called Iktis (probably St. Michael's Mount). During the recess of the tide the intervening space is left dry, and they carry over abundance of tin to this place on their carts; and it is something peculiar that happens to the islands in these parts lying between Europe and Britain, for at full tide, the intervening passage being overflowed, they appear islands, but when the sea returns a large space is left dry, and they are seen as peninsulas. From hence, then, the traders purchase the tin of the natives and transport it into Gaul, and finally, travelling through Gaul on foot, in about thirty days they bring their burdens on horseback to the river Rhone.' Thus we learn from an authentic source how the tin of Cornwall found its way to Italy in the times of the first Roman Emperors; but long before that period the wild inhabitants of Cornwall must have discovered the use of the metallic treasures of their barren soil and the way to barter them for the commodities of the rude tribes of their own island or of the neighbouring nations of Gaul.

When we consider the various and important uses to which tin may be applied, we cannot wonder at its importance in the commerce of the ancient world. The discovery of bronze marks one of the great epochs in the progress of human civilization, and the nations that could command its use became at once superior, in peace and war, to the tribes who had only flint spear-heads for their defence, or flint hatchets for the construction of their huts. At a later period, when iron gradually supplanted the use of bronze for many purposes, tin still continued to be highly esteemed for its many excellent qualities. Possessing a lustre but little inferior to that of silver, it is not soon tarnished, and not only retains its metallic brilliancy a long time, but when lost easily recovers it. Under the hammer it is extended into leaves called tin-foil, which are about one-thousandth of an inch thick, and might easily be beaten into one-half that thickness if the purposes of trade required it. The application of tin to the coating of other metals has been carried to great perfection, and it forms the chief ingredient in various kinds of pewter and other white metallic alloys, such as Britannia metal, which are manufactured into domestic utensils by .casting, stamping, and other ingenious processes. Tin is the substance which, coated with quicksilver, makes the reflecting surface of glass mirrors. It is also very important in dyeing processes, as its solutions in nitric, muriatic, and other acids gives a degree of permanency and brilliancy to several colours not to be obtained by the use of other mordants. A compound of tin with gold gives the fine crimson and purple colours to stained glass and artificial gems, and enamel is produced by the fusion of oxide of tin with the materials of plate glass.

There are only two ores of tin—the peroxide, or tinstone, and the pyrites, or stannine. The former alone occurs in sufficient abundance for metallurgic purposes, and has been found in few countries in a workable quantity. In Asia its richest deposits occur in Sumatra, the peninsula of Malacca, and some smaller islands, particularly Banca and Billiton. The stanniferous region in this part of Asia extends from 20° N. Lat. to 5° S. Lat., and in many places the ore is found in such quantities in the alluvial grounds as to be separated in the easiest manner, by washing or 'streaming,' from the gravel or sand with which it is mixed. The facility with which it is obtained renders its cost of extraction so small that large quantities find their way to the European markets. In 1873 the Dutch mines of Banca furnished 4,480 tons of tin, and those of Billiton 3,260 tons of the same metal, but England still continues to be the first tin-producing country of the world, as the mines and streams of Cornwall and Devonshire yielded in the same year 14,884 tons of ore,[*] from which 9,972 tons of metal were extracted. Since the last year New South Wales and Queensland have also entered the list of tin-producing countries, and are rapidly increasing in importance. A non-official return, but which is thought to be near the truth, gives as follows

[*] 'Mineral Statistics of the United Kingdom for 1873,' by R. Hunt, F.R.S., Keeper of Mining Records.

the quantities of Australian tin (reduced to pure metal, the ore being estimated at 70 per cent.) imported into London during the last three years:—

1872.	1873.	1874.
151 tons.	2,472 tons.	5,067 tons.

Besides its own enormous produce Great Britain imports vast quantities of foreign tin, amounting in 1873 to 5,612 tons of ore, and to 7,791 tons of metallic tin. The exports of British tin amounted in 1873 to 115,946 cwt., and of foreign and colonial production to 28,869 cwt.

Both Cornwall and Devon possess tin mines, which, however, are most important in the county

> ' Where England, stretched towards the setting sun,
> Narrow and long, o'erlooks the western wave.'—COWPER.

The undulating surface of this arid peninsula, which, being remarkable neither for agricultural nor foreign commerce, has been celebrated since the remotest ages for the mineral riches concealed beneath its barren soil, consists almost exclusively of slaty transition rocks or *killas*, traversed or intersected by a central granitic range and by dykes of porphyry or *elvan* which cut the slate and granite, occasionally traversing both in one continuous body of rock, somewhat in the manner of trap-dykes, and evidently of a later formation. The lodes or mineral veins traverse the granite, the slate, and the elvan indiscriminately, but they occur more especially at the junction of granite and slate. They have commonly one prevailing direction, but they invariably throw off into the containing rock 'shoots, strings, and branches,' often in such abundance that, instead of one main lode, called a champion lode, the whole is an irregular network of veins. It is not at all certain that the same lode has ever been traced for more than a mile in length. Very often the lode first discovered dwindles to a mere line, whilst some of its offshoots swell out, enlarge, and rival, or even surpass, both in size and richness, the veins from which they have separated.

The metalliferous or valuable contents of a lode generally bear but a small proportion to its unprofitable parts. Instead of forming uniform lines of metal or pure ore, running throughout the whole extent of the vein, they generally occur in what the miners term *bunches*, or in patches of various sizes and shapes. These very rarely occupy the whole space between the walls or containing sides of the lode, but they are mixed up with a variety of other substances, the chief of which is quartz.

Sometimes a lode is filled with a compact and perfectly .solid mass; at other times it abounds in cavities which may occur in any one of the ingredients and also of any size, from those of the hollows of a honeycomb to hollows of several fathoms in length and depth.

In many lodes tin is found associated with copper, and frequently above the latter, so that the upper part of many a copper lode has been worked as a tin lode.

The veins of Cornwall have no determinate size, being sometimes very narrow, and at others exceeding several fathoms in width; sometimes they extend to a great length and depth, at others they end after a short course. They vary so in breadth that in the same lode one part may consist of a mere line between the opposing walls, while another swells to a width of from thirty to forty feet. These great changes, however, seldom happen within several fathoms of each other. Lodes which yield both tin and copper in mixture are considerably larger than those which yield each metal singly. It is also a general fact that the lodes diminish in breadth in proportion to their depth. The richest tin ores are more commonly found between forty and sixty fathoms deep; but in some instances, as in Dolcoath mine, the depth of 200 fathoms has been attained without exhausting the supply, and Tresavean mine has been worked to great profit at more than 320 fathoms from the surface. Tin veins are considered to be good working when only three inches wide, provided the ore be good for its width.

Besides being contained in lodes, tin is also found in alluvial beds, probably resulting from the disintegration of the former during a long series of ages. This stream-tin, as it is called, is met with either in a pulverised sandy state in separate stones, called *shodes*, or in a continued course of stones, which are sometimes found together in large numbers, and occur at depths varying from one to fifty feet. This course is called a stream, and when rich in ore was formerly called Beauheyl, which is a Cornish word signifying a 'living stream.' In the same figurative style, when the stone was but lightly impregnated with tin, it was said to be 'just alive,' and dead when it contained no metal.

Tin streams of irregular breadth, though seldom less than a fathom, are often scattered in different quantities over the whole breadths of the moor bottoms or valleys in which they are found. As the confluence of rivers makes a flood, so the meeting of tin-streams makes what is called a rich 'floor of tin.' The ore, being thus disseminated both in the alluvium which covers the gentle slopes of the hills, and in that which fills the valleys winding round their base, is easily obtained by conveying over its bed a stream of water, which, by washing away the lighter matter, leaves the heavy ore to be picked up where the operation has been performed.

There can be no doubt that this was the oldest method of tin-getting, and the abundant traces of ancient stream-works which are to be seen from Dartmoor to the Land's End give proof of the great accumulation which must have been formerly worked out by this method. In the course of ages most of these alluvial deposits have been exhausted, and where, thirty or forty centuries ago, large quantities of tin ore were superficially gathered with little ingenuity and labour, the miner is now obliged to descend many a fathom deep into the bowels of the earth and work his slow way through the hard rock.

Yet, after so many centuries of research and extraction, tin-streaming is still carried on in several places, as, for instance, at Carnon, north of Falmouth, where a long line of stream-works extends down the valley. The ore, mixed with rounded pieces of slate, granite, and quartz, lies buried about fifty feet from the surface, beneath the bottom of an estuary, where trees are discovered in their place of growth, together with human skulls and the remains of deer, amidst the vegetable accumulation which immediately covers the stanniferous beds. Thus ruins are here piled up above ruins.

In 1873 the number of tin mines in activity amounted to 159 in Cornwall and to 11 in Devonshire. The most important works are at present Dolcoath, Tincroft, Carn Brea, Pool East, Vor Great Huel, Phœnix, and Basset Huel; but, as nothing is more fluctuating than the fortunes of mines, others will probably soon take the lead.

Wheal Vor, in the parish of Breage, three miles from Helston, may be cited as a conspicuous example of the changes of fortune so frequent in the annals of Cornish mining. Twenty-five years ago it was considered the richest tin mine in Cornwall. More than 200,000*l.* profit had been divided among the shareholders. In 1843 there were fifteen engines at work on this extensive sett, which had the appearance of a town, and the machinery was valued at 100,000*l.* Here was put up the first steam-engine ever erected in Cornwall, between the years 1710 and 1714. The lode from which the chief part of the ore was raised was still productive in 1843, when the mine employed 1,200 persons, and the monthly cost of working had been, some years before 1843, about 12,000*l.* per month. The mine, however, became less profitable, and, wearing out by degrees, finally stopped.

There was formerly a blacksmith's forge at the bottom of this mine, in full operation at 1,470 feet below the surface of the earth. All the miners' tools were steeled, sharpened, and repaired, and bucket-rods cut and welded in this subterranean smithy, which was clear and free from dust, smoke, and sulphur, and did not in the least annoy the miners. Within the last few years the mine has been resuscitated with a capital of 200,000*l.*; but, as the shares have fallen from 40*l.* to 8*l.*, the attempt seems to have been far from profitable.

The history of the rise and fall of Huel Wherry,* a tin mine which was opened, more than a century ago, in the midst of the sea, near the town of Penzance, is too interesting to be passed over in silence. In this place a gravelly bottom was left bare at low water. Here a multitude of small veins of tin ore crossed each other in every direction through elvan rocks, and were worked whenever the sea, the tide, and the season would permit, until the depth became unmanageable. About the year 1778 Thomas Curtis, a poor miner, was bold enough to renew the attempt. The distance of the shoal from the neighbouring beach at high water is about 120 fathoms, and this distance, in consequence of the shallowness of the beach, is not materially lessened at low water. It is calculated that the surface of the rock is covered about ten months in the year, and that the depth of the water upon it

* 'Cornwall, its Mines and Miners.'

at spring tide is nineteen feet. A very great surf is caused, even in the summer, by the prevailing winds, while in winter the sea bursts over the rock in such a manner as to render useless all attempts to carry on mining operations. Yet all these difficulties had to be overcome by a poor uneducated man. As the work could be prosecuted only during the short time the rock appeared above water—a time still further abridged by the necessity of previously emptying the excavation already made—three summers were spent in sinking the pump-shaft, which was a work of mere bodily labour. A frame of boards, made watertight by pitch and oakum, and carried up to a sufficient height above the spring-tide, was then applied to the mouth of the shaft. To support this boarded turret—which was twenty feet high above the rock, and two feet one inch square—against the violence of the surge, eight stout bars were applied, in an inclined direction, to its sides, four of them below, and four, of an extraordinary length and thickness, above. A platform of boards was then lashed round the top of the turret, supported by four poles, which were firmly connected with these rods. Lastly, upon this platform was fixed a windlass for four men.

By such an erection it was expected that the miners would be enabled to pursue their operations at all times, even during the winter months, whenever the weather was not particularly unfavourable. But as soon as the excavation was carried to some extent, in a lateral direction, the hope was disappointed, for the sea water penetrated through the fissures of the rock, and, in proportion as the workings became enlarged, the labour of raising the ore to the mouth of the shaft increased. To add to all this, it was found impossible to prevent the water from forcing its way through the shaft during the winter months, or, on account of the swell and surf, to remove the tin-stone from the rock to the beach opposite. Hence the whole winter was a period of inaction, and the regular working of the mine could not be resumed before April. Nevertheless, the short interval which was still allowed for labour below ground was sufficient to reward the bold and persevering projector.

The close of this wonderful mine, from which many thousand

pounds worth of tin was raised, was as romantic as its commencement. An American vessel broke from its anchorage in Gwavus Lake, and, striking against the stage, demolished the machinery, and thus put an end to an adventure which, both in ingenuity and success, was in all probability unequalled in any country.

This wonderful mine was worked again a few years since; but, although a very large sum of money was expended, and it had all the advantage of improved machinery, yet it failed to be a profitable adventure, and was eventually abandoned.

The Carclaze tin mine, near the town of St. Austell, though unimportant with regard to its produce, deserves to be noticed for its picturesque appearance and the manner in which it is worked. It consists of a large open excavation, of a mile in circuit and from twenty to thirty fathoms in depth, looking more like a vast natural crater than a hollow made by human hands; and for hours the visitor might traverse the dreary and barren hilly common in which it is situated without suspecting that a mine is close at hand. No engine-house and chimney towering aloft announces it from afar; the whole business is confined to the interior of the punch-bowl hollow. Every detail of the works is here exposed to view, and it would seem as if a complete mine had been turned inside out for the benefit of timid travellers, who would wish to see the various operations of mining without the risk of a descent below the surface.

The walls of this vast hollow or crater are almost perpendicular, and the view from the ridge of the precipice, into which but few footpaths descend, is rendered interesting by the fantastic shape of the rocks, worn or hewn into a thousand grotesque forms by the action of the waters or the pickaxe of the miner; by the enormous number of holes and hollows resulting from ancient excavations; by the white colour of the granite, veined with the darker metalliferous streaks; by the water-wheels at the bottom, which, worked by streams from the neighbouring commons, propel the machinery for crushing the stones, loosened by the water as it flows down the sides of the cavity; and by the men, women, and children, scattered over the works. The ore is obtained without much difficulty, and is easily separated from the friable and decomposed granite, in which it is embedded, by repeated washings in the streams that are made to flow out at the bottom of the mine through a channel or tunnel, and which carry away the soft granite by their rapid current, while the heavier metalliferous substances are precipitated.

As the ores are very poor, not even containing one per cent. of tin, Carclaze, which has been already worked for many centuries, would long since have been abandoned but for the abundance of the ores and the comparatively small expense of their extraction.

The dressing of the tin ores, or the process by which they are separated as far as possible from the earthy impurities which are mixed up with them, and are generally much lighter, begins with cleaning and sorting, and then goes on to washing and stamping, and finally to calcination in the burning-house and to smelting.

The tin ores of Cornwall and Devonshire are all reduced within the counties where they are mined, as the law prohibits their exportation—a most absurd and antiquated regulation, which, however, in this case is not injurious to private interests, as the vessels which bring the fuel from Wales for the smelting furnaces return to Swansea and Neath laden with copper ores. The smelting works, not exceeding seven or eight in number, belong generally not to the proprietors of the mines, but to other parties, who purchase the ore from the proprietors.

The smelting is effected by two different methods, which may be briefly described by stating that, by the first and most common, the ore, mixed with culm, is exposed to heat upon the hearth of a reverberating furnace, in which pit coal is used as fuel; while by the second method, which is applied merely to stream-tin, and which is followed in order to obtain tin of the finest quality, the ore is fused in a blast-furnace called a blowing-house, in which wood fuel or charcoal is used. The melted tin runs off from the furnace into an open basin, whence it flows into a large vessel, where it is allowed to settle. The scoriæ are skimmed off, and the subsequent operations consist of refining by allowing the mass of the metal to rest, then submitting the upper and pure portion to the refining basin, and remelting the lower part. In order

to convert the blocks into *grain-tin*, they are heated until they become brittle, and made to fall from a considerable height in a semi-fluid state, thus producing an agglomerated mass of elongated grains.

The number of persons that find occupation in and about the Cornish and Devonian tin mines may amount to about 20,000. The wages are, on an average, much inferior to those of the pitmen and pitlads in the northern coal-fields; but, on the other hand, the Cornish miner is exempt from many evils to which the northern miners are subject. He has not to fear the fatal fire-damp, and can sit at ease and hear or read of explosions that have destroyed hundreds in a few minutes.

His intellectual superiority to the agricultural labourer may be at once inferred from the nature of his pursuits. The latter plods on through life like a mere human machine, and, as he is never thrown on his own resources in the progress of his monotonous occupations, his stock of ideas remains scanty and confined. But the Cornish miner is the reverse of all this. He is engaged mostly in work requiring the exercise of the mind. He is constantly taking a new 'pitch'* in a new situation, where his judgment is called into action. His wages are not the stinted recompense of half-emancipated serfdom; but they arise from contract, and depend upon some degree of skill and knowledge. In fact the chances of the lode keep expectation constantly awake, and thus :—

> 'Hope reigns triumphant in the *miner's* breast,
> Who never is, but always to be blest.'

If he is at all imaginative, golden dreams enliven the darkness of his underground labour. He is in fact a kind of subterranean stock-jobber, and 'settling day' is as anxious a

---

* The lodes in the Cornish tin and copper mines are divided by shafts and galleries into rectangular compartments, called 'pitches.' These are open to the inspection of all the labouring miners in the county, and, by an admirable system, each 'pitch' is let by public competition, for two months, to two or four or more miners, who may work it as they choose. These men agree to break the ores, wheel them, raise them to the surface, and bring them (if desired) into a fit condition for the market. The ores so raised are sold every week, and the miner immediately receives his *tribute*, or percentage for which he agreed to work. The sinking of shafts and the driving of levels is paid by tut-work, or task-work, at much per fathom.

time for the humble tributer at the Land's End as for the bold speculator of the Stock Exchange.

When not exhausted after his hard day's labour, the miner frequently cultivates a small patch of land. Many have tolerable gardens, and some are able to perform their own carpentry, while, if near the coast, others are expert fishermen.

CHAPTER XXVIII.

IRON.

Iron the most valuable of Metals—Its wide diffusion over the Earth—Meteoric
Iron—Iron very anciently known—Extension of its Uses in Modern Times—
British Iron Production—Causes of its Rise—HotBlast—Puddling—Coal-smelt-
ing—The Cleveland District—Rapid Rise of Middlesborough—British Iron Ores
—Production of Foreign Countries—The Magnetic Mountain in Russia—The
Eisenerz Mountain in Styria—Dannemora—Elba—The United States—The Pilot
Knob—The Cerro del Mercado.

AS an instrument of civilisation iron is the most valuable
and indispensable of all mineral substances. Even
coal is of inferior importance to the welfare of mankind,
for iron may be obtained without its aid, while coal could
not possibly be extracted from the bowels of the earth
without the assistance of iron. Hard and malleable, tenacious and ductile, endowed with the singular property of
welding, which is found in no other metal except platinum,
and acquiring new qualities by its conversion into steel, it
accommodates itself to all our wants and even to our
caprices, so that no other metal has such various and extensive uses. It clothes our war ships with a case of
impenetrable armour, and sets the finest watch in motion; it
provides the sempstress with her needle, and guides the
mariner over the ocean; it furnishes the husbandman with
his ploughshare, and the soldier with his sword; it concen-
trates in the steam-engine the sinews of a thousand horses,
and mocks on the railroad the fleetness of the swiftest
courser. It is, in one word, the embodiment of power, the
chief agent of all social progress.

' Were the use of iron lost among us,' says the illustrious
Locke, ' we should, in a few ages, be unavoidably reduced to
the wants and ignorance of the ancient savage Americans;'

nor will this view be deemed extravagant if we reflect that, but for iron, man would be *virtually without* tools, since it is almost the only metal capable of taking a sharp edge and keeping it.

The bounty of the Creator, which bestowed on man this inestimable gift, has also provided for its wider diffusion over the earth than is the case with any other of the useful metals. Few mineral substances or stones are without an admixture of it. Sands, clays, the waters of rivers or springs, are scarcely ever perfectly free from iron, while animal and vegetable substances likewise afford it in the residues which they leave after incineration. Its mines may truly be said to be inexhaustible; in some its ores occur in compact masses of extraordinary magnitude, in others they spread in vast strata or extend in veins of a prodigious length.

Yet, in spite of its wide diffusion, the extraction of iron from its ores generally requires so much metallurgic skill that its use would probably have remained undiscovered by the ancients if Providence had not in a wonderful manner revealed, as it were, its existence to mankind.

All iron of a *terrestrial* origin is combined with other substances, which conceal its true nature from the uninitiated eye, and from which it is with difficulty separated; but here and there, scattered over the surface of the earth, are found solitary masses of metallic iron, which, having fallen from the skies, may truly be called erratic boulders from another world. The idea of their having dropped from the clouds was long ridiculed by the learned; but their fall has been so often observed, and so circumstantially recorded in the annals of almost every age, that scepticism has been obliged to yield to the weight of accumulated evidence, and science no longer doubts their meteoric origin. Nothing can be more interesting than these mysterious heralds from the distant fields of ether, which, after wandering through space for countless ages, have at length been brought within the sphere of attraction of our planet, and, alighting on its surface, afford us tangible proofs that many of the substances of which our earth is composed—iron, nickel, silex, &c., &c.—exist beyond its limits, and that most probably our whole solar system is constructed of the same materials as our globe.

But meteoric iron—which sometimes occurs in enormous masses *—is more than a mere object of curiosity, for it has had a most important influence on the progress of the human race. On such a mass undoubtedly the *first smith* exercised his skill, and it was this which first made mankind acquainted with a metal more valuable than copper or gold.

As we see from the example of the Esquimaux, whom Captain Ross (1819) found in possession of knives and harpoons which they had made from masses of meteoric iron, the discovery was probably made at a very remote period, while man was still in the savage state; but iron having once become known, the desire to obtain it in larger quantities from other sources naturally grew with the progress of civilisation, and gradually led to the knowledge of its ores and of the art of utilising them. Thus there can hardly be a doubt that iron-smelting was practised long before historic times. In India and China the origin of its use loses itself in the remotest antiquity; and the imposing monuments of ancient Egypt, many of which are at least five thousand years old, could not possibly have been erected without the aid of iron. In the Book of Deuteronomy (iv. 20) the land of Egypt is compared to an iron-furnace—a figurative expression which shows that even at that early period iron-smelting must have been a well-known branch of industry.

The iron weapons found in the lacustrine dwellings of Switzerland likewise point to a very ancient use of iron in Central Europe, no less than the fact mentioned by Cæsar, that during the siege of Avaricum (Bourges) the works erected by the Romans for taking the town were repeatedly destroyed by the subterranean galleries of the besieged, who, as the conqueror relates, were accustomed to such underground labour from their habitually working in iron-mines, an industry which, to judge from this passage, must even then have been of ancient date in Gaul.

* The weight of the mass found at Otumpa, in the Gran Chaco Gualamba, in South America, by Don Rubin de Celis (1783) was estimated at about fifteen tons. A piece from this mass, weighing 1,400 pounds, is now in the British Museum.

For many ages the uses of iron remained chiefly confined to the instruments of agriculture and of war—to the plough-share and the sword. With the progress of civilisation its employment extended to many purposes unknown before; and in our times the construction of ships and buildings, of railroads and bridges, absorbs quantities which would have appeared incredible almost within the memory of living man. Hence the manufacture of iron has made more rapid progress since the beginning of the present century than in any former period of the world's history, and even the present immense production scarcely keeps pace with a demand to which it is not easy to assign a limit.

Among the iron-*producing* countries of the globe Great Britain occupies by far the first rank, and there is every reason to believe that it will long continue to maintain it. The British ores, indeed, are generally poor, as clay, silica, phosphorus, and a variety of impurities which are with difficulty separated from the metal, enter into the composition of those which supply the greatest part of our iron; but this deficiency is more than counterbalanced by many advantages. Most of the British iron mines are situated in those districts where coal is cheapest, the ore being often even raised from the same pit as that from which the coal is extracted. Limestone (the necessary flux) is at hand, while fire-clay—no unimportant article in the building of the furnaces, on whose long-continued working so much depends—is found in the same ground as the ore itself. The largest and most complete manufactories have long been established in the most convenient places. With an almost unlimited amount of capital, the most perfect and the cheapest communication by water is open to all parts of the world; and the further processes which the metal has to undergo are performed at once on the spot in the best manner and at the smallest possible expense. No other land can boast of equal or greater facilities for the production of an unlimited quantity of cheap iron; so that, even with the assistance of heavy protective duties, most of the iron-producing countries of Europe find it difficult to compete in their own markets with the produce of Great Britain.

The art of making iron in this country is of very ancient,

though of unascertained date.   It was probably found by the
Romans in a far advanced state.   It certainly was carried on
by them subsequently to a great extent—a fact proved by the
immense beds of iron cinders discovered in the Forest of Dean ;
nor has it ever been discontinued by the other races who
in succession have held sway in the island.   But, though of
such ancient origin, and enjoying so many natural advantages, our iron manufacture remained confined within very
narrow limits so long as the ore was exclusively smelted by
means of charcoal made from wood.   The manufacture was
even for some time partially prohibited in England, the consumption of wood-charcoal in the process of smelting being
so great as to create apprehensions that, if care were not
taken of the remaining forests, enough timber would not be
left to supply the wants of the navy.   It seems almost incredible in our days that Acts were passed in the reigns of
Elizabeth and James forbidding the felling of timber for the
smelting of iron, except in certain districts of Kent, Sussex,
and Surrey, then the principal seats of the manufacture,
and even there the erection of new works was expressly forbidden.

Attention was then directed to the smelting of ironstone
by means of pit coal.   Amongst others, Lord Dudley gal-
lantly struggled to establish a manufactory in the neighbourhood of Stourbridge, and partially succeeded ; but what with
riots among the ironworkers, who destroyed his works, and
the wars of the Great Rebellion, which ruined his fortune, he
reaped no advantage from his enterprise.   Nothing contri-
buted to arrest the decline in this branch of trade, and towards
the middle of last century the number of furnaces, which in
the reign of James I. had amounted to 300, fell off to 59, the
total make of which amounted to not more than 17,350 tons,
being an average of 294 tons per annum for each furnace—a
quantity very little exceeding that sometimes made in a
single week in some of the huge furnaces in Wales in the
present day.   The partial use of pit coal in the process of
smelting was revived in Coalbrookdale, in Shropshire, about
1713.   The chief difficulty was to keep the coal in a state of
combustion sufficiently intense for the purpose of smelting
the ore ; the hand-worked bellows, or the more powerful

water movement, which produced blast enough for charcoal, having comparatively little effect upon coal.

This obstacle was finally overcome through the perseverance and enterprise of Dr. John Roebuck (a physician in Birmingham, and grandfather of the late distinguished member for Sheffield), who, seeking for more economical methods of smelting iron ore than those then in use, founded in 1759 the now celebrated Carron Works, where John Smeaton, the illustrious architect of the Eddystone lighthouse, first introduced (1760) a new contrivance for throwing a powerful and constant blast into the furnace. By means of a forcing pump, a large column of air, of triple or quadruple density to that which had been previously obtained, could now be poured into the furnace; and effects equivalent to this great improvement followed. The same smelting oven that formerly yielded ten or twelve tons weekly now sometimes produced forty tons in the same period; and such was the impulse given to the trade by this unexpected success of a powerful blast with pit coal that in 1788 the manufacture of pig-iron in England, Wales, and Scotland amounted to 68,300 tons, being an increase of 50,950 tons on the quantity manufactured previous to the introduction of pit coal.

In 1782 Mr. Cort, after many years of experiment, discovered the means of converting cast or pig-iron into malleable iron by a process which was at once sure, rapid, and economical. The iron is re-melted in a puddling furnace, as it is called, which is heated with raw coal, and there, by a series of operations, the object of which is to give the iron malleability and toughness by expelling the carbon, it is manipulated until it acquires the consistency of a solid white-hot ball. In this shape it is subjected to the action of an enormous hammer, by which the coarser parts are beaten from it, and it is formed into the shape of thick short bars, called blooms or slabs. While still red-hot it is passed through a series of grooved rollers, till it is drawn out into a long bar, the exact dimensions of which are regulated by the requirements of the manufacture for which it is destined. The bars thus made are technically called puddled bars, and considered as *half-manufactured* iron. To refine them into *merchant*-iron, they are again submitted to the action of fire, and, when hot enough, are welded together and formed into the various denominations of bars, rods, hoops, sheets, or plates. Mr. Cort's discovery, though of immense importance, would yet have proved of comparatively small value without the aid of the double-power steam-engine, which was about the same time invented by James Watt, and supplied the power which was needed to give our iron-works their full development.

Hitherto the 'top measures' only of the mineral had been worked, and generally on 'the rise of the mine,' where the water would not lie, or those strata favourably situated on the side of a hill where levels could be driven in and the water released. Water was the great enemy in the pits, and even in shallow workings it often accumulated faster than a gin turned by horse-power could bring it to the surface. By the new agency of steam the deepest pits were drained, and materials were drawn up from the bowels of the earth in a quantity and with a rapidity and security hitherto unknown. By the same means that prodigious blast was obtained for the furnaces to which all subsequent improvements of the manufacture owe their origin. Instead of the rude machinery of waterwheels and bellows, huge engines of enormous power forced an immense volume of air through several small *tuyeres* or tubes so disposed at the lower part of the furnaces that in each portion of the ignited mass an equally diffused blast might raise an equal intensity of heat. Furnaces of greater height and much larger capacity than any hitherto known were erected, and in its general aspect the iron manufacture assumed very much the appearance which it maintains at the present day.

Most readers are aware* that the flaming towers which give such an unearthly effect at night to what is called the Black Country, round Wolverhampton, are iron furnaces, and that the projecting circular galleries which surround their tops are contrived for pouring down their capacious throats, by apertures placed at equal distances, an equable and regular supply of the materials with which they are fed. Besides the iron-stone and the fuel, there is needed a third substance, which is called 'a flux,' because it forms a fusible compound

* 'Quarterly Review,' vol. cix. p. 114.

with the earthy matter of the mineral. When we are acquainted with the foreign matter in combination with the
ore, chemistry tells us what substance we ought to add for

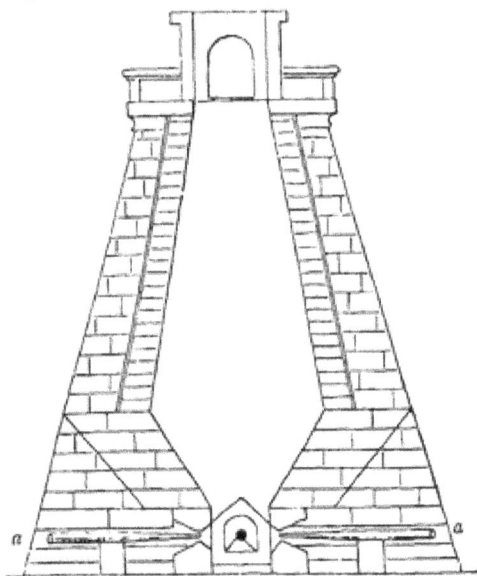

BLAST FURNACE.
a. Tuyeres.

the purpose of eliminating the metal. Among the wonderful
provisions of nature for the convenience of man, none is
more remarkable than that by which many substances are
fusible in conjunction at a temperature which either could
resist separately. The British ores are for the most part
argillaceous, that is to say, they are combined with what, in
its general character and appearance, resembles clay. To
all such limestone in due proportion must be added; but if
the earthy matter consists of lime, clay is the proper flux.
In either case the foreign matter and the flux are fused into one
substance. The liberated iron sinks downwards, and having
now itself become fusible by the combination of carbon, with
which it has been impregnated by the fuel, it melts as it
reaches the point of fusion, and settles down in the lowest part
of the furnace, otherwise called the hearth. It is followed by
the slag, or 'cinder' (as it is always called in the trade),
composed of the flux, the foreign matter of the ores, and
the ashes of the fuel, which are now in a vitrified state;

and this artificial lava, being of much less specific gravity, rests on the surface of the iron and protects it from the action of the blast. The furnace is 'continued in blast,' that is to say, in full operation, and must be fed equably and constantly night and day till the manufacturer thinks fit to blow it out, either for the purpose of repairing it, or of reducing his make of iron. At certain intervals, generally twice in the twenty-four hours, the furnace is tapped; that is to say, the stoppage of sand which closes an orifice at the bottom is knocked away, the liquefied metal rushes out, and is guided successively into moulds of sand in the form of short thick bars, which, by a rude metaphor, as old as the invention of casting, are called 'pigs,' while the main channel down which the red-hot torrent flows is called the 'sow.'

The invention of the hot blast gave a new and mighty impulse to the production of iron. Though Mr. Scrivenor [*] mentions the remarkable fact that in the furnaces of Peru a contrivance has been noticed for letting the air pass over hot coals, and thus become heated in its passage to the fire, yet it was personal observation, and not archæological research, that, in 1829, suggested to Mr. Neilson, of the Clyde Iron Works, the possibility of economising fuel by substituting hot for cold air in blowing his furnaces. Before this important discovery more than eight tons of coke had been required to produce one ton of pig-iron; but on heating the blast, previously to its entering the smelting-oven, to a temperature of 300° F., it was found that a saving of two and a half tons of coal could be made on every ton of iron, and on raising it to the temperature of 600° F.—a heat somewhat more than sufficient to melt lead—a still more considerable saving of fuel was effected, while at the same time the important discovery was made that at this high temperature bituminous and even anthracitic coal might be used instead of coke. Another advantage was, that the same steam power now sufficed for applying the blast to four furnaces which had formerly been required for three; and the total result of the improvement was a saving of 72 per cent. of fuel. Thus we have here another instance of the important results that may be gained

* 'History of the Iron Trade.'

A A

from a single good idea when worked out by clever practical men, for the hot blast has most assuredly increased the wealth of England by many millions a year!

Another circumstance likewise tended considerably to increase the production of pig-iron. It was found that the hot blast not only had power sufficient to produce in the raw coal the requisite intensity of heat, but also to expel from it, to a certain extent, the sulphur, which injured the quality of the iron, and thus a great economy in labour as well as in the quantity of fuel was effected. Since then the black-band, an iron-stone found in great quantities in Scotland (and also, to a less extent, in Wales), but not readily convertible into iron by the old methods, and also the Northamptonshire and the Cleveland ores—discoveries of a later date and of an incalculable extent—have been made by the hot blast to yield their iron in great abundance.

The power of using the black-band alone in the furnace—and not, as before the introduction of the hot blast, in small quantities only, and combined with other ores—constituted a new era in the manufacture of iron, and gave to Scotland, till then an iron-making district of little importance,[*] the pre-eminence over all others for the production of soft fluid iron, best suited to ordinary founding purposes.

The Cleveland district, formerly unknown in metallurgy, is now the seat of a vast industry, keeping more than a hundred furnaces in blast.

The head-quarters of this new iron-country are established at Middlesborough, on whose site there existed but one house in 1829, but which in 1861 had grown into a town of 24,000 inhabitants, and still increases at the rate of 1,000 a year. Branch railways bring the stone here for smelting from all the neighbouring quarries; and the dense cloud of smoke that hangs over the place serves as a land-mark, not only from the high ground of Yorkshire, but even from some parts of Westmoreland.

---

[*] The metal was formerly so scarce in their country that in the times of the Edwards the Scotch were accustomed to make predatory incursions into England for the sake of the iron they could carry off. Now they not only manufacture sufficient for their own use, but actually export above half a million tons.

But Middlesborough, 'the youngest child of England's enterprise,' as it has been called by Mr. Gladstone, is by no means one of the loveliest of her offspring. Scarcely a blade of grass, and not a single tree, relieves the dull monotony of its dreary streets of small houses, darkened by perpetual smoke, which, as the wind sways it, affords, at rare intervals, glimpses of distant hills or of the Tees, serving only to make this prison of a town more gloomy. Mines and furnaces have also been established in other parts of the district—in Rhosdale, at Grosmont near Whitby, and elsewhere—and not a year passes without the opening of new veins and the rising of new smoke-clouds amid the lovely dales of north-western Yorkshire. The iron which eventually finds its way to Middlesborough is sent thence to every part of the world. Its quality is essentially inferior to that derived from the coal-measures; but for ordinary purposes, and for mixing with the finer-classes, it is of great value. Looking to the future, we cannot doubt that the Middlesborough district is destined to have no rival in any part of the world, for even now its works compete in magnitude with those of our old iron-fields.

A material which had hitherto been thrown away was also, by the agency of the hot blast, made available for the purposes of the iron master. The 'tap cinder,' or refuse of the puddling furnace (which is not to be confounded with the cinder of the blast furnace), contains a considerable percentage of metal, and when thrown again into the furnace greatly increases the yield, though it proportionally deteriorates the quality, of the iron. The results of all these successive discoveries and innovations, aided by the facilities of transport afforded by canals and railroads, are truly astonishing.

The make of iron which, on the introduction of steam, had suddenly risen to nearly 50,000 tons per annum, reached 125,000 in 1796, and in 1806 had advanced to nearly 260,000. In 1825 the make was nearly 600,000 tons; in 1840 it amounted to 1,300,000 tons; and in 1854 to 2,700,000 tons. In 1865 it reached the enormous figure of 4,819,254 tons, and in 1873 no less than 5,566,451 tons of pig-iron were produced—a colossal expansion without a parallel in the annals of metallurgic industry. The number of furnaces in blast in 1873 was in England 448¼, distributed over 164 iron works; in Wales 109, distributed over 34 works; and in Scotland 126, over 27 works. To supply these furnaces there were raised 15,577,499 tons of ore, the estimated value of which at the place of production was 7,573,676l., and imported 1,242,536 tons of foreign ore. The value of the pig-iron, at the mean average cost at the place of production, was about 40,000,000l.

On reviewing the chief districts which furnished this incredible mass of iron, we find Yorkshire taking the lead with 5,750,029 tons of argillaceous carbonate, and Northamptonshire following next with 1,412,255 tons of hydrated oxide. Cumberland and Lancashire respectively produced 1,229,826 and 926,497 tons of red hematite, which, consisting almost entirely of red peroxide of iron, is reckoned among our best ores. Staffordshire, North and South, yielded 1,013,530 tons of argillaceous carbonate, and Shropshire and Lincolnshire 430,725 and 420,281 tons of argillaceous carbonate and hydrated oxide; Derbyshire (365,127 tons), Gloucestershire (199,342 tons), Wiltshire (140,139 tons), Northumberland and Durham (123,282 tons), likewise yielded large quantities of iron ore which a hundred years ago would have seemed prodigious. The counties of Cornwall and Devon, so rich in tin and copper, produced only 31,455 and 9,514 tons of iron ore. The production of Scotland amounted to 1,986,000 tons of argillaceous carbonate and black band, and that of South Wales and Monmouthshire to 943,926 tons of argillaceous carbonate and brown hematite.*

The finest iron ores—such as the black oxide or magnetite, specular iron, and spathose iron, or sphærosiderite, which furnish the best kinds of iron—are unfortunately but of rare occurrence in Great Britain.

As we see by the following tablet† of the production of cast iron in the chief European mining States—

* 'Mineral Statistics of the United Kingdom for the Year 1873,' by R. Hunt, Keeper of Mining Records.

† From the official reports of the International Jury of the Universal Exhibition of 1867 in Paris.

Countries	Years	Tons	Years	Tons
Great Britain . .	1840	1,400,000	1865	4,527,000
France . . .	1845	271,000	1864	1,213,000
Prussia . . .	1836	32,800	1865	772,000
Belgium . . .	1845	134,500	1864	450,000
Russia . . .	1838	171,000	1865	278,000
Austria . . .	1835	200,000	1865	250,000
Sweden . . .	1835	97,500	1865	227,000
Spain	1850	26,000	1865	48,000
Italy	1838	20,000	1865	27,000

France occupied the next rank to Great Britain in 1864; but in 1869 the German Zollverein produced 1,400,000 tons, and France only 1,350,000 tons, from which the production of Elsass-Lorraine, amounting to 250,000 tons, must now be deducted, and added to that of the German Empire. Nassau possesses inexhaustible supplies of specular iron ore of a remarkable purity, which not only feed the blast furnaces of Westphalia, but are also largely exported to England.

In proportion to the smallness of her territory Belgium (440,000 tons) rivals Great Britain in the production of iron, and surpasses the vast empire of the Czar (400,000 tons).

But Russia has the advantage over all the countries previously mentioned of possessing inexhaustible deposits of magnetic iron ore (magnetic loadstone—magnetite) which affords bar-iron of the very best quality; and though hitherto the immense distances which separate the mines from the larger centres of consumption have retarded the progress of the iron manufacture, the construction of railroads is gradually overcoming these obstacles; and possibly even the now unworked Siberian mines of the Altaï and of Transbaikalin, where coal is found along with iron, may acquire importance at a not far distant time. At present the chain of the Oural (Permia and Orenburg) furnishes nine-tenths of all the iron produced in the empire. The most remarkable of the Ouralian mines is the famous magnetic mountain Wissokaja Gora, in the neighbourhood of Nishne-Tagilsk, which Peter the Great bestowed in 1702 on the armourer Nikita Demidoff of Tula, along with a vast extent of forests and arable land. This magnetic mountain or hill, which is 300 fathoms long, 250 broad, and 240 feet high, rises from the midst of a plain, in the form of a broad, flat eminence. It consists

almost entirely of *pure magnetic iron ore*, and is worked
like an open quarry; but, on account of its hardness,
the ore requires to be blasted with powder. Although
many millions of tons have been extracted from it since
it first came into the possession of the Demidoff family, it
may easily be imagined that a mass of at least 600,000,000
cubic feet of iron-ore is not easily removed, and will outlast
the labours of many generations. The quantity of cast-iron
annually produced amounts to 25,500 tons, which is converted
(partly at Nishne-Tagilsk and partly in the neighbouring
forges) into bar-iron, anchors, kettles scythes, nails, wire, &c.
The excellent quality of the iron allows it to be rolled into
very thin plates, which are frequently made use of in Russia
for the roofing of houses, and are also manufactured at
Nishne-Tagilsk into lacquered wares, which find a ready sale
throughout the whole of European and Asiatic Russia.

The iron industry (350,000 tons) of Austria has its chief
seats in Styria, Carinthia, and Bohemia, and, though out
of proportion with the vast natural resources of the empire,
has of late made rapid progress. The ores, which are of an
excellent quality, are mostly smelted with charcoal, as they
are generally situated at a great distance from the coal
mines. The Noric or Styrian iron has enjoyed an excellent
reputation ever since the time of the Romans, when the
imperial manufactory of Lauraceum on the Danube supplied
the legions with swords and javelins.

In a pass of the Styrian Alps, between the valleys of the
Mur and the Enns, lies one of the most remarkable iron mines
in the world—the famous Erzberg or iron-mountain, which
rises to a height of 3,000 feet, and whose summit and sides
are almost everywhere coated with a thick mantle of the
richest ore. Authentic records show that it has been worked
ever since the year 712; and probably the Romans derived a
part of their Noric iron from this source, as the ore is not
here concealed in the bowels of the earth, but crops out on
the surface, near a mountain pass which was undoubtedly
known to them. As at Nishne-Tagilsk, the ore thus conveniently situated is quarried from the mountain, and thus
in course of time extensive excavations or grottoes have
been formed all over its surface, affording a most interesting

spectacle. The bottom of these iron-stone pits is irregularly strewn with large blocks of ore, through which wind narrow footpaths. Roads lead from one pit to the other, and close by are situated the small huts in each of which ten or twenty of the miners sleep as long as the working season lasts, for their families generally live lower down in the deeper valleys. On the top of the mountain stands a colossal crucifix of iron, near which an annual thanksgiving feast is celebrated. Though the mines are easily worked, the conveyance of their produce to the smelting-ovens of Eisenerz and Vordernberg is a matter of much greater difficulty, and requires numerous tram-roads, galleries, and shafts, through which the ore is precipitated from a higher to a lower level. At the foot of the mountain they all unite in one main shaft, which leads into a gallery ornamented with a monumental gate. Thus the whole of the Erzberg is covered with various machines, pits, horizontal and vertical galleries, tunnels, and roads, and represents as it were a *mine turned inside out*, where all the operations which are elsewhere performed underground are (as at the Carclaze tin-mine in Cornwall) exposed to the open day. Here, instead of dirty and dangerous ladders, convenient footpaths, bordered with trees and illumined by the sun, lead from gallery to gallery and from pit to pit; and instead of being confined in dismal passages, the miner enjoys magnificent views of the grandest Alpine scenery. The annual produce of the Erzberg amounts to 25,000 tons of excellent iron, as the ore (sparry iron, carbonate of iron, sphærosiderite) is nearly as pure as the magnetic iron-stone of Russia, and affords natural steel with the greatest facility. The railroad now being made in Styria will no doubt greatly increase this production, which might be continued for many thousand years without exhausting the vast mineral wealth which Nature has here deposited.

Though Sweden furnishes but a small quantity of iron (320,000 tons) when measured by an English scale, yet the quality of its produce is unrivalled in the world. The purest iron-ores (magnetic iron-stone) abound in the mountain chains which traverse the kingdom, and immense forests afford almost inexhaustible supplies of charcoal for their smelting, so that hitherto the want of roads has alone prevented the production of iron from attaining dimensions equal to the natural resources of the country. Many of the richest mines, particularly in the more northern provinces, have never yet been worked, as for instance the enormous mounds of magnetic iron-ore at Gellivara (67° N. lat.) in Swedish Lapland, beyond the Arctic circle, which, from their situation in a polar desert, have hitherto been totally useless. In 1865 an attempt was made to utilise them by means of a railroad and the canalisation of the river Luleo, but after a heavy expense the works were finally abandoned.

Thus the manufacture of iron, which, under more favourable circumstances, would reach as far as its ores extend, is confined to Dalecarlia, the more central provinces of Kopperberg, Wermland, and Upsala, where the celebrated mines of Dannemora, which furnish the fine Oeregrund iron, largely imported into England for the manufacture of steel, deserve our particular notice, both for their ancient renown and their wild and colossal grandeur. An excellent road leads from the famous university town of Upsala to Old Upsala, old and hoary in the fullest sense of the word, for its church dates from the Pagan times, and close by rise three tumuli which, according to popular tradition, contain the remains of no less important personages than Odin, Thor, and the divine Freya.

Further on towards the north-east, six geographical miles from Upsala, lie the iron-works of Oesterby, remarkable for their beautiful situation in a natural park of forests and lakes, and thence half-an-hour's walk over the plain brings one to the far-famed pit of Dannemora.

The country around is perfectly level—a succession of pine-woods and open grounds—and no sign announces the vicinity of the mine, until at length the traveller sees a few huts and some machines for lifting the ore, and then suddenly stands on the brink of an enormous pit, or rather of a vast crater, whose black and precipitous walls inclose an area of at least a mile in circumference. On looking down into the abyss, which descends to a depth of 450 feet, and is here and there enlivened with patches of perennial snow, he perceives along its black walls the still blacker entrances to labyrinthine caves, fringed in some parts with long stalactites of ice of aquamarine transparency and colour. From some of these hollows the flames of piles of fir-wood are seen to creep along the hard rock which they are to soften, and the deep gulf is animated by troops of miners, the distant clang of whose hammers, closely resembling the clicking of a number of clocks, forms a strange concert with the creaking noise of the machinery. Attracted by the novelty of the interesting sight, the eye wanders from one object to another, and time steals on rapidly and unperceived—when suddenly a bell tolls, and the scene as suddenly changes.

It is noon, and the tuns, which before were hauled up from the deep laden with ore, are now seen ascending with a living freight—men, women, and children—standing quite unconcernedly on the narrow edge of the tub, and holding with one hand the chain to which it is attached.

Soon a deathlike silence reigns in the pit—a striking contrast with the noise and life it erewhile displayed; and now loud shouts are heard, warning all those who may have remained behind that the battery prepared during the previous working hours is about to explode. Again a profound silence—and then loud thunder bursts forth, with many an echo, from the depth of the abyss.

For several minutes the whole neighbourhood trembles as if shaken by an earthquake. Through the black clouds of smoke which ascend from the gulf, pieces of stone or ore are hurled upwards, frequently far beyond the brink of the pit, and most of the detonations are followed by the crash of the falling fragments rent by the explosion from the mother-rock.

For many centuries this remarkable mine has afforded employment to many hundreds of workmen, without showing any signs of exhaustion; for its mighty mass of magnetic iron-ore descends to an unknown depth, and seems to be practically inexhaustible.

Though the mineral resources of Spain are immense, yet its iron-industry is so little developed that more than two-thirds of the excellent ores of Biscay, instead of being smelted in the country, are exported to France and England.

In Italy the red oxide and magnetic iron-stone mines of

Elba have been celebrated since the remotest antiquity; but, from the want of fuel on the island, their entire produce, which amounts to about 100,000 tons, is exported to the coast of Italy, to France, and to England. The principal mines are situated on the slope of a steep mountain fronting the sea, and are divided by horizontal terraces into five stories or huge · steps communicating with each other by means of oblique roads, on which carts convey the ores to the shore. Though worked for more than 2,000 years, the mines, which occupy about 700 workmen, are apparently able to supply the wants of the remotest posterity.

On turning to America we find the United States making gigantic strides in the extension of their iron manufacture, which has risen from 347,000 tons in 1840 to 2,000,000 tons in 1869; and as here none of the elements of progress are wanting—a boundless mineral wealth, liberal institutions, which allow the freest scope to individual energy, and an unrivalled spirit of enterprise—there can hardly be a doubt that finally the United States must become the first iron-country in the world. The masses of magnetic iron-stone and red oxide which extend along Lake Superior, over a length of 120 miles and a breadth of from five to thirty, would alone suffice to provide for the wants of the whole of the human race for many thousands of years. They only began to be worked in 1849; and in 1866 the railroad which leads from the mines transported 204,454 tons of ore.

The thriving town of Marquette, the central point of this new seat of industry, was, scarcely twenty years ago, a complete wilderness, where the Red Indian pursued the beasts of the forest, unconscious that the treasures concealed beneath his natal soil would one day be the cause of his expulsion from the hunting-grounds of his fathers.

The State of Missouri possesses two 'iron-mountains' similar to the magnetic mountain of the Demidoff: one of them, called Pilot Knob, is 600 feet high, the other 220. An immense mass of magnetic oxide has also been discovered in California, near the northern extremity of the Sierra Nevada. But though iron is found in abundance in many parts of the Union, (Tennessee, North and South Carolina, Georgia, &c.), the

States of New York and Pennsylvania produce by far the greatest quantity.

Brazil and the island of Cuba likewise contain vast deposits of the richest iron-ores; and in Mexico we find the famous Cerro del Mercado, an iron-mountain 633 feet high, which rises in grotesque form from the valley of Durango. This wonderful mound has been calculated to contain 3,244,000,000 cubic feet, or 454,000,000 tons, of magnetic iron ore, capable of yielding 290,000,000 tons of cast iron, or more than fifty times the annual production of Great Britain! A more industrious and civilised race would here find a boundless field for profitable employment; but the indolent Mexican, steeped in ignorance and falling from one revolution into another, still leaves these treasures almost untouched, and, neglecting the vast resources of his country, draws nearly his whole supply of iron from the distant forges of Great Britain.

CHAPTER XXIX.

LEAD.

Its Properties and extensive Uses—Alston Moor—Belgian Lead Mines—Galena in America—Extraction of Silver from Lead Ores—Pattinson's Process—A great part of our wealth is due to the laboratory.

LEAD was but little prized by the ancients, who, unacquainted with gunpowder, needed no bullets for war or for the chase. The history of its first discovery is lost in obscurity, but it probably became known much later than copper or tin, as less metallurgic skill is required for the smelting of the cupreous or stanniferous ores than for the reduction of galena or sulphide of lead, which is the most abundant of the plumbiferous ores, and may indeed be regarded as the only commercial ore of any value, if we except the carbonates, which are probably formed by its decomposition.

Lead, however, is mentioned both in the Book of Job and in the fourth book of Moses. 'Oh! that my words were graven with an iron pen on tablets of lead,' exclaims the long-suffering patriarch; and the legislator of the Jews commands his people to 'make go through the fire gold, silver, brass, tin, and lead, and everything that may abide it.' The Phœnicians, who provided Greece and Egypt with Spanish lead, frequently made use of this metal to increase the weight of their anchors; and Herodotus, describing a bridge in Babylon, mentions that its stones were fastened with clasps of iron soldered with lead. The physical properties of this metal qualify it for a great variety of uses. As it is but little altered by exposure to air or water, it makes excellent pump-tubes and rain-gutters; while its considerable weight, its softness, its flexibility, and the facility with which it melts at .a comparatively low temperature, render it an invaluable material for the soldier's bullet or the huntsman's shot. Combined with oxygen it forms the pigment called red-lead or minium, and united with carbonic acid white-lead or ceruse, which is still more frequently used in painting.

In the manufacture of glass and crystal it plays an important part, as it forms one of the chief ingredients of flint-glass and crown-glass, of which, as is well known, the achromatic lenses are made which have so wonderfully improved the distinctness of our telescopes. United with tin it forms an alloy which is more fusible than either metal alone, and which is consequently used as a solder by the plumbers, while with antimony it combines into a hard mass which serves to make letters for the printing-press. All these various uses absorb immense quantities of lead, and render it one of the most valuable products of the subterranean world.

Among the lead mines of Europe we find the first rank occupied by those of Spain, which in 1863 furnished the enormous mass of 309,940 tons of galena. The principal mines are situated in Guipuzcoa, Catalonia, Arragon, the Sierra Morena, and above all, in the mountain chain of the Alpujarras, where the rocks of the Sierra Gador are everywhere traversed by veins of galena. The production of North Germany is also very considerable. In 1865 Prussia furnished 57,808 tons of galena, and the Hanoverian Hartz Mountains, which produced 101,411 tons in 1864, and now belong to the same monarchy, have raised its previous production to a threefold amount.

The chief lead mines of England are situated near Alston Moor, where the three counties of Northumberland, Durham, and Cumberland meet together. The lofty hills of the district, bare of wood and almost wholly covered with marshy heaths, are intersected by the valleys through which run the Tyne, the Wear, and the Tees, with their numerous branches. The country, though little frequented by tourists, is wild and picturesque ; but the deep gorges with which it is furrowed have more than a mere romantic interest, for they lay open to view numerous veins of ore, and direct the operations of the miner to the places where it is sufficiently abundant to reward his toil. The town of Alston, the mining centre of the district, is beautifully situated close to the river Tyne, which, about five miles above it, ascends, between lofty hills, to the foot of Cross Fell, this picturesque mountain giving a character of considerable grandeur to the surrounding scenery. The mines of the Alston Moor district produced in 1873, 20,359 tons of lead. The waters are drawn off by long adit levels, and the ores are dragged out by horses to the day. This region extends southward to the Yorkshire valleys of Swaledale, Arkendale, and to Grassington, where numerous lead mines are worked under very similar circumstances.*

Another important group of veins of lead occurs in the slaty rocks of Cardiganshire and Montgomeryshire, all of which have an E.W. direction, although so far from parallel that they often meet, and frequently form at such points of intersection 'courses of ore.' Some of these mines have been already profitably worked, in the seventeenth century. In 1873 they produced 10,061 tons of lead.

Shropshire, Derbyshire, Cornwall, and Denbighshire follow next with a yield of 5,623, 3,116, 2,923, and 2,242 tons. Flintshire produced 1,470 tons, and Carnarvonshire 738.

There are also considerable lead mines in the south of Scotland, at Wanlockhead and Leadhills in Lanarkshire; and those of Strontian, in Argyleshire, likewise deserve to be noticed. The Isle of Man has two important lead mines, the Foxdale and Laxey, the former remarkable for the great size of its main lode. The elevated tracts of Wicklow likewise contain some valuable lead mines, at Luganure and Glenmalure.

The island of Sardinia, already renowned among the

---

* In 1873 the total produce of our 206 lead mines amounted to 73,500 tons of ore, which yielded 54,235 tons of lead. In the same year we imported 11,088 tons of ore and 62,563 tons of pig and sheet lead, chiefly from Spain, Greece, and Italy. Our principal mines are Van in Montgomeryshire (5,740 tons of ore in 1873, besides 48,824 ounces of silver); Allendale in Durham (8,370); Frank Mills (3,557) and Exmouth (2,631) in Devonshire; Stonecroft and Greyside (2,935) in Weardale; Roman Gravels in Shropshire (2,550); Chivorton West in Cornwall (2,224); and Minera in Denbighshire (2,913). The produce of some of the mines is ludicrously small. Budnick Consols in Cornwall produced in 1873, 14 cwt. of lead ore, value 6*l.* 12*s.*, and Crenver 5 cwt., value 2*l.* 11*s.*

ancients for its rich lead mines, produces about 15,000 tons, or nearly as much as France, where, however, the extraction of galena has of late years made considerable progress.

Belgium, which in 1841 produced no more than 34 tons, raised in 1864 no less than 16,780 tons, chiefly from the mines of Bleyberg-à-Montzen, situated in the carboniferous limestone near Verviers. To render these rich deposits available, vast difficulties had to be surmounted by the united powers of enterprise, capital, and engineering skill. Rivers and brooks, diverted from their ancient course, were made to flow through new water-tight channels; and such is the amount of drainage required in that aquiferous region, that engines of two thousand horse-power have to raise from a depth of 360 feet a daily quantity of 800,000 cubic feet of water.

In Greece the immense mounds of scoriæ accumulated near the ancient silver mines of Laurium have been found to contain about ten per cent. of lead. Their total mass is estimated at no less than 3,000,000 tons, and they afford a convincing proof both of the importance of the ancient silver production of Attica and of the imperfection of the old Athenian mining operations. A French company has lately been formed for smelting this prodigious accumulation of scoriæ, once cast aside as rubbish.

In Siberia the famous lead mines of Nertschinsk, where many an unfortunate exile is doomed to end his days, are worked merely for the silver contained in them.

In South America, the Chilian lead mine of Mina Grande, near Coquimbo, is renowned for its extreme richness, and Brazil has considerable veins of galena, in the province of Minas Geraes; but probably the United States of North America (Wisconsin, Arkansas, Iowa, Illinois), possess the largest galena deposits in the world. In Wisconsin they extend all over a vast territory of more than 4,000 square miles. As yet the works are conducted in the most negligent manner, by a crowd of adventurers. In winter, when the air in the pits is more salubrious, and agricultural labour ceases, needy farmers, bankrupt traders, and thriftless artisans, flock from all parts to the lead country, for the purpose of repairing their broken fortunes. In summer, when malaria

renders the pits extremely unhealthy, this nomadic population
melts away like chaff before the wind. Yet, in spite of the
rough mode of extraction which prevails in the American
lead country, the mines of Iowa, Wisconsin, and Illinois,
yielded about 20,000 tons of lead in 1866, and their produce
is constantly increasing.

In all lead mines the galena often occurs in pieces so
large that they do not require to be separated from the vein-
stone by the processes of stamping and washing. They are
then called *pure ores*, and the most simple preparation is
sufficient to prepare them for the smelting furnace. When
the ore has been picked and so far prepared, it is first
roasted or heated in a reverberatory furnace—an operation
which causes the oxygen of the air to combine with the
two elementary bodies of which galena is composed. After
undergoing this chemical change, the ore is mixed with
coke, charcol, or peat, and reduced by smelting in a small
blast-furnace of a peculiar kind. Under the influence of
heat, the carbon of the coal, uniting with the oxygen of the
ore, flies off in the form of carbonic acid gas, while the
metallic lead, which in the finer ores amounts to 70 or 80
per cent., sinks to the bottom of the furnace. Almost all the
varieties of galena contain a greater or less proportion of
silver, which it is often found worth while to separate from
them. This process is at present effected according to a
most ingenious method, founded on the circumstance first
noticed in the year 1829, by the late H. L. Pattinson, of
Newcastle-on-Tyne, that when lead containing silver is
melted in a suitable vessel, afterwards slowly allowed to cool,
and at the same time kept constantly stirred at a certain
temperature near the melting point of lead, crystals begin
to form. These, as rapidly as they are produced, sink to the
bottom, and on being removed are found to contain much
less silver than the lead originally melted. The still fluid
portion, from which the crystals have been removed, will at
the same time be proportionally enriched. By repeated
meltings and crystallizations in a series of cast-iron pots
with fire-places beneath, the workman is thus able to de-
prive almost entirely of its silver by far the largest portion of
the lead operated upon, while the remainder becomes an

exceedingly rich alloy of both metals, so as to contain fifty times its original proportion of silver. This rich lead is subsequently exposed in a refining furnace to a strong blast of air at a high temperature, fresh supplies of lead being constantly introduced as the operation proceeds. By this means the lead becomes rapidly oxidized and converted into litharge, which partly runs off in the fluid state, and is partly (about 10 per cent.) lost by sublimation, while the silver forms a cake at the bottom of the cupel. The brightening of the silver, which lustrously shines forth at the moment of the separation of the last traces of lead, indicates the precise period at which the operation should be ended, and the blast is then turned off and the fire removed from the grate. Before the introduction of Pattinson's ingenious process, the separation of the silver from the lead was attended with a much greater loss of the latter metal, as greater quantities had to be cupelled to effect the same result. The economy obtained amounts to no less than 98 per cent; for where formerly 100 cwt. of lead were lost by sublimation, the same quantity of silver is now obtained with a loss of no more than 2 cwt., and at the same time with a considerable saving of feul.

Ores very poor in silver, as for instance those of Alston Moor or Derbyshire, which formerly could not be profitably cupelled, are advantageously treated by the Pattinson process, which in its turn is giving way to Parkes' new method of separating the silver from the lead by means of the more powerful affinity of zinc. This is but one example of the valuable practical results which may be obtained from a single scientific discovery. But chemistry has introduced thousands of similar technical improvements in almost all branches of manufacturing industry; and were we to add together the profits thus obtained, we should find that a great part of our wealth is due to the laboratory.

CHAPTER XXX.

MERCURY.

Not considered as a true Metal by the Ancients—Its Properties and Uses—
Almaden—Formerly worked by Convicts—Diseases of the Miners—Idria—Its
Discovery—Conflagration of the Mine—Its Produce—Huancavelica—New
Almaden.

AMONG the metals known to antiquity mercury was the
last discovered. It is mentioned neither in the Bible
nor in Homer, who accurately, though briefly, describes the
characters and uses of all the other ancient metals; but we
learn from the works of Aristotle that its discovery must
have preceded the times of Alexander the Great.

From its always remaining fluid in the temperate climates
of the earth, it was, however, not considered as a true metal;
for the ancients had no means of ascertaining that at the low
temperature of $-39°$ Fahr. it becomes malleable and assumes
the solid form. The Greeks called it Hydrargyros or *water-silver*, from its fluidity and argentine colour; the Romans,
'argentum vivum,' or live silver, from which our 'quick-silver' has been derived; and in the Middle Ages the alche-mists gave it the planetary name of Mercury, which has
been generally adopted in modern scientific language.

At a very early age cinnabar (sulphide of mercury)—the
beautiful scarlet ore from which it is chiefly obtained—was
employed by the ancients as a colouring material for im-parting a florid complexion to triumphant generals or to
guests at the festive board. The extent to which this cos-metic was used may be inferred from the facts mentioned
by Pliny, that the Greeks imported red cinnabar from
Almaden 700 years before the Christian era, and that Rome
in his time received 700,000 pounds from the same mines.

As the alchemists considered quicksilver as the fittest substance for transmutation into gold, it became the subject of innumerable experiments; and though these manipulations had not the desired effect, they accidentally led to the discovery of several of its combinations, which soon became known as powerful medicines. But it was reserved for modern times to appreciate and understand the full importance of mercury, and to extend the field of its utility to a variety of uses unknown to former ages.

Alloyed with tin-foil, it forms the reflecting surface of looking-glasses; and by its ready solution of gold and silver, and subsequent dissipation by a moderate heat, it becomes the great instrument of the arts of gilding and silvering copper and brass. The same property makes it available in extracting these precious metals from their ores. To science it is a substance of paramount value. Its great density, and its regular rate of extension and contraction by increase and diminution of temperature, give it the preference over all liquids for filling barometer and thermometer tubes, so that without mercury we should know but little indeed about the laws of caloric and of atmospherical pressure. In chemistry it furnishes the only means of collecting and manipulating, in the pneumatic trough, such gaseous bodies as are condensible over water. To its aid, in this respect, the modern advancement of chemical discovery is pre-eminently due, and without its assistance many a branch of industry which now greatly adds to the wealth of the nation could never have existed.

Mercury does not rank among those metals which are copiously disseminated over the surface of the earth. Considerable deposits or veins of even its commonest and most abundant ores (red cinnabar, hepatic cinnabar) are confined to a few spots, and all Europe possesses but two important quicksilver mines—at Almaden in Spain, and Idria in Carniola.

The Sierra Morena or the Black Mountains in the Iberian Peninsula, so well known as the scene of the exploits of Don Quixote, are no less renowned in the mining world for their subterranean riches, among which the famous quicksilver mines of Almaden del Azogue hold a conspicuous rank. This small town of about 10,000 inhabitants is situated on the northern border of the above-mentioned mountain chain, at the south-western extremity of the province of New Castile, and, unlike most other Spanish cities, affords the pleasing sight of well-built houses and clean and regular streets. As is generally the case in mining districts, the neighbourhood is sterile but picturesque, and from the neighbouring mountains magnificent prospects open on every side.

The mines of Almaden, after having been the property of the religious knights of Calatrava, who had assisted in expelling the Moors, were farmed off to the Fuggers, the celebrated merchant princes of Augsburg, whose descendants still rank among the high aristocracy of Germany. Afterwards, from the date of 1645 till the present time, they have either been worked on Government account or farmed off to private companies. A visit to these celebrated mines is highly interesting. A spacious tunnel or gallery, completely walled with solid masonry, leads into the bosom of the mountain, and branches out at its extremity into several galleries hewn in the slate which forms the matrix of the vein. One of these galleries conducts to the Boveda de Santa Clara, a vast circular hall in which formerly stood a horse-gin for raising the ore to the surface. At present, however, this work is performed through a shaft which descends to the lowest level of the mine. From another gallery convenient stairs lead down to the first working level of the mine, and thence short ladders to the deeper stories. The shafts are wide, the galleries high enough to allow one to walk upright through them. The upper stories are almost thoroughly dry, the lower ones humid; the water, however, is easily removed by hand-pumps, which raise it from story to story into a large subterranean reservoir, emptied once a week by a steam-engine of fifty-four horse-power. The veins are so extremely rich that, though they have been worked pretty constantly during so many centuries, the mines have hardly reached the depth of 1,140 feet. The lode actually under exploration is from fourteen to sixteen yards thick, and it becomes thicker still at the crossing of the veins. Its present annual produce amounts to about 20,000 cwt. of pure mercury. The ore presents a beautiful sight in the galleries where it is worked, on account of the dark red glittering colour of the cinnabar, which is sometimes earthy and sometimes forms compact crystalline masses, or fine crystals mixed with calcareous spar. Often when a hewer detaches a block of ore with his pickaxe, quicksilver masses of the size of a pigeon's egg roll from a crevice in which they were lodged, and, leaping along the floor, divide into thousands of small drops. This, however, is no loss, for all the rubbish which accumulates in the galleries is carefully collected.

The ores are treated in thirteen double distillatory furnaces, called alodels, and yield only ten per cent. upon an average. But the analysis of the ores leaves no doubt that the barbarous apparatus employed in its sublimation causes a loss of nearly one-half of the quicksilver, which is dispersed in the air, to the great injury of the workmen's health. This apparatus has remained without any material change for the better since the days of the Moorish dominion in Spain. The furnaces are heated with brushwood, particularly with the resinous branches of the *Cistus ladaniferus*, which grows over the greatest part of the surrounding mountains. This dark evergreen shrub often extends over many miles of ground, and when in flower covers the hills with a beautiful snow-white carpet.

Formerly only criminals condemned to hard labour for life were employed to work in the mines of Almaden. At sunrise they were conducted from a prison (which still exists), through a subterranean passage into the mines, where they were obliged to toil till evening, when they were led back again to their dungeons, so that they never saw the light of day. After a few years these poor wretches generally died from inhaling the poisonous vapours of the mercury. Reduced to despair, they set fire, at the beginning of the last century, to the galleries, which were then constructed of wood, and thus rendered the mining operations impossible for many years. Since then only free miners are employed, who are well paid, and not allowed to work longer than six hours a day. Most of them, however, die at an early age (between thirty and forty), and those who live longer are affected by a spasmodic trembling, so as to be unable to keep a limb quiet.

After the mines of Almaden those of Idria in Carniola are

the richest in Europe. This neat little town lies in one of the largest and most picturesqe basins of the Julian Alps. The gigantic and naked rock walls which inclose the sequestered valley are only partially decked along their summits with clumps of firs ; but lower down the slopes are covered with beautiful meadows and forests, and here and there on the projecting spurs of the mountains stand picturesque chapels, which serve to heighten the beauties of the magnificent panorama.

Chance—to which man owes so much both of good and evil—also led to the discovery of the subterranean treasures of Idria. In the year 1497 a peasant found, in a tub which he had placed under a brook, some drops of a heavy liquid shining like silver. Although quite ignorant of the value of his discovery, he still was cunning enough to carry them to a goldsmith without mentioning the place where he had found them. At length a man named Anderlein, having promised him a handsome reward, became master of the secret, and associating himself with several wealthy persons, began to work the mine. After several years the property passed into the hands of a company of goldsmiths of Salzburg and Augsburg, which derived such profits from the mine as to excite the greed of the neighbouring Venetians, who in the year 1510 forcibly drove away the Germans, but were soon after expelled by the Emperor Maximilian, who, as soon as he heard of this insolent attack, sent some troops into the forest of Idria and restored it to its rightful owners. In 1578 the mine was incorporated among the domains of the State, and began to be worked with greater regularity.

The entrance lies to the south of the town, on the slope of a small hill projecting from the circular zone of mountains which gird the basin. The visitor may either descend, in less than five minutes, to the bottom of the pit, through a perpendicular shaft, in one of the large tubs which serve to raise the ore, or he may make use of convenient stairs. Here and there are landing-places, whence galleries lead to the various fields or stories of the mine, the lowest of which is 145 fathoms under ground. The vein, however, descends much further, to an unknown depth, and its horizontal extent has likewise not yet been measured.

As the limestone in which the ores are embedded does not form a solid compact mass, but is of a loose nature, most of the galleries had, from the very beginning of the mining operations, to be artificially propped. Until the end of the last century wooden beams were chiefly used for this purpose —a construction which frequently gave rise to fearful fires. Thus on March 16, 1803, the workmen saw a thick smoke issuing from several of the deepest galleries. It rose higher and higher, and spread through the upper galleries of the mine. No fire was to be seen, no sound of flames was to be heard, yet there could be no doubt that the mine was burning. Some of the workmen boldly undertook to descend to the scene of the fire, but in vain; they were obliged to flee before their enemy, for the smoke was not only dense and suffocating, but so impregnated with poisonous vapours that no living being could exist within its reach. An attempt was now made to smother the fire by shutting out the air. All the galleries were blocked as near as possible to the supposed seat of the fire, as well as the two shafts which led to the upper world. The mine remained thus closed during five weeks, but to no purpose; for when it was re-opened, the fire burst forth more furiously than before. The flames now howled fearfully from the bottom of the pit, and the mercurial and sulphurous fumes which rose from it threatened instant death to anyone who should be rash enough to approach them. The director of the mine now resolved to flood the works, as a last resource against the ravages of fire. A river was turned into the perpendicular shaft, and ran during two days and three nights into the pit. On the first day there was no perceptible effect, but on the second—whether it were that the vapours produced by the meeting of the antagonistic elements were striving for an outlet, or that new inflammable gases had been formed—a most terrific explosion took place, which made the whole mountain tremble as if shaken by an earthquake. The huts close to the opening of the shaft were rent to pieces, the stone houses at the foot or on the slopes of the hill fell in with a tremendous crash, and the inhabitants of Idria, fancying the day of judgment had arrived, fled in terror to the hills.

In the mine itself the explosion was afterwards found to

have torn up the galleries, to have burst the vaulted roofs, and to have hurled away the stairs. But the victory was now won. The vapour gradually drew off, and after a few weeks it was possible once more to descend into the pit. Two full years, however, passed before the water was fully pumped out into the Idriza, where it poisoned all the fishes, with the exception of the eels, who, it seems, are proof against everything except roasting or boiling. Even after all the water had been removed, it was still found impossible to work in the mines, partly from the heat, but chiefly on account of the venomous fumes, which soon produced all the symptoms of mercurial disease.

In order to stimulate almost superhuman exertions, an exorbitant salary was offered to all those who should venture to explore the most dangerous passages, and gather the quicksilver, which in some places had collected in considerable masses. But many paid for their greed with their lives, and for many months afterwards the air remained so noxious that the ordinary mining works could not possibly be carried on. To prevent similar accidents for the future, and also on account of the increasing scarcity of wood, the galleries have since then been walled with stone; but when we consider that the whole length of the subterranean passages amounts to no less than fifty miles, we cannot wonder that many of them are still propped with wood, and that as recently as 1846 fire raged in the mine, which was again quenched by putting part of it under water.

The stone galleries are vaulted, and of a masterly construction, seven feet high and six feet broad. The cost is less than one would suppose, as the progress of the mining operations furnishes the necessary materials.

The temperature in many galleries is equal to that of a conservatory; and if a floral hall, bathed in light and filled with delicious odours, is felt to be disagreeably hot, the warmth of the air is naturally far more intolerable in dismal excavations, where the workman pursues his laborious task by the weak glimmering of a lamp, and in an atmosphere full of deadly vapours.

Here, as in Almaden, a premature old age is the lot of the unhappy miners, who while young tremble like old men. Yet some attain a tolerably good old age, and he who reaches his forty-fifth year with trembling limbs is said to get accustomed to the effects of the poison, and may then live, or rather vegetate, till sixty or seventy.

Scarcely any animals live in the mines of Idria. Even spiders cannot long resist the noxious atmosphere. Rats, however, formerly existed there in considerable numbers, but have almost entirely disappeared since the last fire, which proved too much even for them. In some parts of the mine the mercury is inclosed in the clay-slate in extremely fine globules, so as scarcely to be visible to the naked eye; but on removing the ore many of them fall out and collect on the floor, in larger drops or small pools, which are carefully gathered in leather pouches or bags. By far the greatest part of the mercury of Idria, however, is combined with sulphur, and is obtained by distilling the ore in vast furnaces.

It may easily be imagined that the carriage of a liquid body of the weight of mercury requires the greatest care. The old mode of packing, still partly in use, is in sheep-skins, which can acquire the necessary firmness only by being tanned with alum, and are attentively examined before being used. After having been filled, the sack is first tied on a wooden table, and, having successfully stood the ordeal of severe pressing and punching, is enclosed in a second skin. Two of these packages, each containing forty-one pounds and a few ounces, are then placed in a small cask, and three of these in a square box. But as, in spite of all these precautions, the sacks will sometimes burst, the mercury is now frequently transported in large iron bottles, the stoppers of which are firmly screwed down by means of a machine invented for the purpose. All the quicksilver packed up in this manner is sent to Trieste, and thence chiefly to England, while that which is destined for Vienna and Germany is exported in sacks.

In the years 1856, 1857, and 1858, 4,570, 7,178, and 4,331 hundredweight of mercury were produced in Idria, and sold at an average price of one hundred florins. In 1850 mercury was worth two hundred and fifty florins the hundredweight, and thus we see how detrimental the competition of California has been to the Austrian treasury, which, in its chronic state of atrophy, is little able to bear any diminution of revenue.

A great part of the produce of the mines—about 1,000 hundredweight—is manufactured into cinnabar in Idria, which supplies almost all Europe with this splendid red colouring matter. All the other European quicksilver veins, in Tyrol, Bohemia, Hungary and the Bavarian Palatinate, are utterly insignificant when compared with Almaden and Idria, as none of them produce more than a few hundredweight.

But even Almaden and Idria have lost much of their former importance since the discovery of the inexhaustible quicksilver mines of California, whose total product for the years 1869, 1870, and 1871 has been estimated at 36,600, 29,546, and 31,881 flasks, each containing $76\frac{1}{2}$ pounds of mercury. In 1871 New Almaden, in Santa Clara County, by far the most prolific of the mines, yielded 18,763 flasks, the New Idria mine in Fresno County, 9,227 flasks, and the Redington mine, near Knoxville, Lake County, 2,128 flasks. Since 1872 many promising lodes have been discovered, warranting the belief that the future productions of this valuable mineral will be greatly increased, for in no other part of the world has cinnabar been found so widely disseminated as in California. As the uses of mercury are few, a great falling off of price has been the consequence, but the greater cheapness of this metal has had a most favourable influence on the production of silver in Mexico, Peru, Bolivia, &c., as many of the poorer ores can now be profitably worked, and thus California, by lowering the price of silver, has greatly disturbed the relative value of the two precious metals.

It is a remarkable circumstance that, while Europe has for the last three centuries received almost all its silver from America, Mexico and Peru were all the time dependent upon the old world for the mercury without which Potosi and Guanaxuato would have been comparatively unproductive. Quicksilver, it is true, had been found here and there, but the only mine of importance was that of Huancavelica in Peru, the discovery of which in the year 1567 is attributed to the Indian Gonzalo Navincopa; though, according to Humboldt, it was already known to the Incas, who made use

of cinnabar to paint their cheeks, as Roman senators and
Athenian archons had done before them.   Here, at a height
surpassing that of the Peak of Teneriffe by 1,500 feet, from
4,000 to 6,000 cwt. of silver were annually obtained, until
the folly of a director ruined the chief mine.   Ever since 1780
·Huancavelica had with difficulty supplied the growing wants
of the Peruvian silver mines, for at a greater depth the ore
was found to be mixed with sulphuret of arsenic, which greatly
deteriorated its quality.   As the lode forms an enormous mass,
strong pillars had been left standing to support the roof, and
these props the above-mentioned director had the improvident
temerity to remove, in order to increase the produce of the
mine.   What anyone with a little experience or common-
sense might have foreseen, took place.   The rock, deprived
of its supports, gave way; the roof fell in with a tremendous
crash, and the mine was ruined—a memorable warning
against the greed which, snapping at a shadow, loses the
substance.

CHAPTER XXXI.

THE NEW METALS.

Zinc—the Ores, but not the Metal, known to the Ancients—Rapid Increase of its
Production—Chief Zinc-producing Countries—Platinum—Antimony—Bismuth
—Cobalt and Nickel—Wolfram—Arsenic—Chrome—Manganese—Cadmium—
Titanium—Molybdenum—Aluminium—Aluminium Bronze—Magnesium—So-
dium—Palladium—Rhodium—Thallium.

THE metals known to the ancients were either such as
occur in a native state, and whose lustre must attract
even the attention of the savage, or such as are easily extracted from their ores by the simple agency of fire and
carbon, and consequently require no complicated metallurgic
treatment. Their number is limited to the seven substances
described in the preceding chapters; but the art of the
modern chemist has greatly extended our knowledge of
metals, and revealed to us the existence of no less than fifty-
six of these elementary bodies.

Some have been found to lurk under the obscure disguise
of alkaline and earthy matters, such as clay and chalk, magnesia and sand, soda and potash; others have been discovered
in the water of mineral springs, or under the brilliant mask
of precious stones. Most of these were unknown before the
beginning of the present century, nor can there be a doubt
that future researches will make us acquainted with many
metals whose existence is still a secret to mankind.

Most of these new metals are as yet mere objects of
curiosity, either from their rarity or the great difficulty and
cost of their production. But some of them are already of considerable use, and within the last fifty years zinc has obtained
a rank among the most important products of the mineral
world. Calamine, the chief ore which provides us with this
metal, was indeed known to the ancients, who by smelting it

with copper ores obtained an alloy similar to our brass; *
but the metal itself seems to have been first discovered by
the famous alchemist Bombastus Paracelsus, who flourished
towards the end of the fifteenth century. Zinc, however,
remained unnoticed as a useful metal until the year 1805,
when Hobson and Sylvester's discovery that it is malleable
at a temperature of 300° F., and can then be worked to
any shape with great facility, caused it to replace lead for
many purposes, in which its hardness and other valuable
qualities render it superior. As it is very easily extended
into thin sheets, and combines the advantages of lightness,
salubrity, and durability, it is frequently used for the roofing
of houses and for the sheathing of ships. Many of our
domestic utensils, particularly those which serve for the
holding of liquids, are now made of zinc. Large quantities
are moulded into architectural ornaments; and the splendid
white colour of the oxide of zinc has made it a triumphant
rival of ceruse, or white-lead. To provide for so many uses,
the production of zinc has in a short time made strides
without a parallel in the history of metals. While before
1808 from 150 to 200 tons sufficed for the annual consumption of Europe, more than 110,000 tons are now required; so
that in little more than half a century the demand has increased more than five hundred times, and a metal previously almost unnoticed is now produced in masses worth
several millions of pounds.

The chief zinc-producing countries of Europe are Prussia
and Belgium. The Prussian mines, which in 1870 yielded
1,278,388 hundredweight, or about 64,000 tons, are situated
in Silesia, Westphalia, and the Rhenish provinces. In the
same year Belgium produced 35,500 tons, chiefly from the
mines of the Vieille Montagne, near Aix-la-Chapelle, where
calamine occurs in a large mass, imbedded in chalk, and is
worked like an open quarry.

In England calamine is, next to galena, the most important
ore obtained from the Derbyshire mines, and of late years
large quantities of blende or sulphuret of zinc—an ore which,

---

* A coin of Nero, analysed by Arthur Phillips, was found to consist of 81·07
per cent. copper, 1·06 tin, and 17·73 zinc; another, of Hadrian, of 85·78 copper,
1·19 tin, 1·81 lead, 0·43 zinc, and 0·74 iron.

on account of the special difficulties offered by its treatment, had hitherto been neglected—are likewise furnished by the Isle of Man, Denbighshire, Flintshire, and Cornwall.

In 1873 our entire production of zinc ores amounted to no more than 15,968 tons, which yielded 4,471 tons of metallic zinc. Besides our own produce, 30,087 tons of foreign ores, imported chiefly from Italy (21,693), Spain (5,129), France (1,486), and Norway (1,114), were smelted in the zinc works at Swansea.

For many years the United States depended upon Europe for their whole supply of zinc; but as nature seems to have denied none of her mineral riches to the great republic, the discovery of immense deposits of calamine and blende in the state of Tennessee has enabled them to compete successfully with foreign produce, and the works of Leehigh and Lasalle now furnish a large proportion of the zinc consumed in the country.

Platinum, the heaviest body in nature, was first discovered by the Spaniards, in the gold mines of Darien, probably in the first half of the sixteenth century; * but as it remained infusible in the strongest heat, and no method was known for purifying its ore, in which it is remarkably combined with six or even seven other metals, it continued for a long time to be a mere object of curiosity. In 1772 Count Sickingen discovered that it can be welded like iron when urged to a white heat, and first succeeded in producing platinum wire and platinum leaves. A few years after the celebrated Swedish Chemist, Bergmann, isolated it from the metallic substances associated with its ore, and proved it to be a peculiar metal.

Platinum is found in almost all the auriferous districts of the globe, but generally in such small quantities as not to be worth the collecting. Kuschwa Goroblagdat and Nishne-Tagilsk, in the Oural, furnished in 1871, 125 pud (5,000 lbs.), which is nearly ten times the amount from Brazil, Columbia, St. Domingo, and Borneo. But, in spite of this scanty production, its discovery must be considered as one of the most important conquests which science has made in the material world, as its perfect infusibility, its hardness,

---

* Kopp, 'Geschichte der Chemie,' vol. iv. p. 221.

its unalterability by air and water, and its property of with-standing the action of the most corrosive simple acids, render it an invaluable material for the fabrication of various chemical vessels, without whose assistance many important discoveries could not possibly have been made. To the manufacturers of sulphuric acid large retorts of platinum are indispensable for concentrating this highly corrosive fluid, which devours every other metallic vase with which it comes in contact. The price of platinum is intermediate between that of gold and silver.

The ores of Antimony played a great part in the labours of the alchemists, but the metal is first mentioned in the works of Basilius Valentinus, who flourished during the second half of the fifteenth century. It is used chiefly in several important alloys. Combined with lead it constitutes type-metal, and united with lead and tin it is employed for making Britannia metal, and the plates on which music is engraved. Nearly all the antimony of commerce is furnished by the grey sulphuret (stibnite), which occurs in Hungary, Saxony, South America, and Australia. Though Cornwall produces a considerable quantity of antimonial ore, our chief supply is derived from Singapore, the emporium of the various mines of Borneo and other parts of the Malayan Archipelago.

The grey antimony ore was employed by the ancients for colouring the hair and the eyebrows, and for staining the upper and under edges of the eyelids—a practice still in use among Oriental nations for the purpose of increasing the apparent size of the eye. According to Dioscorides, it was prepared for this purpose by inclosing it in a lump of dough, and then burning it in the coals till it was reduced to a cinder. It was then extinguished with milk and wine, and again placed upon coals and blown until it was ignited, after which the heat was discontinued, lest, as Pliny says, 'plumbum fiat'—it become lead. It hence appears that the metal antimony was occasionally seen by the ancients, though not distinguished from lead.

Bismuth, a metal of a dull silver-white colour, inclining to red, is first mentioned in the writings of the alchemists of the Middle Ages. It is almost exclusively furnished by the mines of Schneeberg in Saxony, where it is generally found in a native state. On account of its great fusibility and brittleness it is seldom used alone; but associated with other metals, it forms several valuable alloys.

In the Middle Ages the Saxon and Bohemian miners believed all those ores from which, in spite of their promising appearance, they were unable to extract a useful metal, to be a work of the gnomes mocking the industry of man. Some of these ores they called Kobold -an opprobrious name given to these evil subterranean spirits, who were supposed to be of dwarfish stature and intense ugliness; others Nickel—a name probably of the same meaning as our old Nick. The progress of metallurgic industry has fully exculpated the gnomes of all evil intentions, for the last century succeeded in extracting the metals Cobalt and Nickel from those rebellious ores. Cobalt, though as yet but rarely employed, gives promise of some future importance, as it appears to be extremely tenacious. A wire made of pure cobalt will carry nearly double the weight that an iron wire of the same thickness will do.

The cobalt ores, which impart a magnificent blue colour to glass, have lost much of their importance as pigments since the discovery of artificial ultramarine, while the nickel ores which usually accompany them, and were formerly thrown away, have become valuable, since the metal which they contain has found some important uses. The United States, Germany, Belgium, and Switzerland now use nickel for their small coin, and large quantities are employed in the fabrication of German silver, or Argentine plate, an alloy of copper, nickel, and zinc, which, from its hardness and brilliant white colour, furnishes an excellent material for tablespoons and forks. Both the nickel and cobalt ores are produced chiefly by Germany (10,798 hundredweight in 1870), Sweden, and Norway. In the United States the Camden works (New Jersey) now produce nickel at the rate of 200,000 pounds a year.

Tungsten, a metal discovered in 1783 by two Spanish chemists, the brothers Juan and Fausto d'Elhujar, in a black mineral called wolfram, which frequently occurs along with tin ores in Cornwall (where it is known under the names of cal, or callen, and gossan), Saxony, Austria, &c., is in its isolated state a mere object of scientific curiosity, but when melted with cast steel or even with iron only, in the proportion of from 2 to 5 per cent., it produces a steel which is very hard and fine-grained, and for tenacity and density is superior to any other steel made. Hence wolfram-steel, which is now coming extensively into use in Germany, makes the best knives and razors; but, unfortunately, the rarity and high price of wolfram confine its production within narrow limits. Several of the tungstates, or salts of tungsten, are used as pigments; and the tungstate of soda has the highly valuable property of rendering fabrics uninflammable, and thus furnishes a means for preventing the accidents which constantly occur from the burning of ladies' dresses.

Albert the Great, a famous alchemist of the thirteenth century, is supposed to have been the discoverer of Arsenic, a tin-white metal, which, however, soon loses its brilliancy when exposed to the air, and turns black. From its poisonous qualities it is only used in some unimportant alloys which serve for the manufacture of insignificant articles, such as buttons or buckles. Some of its ores and combinations, which, from their lively yellow, green and red colour, would otherwise have been valuable pigments, are likewise for the same reason seldom used. A great number of copper, nickel, lead, cobalt, zinc, and iron ores contain some arsenic; but this dangerous substance is obtained chiefly from the common arsenical pyrites (Mispickel—sulphuret of iron and arsenic), which occurs in Cornwall and Devonshire. The whole supply of arsenical ores amounted in 1866 to about 2,610 tons, of which England and Prussia furnished the greater part.

The metal Uranium, discovered in 1789 by the celebrated Klaproth, in a black heavy mineral called Pechblende (pitch-blende), occurring in the mines of the Erzgebirge, is not used as such, but is very valuable in porcelain-painting, as it affords a beautiful orange colour in the enamelling fire, and a black colour in that in which the porcelain is baked. A laboratory has been opened at Joachimsthal, where the ore is converted into uranate of soda for this purpose.

Chrome, like cobalt, is used chiefly as a pigment. Several of its salts are splendid yellow colouring matters, and its oxide imparts the finest green tints to porcelain. The metal itself, which was discovered by Vauquelin in 1797, is, as yet, an object of interest only to the chemist, but may one day become important, as in its pure state it is very hard, unalterable by air and water, and even less fusible than platina. Most of its ores belong to the rarer minerals, and but one (chrome-iron) occurs in sufficient abundance for industrial purposes. It is found in Hungary, in Norway (which annually exports about 16,000 tons to Hamburg and Holland), in Siberia, and in large quantities in Maryland and Pennsylvania. The ore employed in England is obtained mostly from Baltimore, Drontheim, and the Shetland Isles, and amounts to about 2,000 tons annually.

Manganese is likewise a metal which has not yet left the domain of the laboratory, but some of its ores are of considerable and increasing importance. The grey and black oxides of manganese are largely used for the manufacture of the chloride of lime—a substance well known for its bleaching properties. They also serve in the fabrication of flint-glass, as a means for correcting the green tinge which it is apt to derive from iron, and are employed in the manufacture of various kinds of steel. The ores of manganese are chiefly provided by the mines of Nassau, which in 1864 yielded 14,460 tons, and of Huelva in Spain, which furnished 24,430 tons in 1865. Our Cornish mines likewise produce considerable quantities, but are still far from being able to supply the wants of our colossal industry, which, in 1866, required the importation of no less than 48,700 tons of oxide of manganese from foreign countries.

Cadmium, which accompanies most of the zinc ores, was discovered by Stromeyer in 1818. Its sulphuret affords a fine yellow pigment; but the metal itself, which has the colour and lustre of tin, and is very fusible and ductile, has no commercial value.

Rutile, a red-brown mineral, occurring in small quantities in the Alps, Norway, and many other localities, where it is generally found in crystals, imbedded in quartz, was found by Klaproth, in 1795, to be the oxide of a peculiar metal which, acording to the old fashion of giving mythological names to new planets and metals, obtained the name of Titanium. The metal, which has a copper-red colour, has not hitherto been applied to use; but rutile is employed as a yellow colour in painting porcelain, and also for giving the requisite tint to artificial teeth.

Like Titanium, the metal Molybdenum, discovered by Hjelm in 1782, is as yet interesting only in a scientific point of view; but one of its salts is used by the cotton-printers as a valuable colouring matter, and another is indispensable as a re-agent in many chemical researches. Thus more than one of the modern metals has already become an important object to the porcelain-painter or the dyer.

Aluminium, the metal which Sir H. Davy discovered in clay or alumina, and of which the purest native oxides are the varieties of corundum (oriental ruby, sapphire, &c.), has of late become of technical importance, and though the cost of its production is very great—as a pound of aluminium is worth about 4l.—yet it already serves for many purposes. Its silvery lustre and perfect unalterability by atmospherical influences render it an excellent material for objects of art and ornament, and from its low specific gravity $(2\tfrac{56}{100})$ it makes excellent tubes for telescopes and opera-glasses, which when composed of any other metal are of a fatiguing weight. Even culinary vases have already been made of aluminium; for, besides its perfect innocuousness, it cools very slowly when heated, and greasy substances do not adhere to it. Its high price is the only obstacle which has hitherto limited its uses. With copper it forms an alloy (aluminium-bronze) discovered by Dr. John Percy, which possesses the hardness, tenacity, and malleability of iron without its liability to rust, and consequently has already found numerous applications. The beautiful gold colour of this alloy makes it a valuable material for the fabrication of the vases and ornaments used in Catholic churches, and a recent decree of the Pope has authorised its employment for this purpose.

Magnesium, the metallic basis of magnesia, a native earth widely disseminated in the mineral kingdom, and forming a constituent part of whole mountain chains, had ever since its discovery by Sir Humphry Davy been a mere object of curiosity, when a few years ago, Mr. Sonstadt, an English chemist, succeeded in producing it in larger quantities. Its silvery brilliancy, hardness, and ductility, its low specific gravity, and unalterability by air and water, are qualities which will probably lead to an extensive employment when a cheaper method of production shall have been discovered; but even now it has found a highly interesting use. It is so easily inflammable that a wire of considerable thickness can be ignited in the flame of a candle, and the light evolved by the combustion is of almost solar intensity. In lighthouses it serves to guide the mariner in his course; it lights up the obscurest recesses of stalactital caverns, and with its assistance the photograhper no longer depends upon the sun, and reveals to us the hidden paintings and sculptures of rock-tombs and temples as distinctly as if they were exposed to the light of day.

Sodium, the metallic basis of soda, was discovered by Sir Humphry Davy in 1807. It is lighter than water, and white and lustrous as silver; but exposure to air almost immediately converts it into soda. Thus it can never become directly useful, like aluminium or magnesium; but being indispensable for reducing the ores of these two metals, it renders important indirect services, and is consequently produced in considerable quantities.

Palladium, one of the hardest and heaviest of metals, is of a steel grey colour, passing into silver white. Its alloy with silver, which has the valuable property of not tarnishing in air, is eminently fitted for the manufacture of delicate scientific instruments. The Wollaston medal, given by the Geological Society, is, in honour of its discoverer, made of palladium, which is considerably dearer than gold.

In 1804 the same eminent philosopher discovered another metal in native platina, to which he gave the name of Rhodium. Mixed with steel in the proportion of 1 to 50, rhodium produces an excellent metal for making the sharpest-cutting instruments, and a mixture of equal parts of rhodium and steel makes the best telescopic mirrors, as it is not liable to be tarnished. It is also employed for making the unalterable nibs of the so-called rhodium pens.

Thallium, though one of the newest metals, as it was discovered by Mr. Crooks as recently as 1861, already bids fair to render some important services. It imparts to optical glasses a considerable density and dispersive power, and

should no other use be found for it this alone would render
it a valuable acquisition.

Such is the brief history of those new metals which have
already found a useful employment in the industrial arts.
It throws a vivid light upon the rapid progress of modern
chemistry, for the very existence of most of them was un-
dreamt of at the beginning of the present century, and their
discovery could be attained only by an amount of analytical
knowledge beyond the scope of any previous age. On wit-
nessing these triumphs of science we may well ask where
they will end, and when the goal will be reached beyond
which it will be impossible for the human intellect to
penetrate?

immense floats of leaves, and now and then some bulky tree
undermined and uprooted by the current. We near the
coast, and now enter the opening of the stream. A scarce
penetrable phalanx of reeds, that attain to the height and
wellnigh to the bulk of forest trees, is ranged on either
hand. The bright and glossy stems seem rodded like Gothic
columns; the pointed leaves stand out green at every joint,
tier above tier, each tier resembling a coronal wreath, or an
ancient crown with the rays turned outwards, and we see
atop what may be either large spikes or catkins. What
strange forms of vegetable life appear in the forest behind!
Can that be a club-moss that raises its slender height for
more than fifty feet from the soil? Or can these tall palm-
like trees be actually ferns, and these spreading branches
mere fronds? And then these gigantic reeds! are they not
mere varieties of the common horsetail of our bogs and
marshes, magnified some sixty or a hundred times? Have
we arrived at some such country as the continent visited by
Gulliver, in which he found thickets of weeds and grass
tall as woods of twenty years' growth, and lost himself amid
a forest of corn fifty feet in height? The lesser vegetation
of our own country, its reeds, mosses, and ferns, seems here
as if viewed through a microscope; the dwarfs have sprung
up into giants, and yet there appears to be no proportional
increase in size among what are unequivocally its trees.
Yonder is a group of what seem to be pines—tall and bulky,
'tis true, but neither taller nor bulkier than the pines of
Norway and America. There is an amazing luxuriance of
growth all around us. Scarce can the current make way
through the thickets of aquatic plants that rise thick from
the muddy bottom; and though the sunshine falls bright on
the upper boughs of the tangled forest beyond, not a ray
penetrates the more than twilight gloom that broods over the
marshy platform below. The rank steam of decaying vegetation forms a thick blue haze, that partially obscures the
underwood. Deadly lakes of carbonic acid gas have accumulated in the hollows; there is a silence all around, uninterrupted save by the sudden splash of some reptile fish that
has risen to the surface in pursuit of its prey, or when a
sudden breeze stirs the hot air, and shakes the fronds of

the giant ferns, or the catkins of the reeds. The wide continent before us is a continent devoid of animal life, save that its pools and rivers abound in fish and mollusca, and that millions and tens of millions of the infusory tribes swarm in the bogs and marshes. Here and there, too, an insect of strange form flutters among the leaves. It · is more than probable that no creature furnished with lungs of the more perfect construction could have breathed the atmosphere of this early period and have lived.'

As coal seams have been discovered as far to the north as Greenland, Melville Island, and Spitzbergen, where now no trees will grow, it has been inferred that, in the primeval ages which witnessed their birth, a tropical climate must have reigned over the whole surface of the earth; but the vegetation of arborescent ferns does not necessarily imply a very warm climate, as such plants are found to flourish in New Zealand, together with many conifers and club-mosses, so that a forest in that temperate country may make a nearer approach to the carboniferous vegetation that any other now existing on the globe. So much is certain, that a very different distribution of sea and land must in those times have mitigated the severity of the Arctic winter; or, perhaps, as Professor Oswald Heer conjectures, our solar system may then have rolled through a space more densely clustered with stars, whose radiant heat gave to our earth the advantage of a mild climate, even at the poles.

The space of time required for the formation of the coal-fields is as immeasurable as the distance that separates us from Sirius. We know by experience how thin the sheet of humus is which the annual leaf-fall of our trees, or the yearly decay of our moor-plants, leaves behind, and how many decenniums must elapse before one single inch of solid mould is formed. But there are many coal strata eight, ten, or even forty or fifty feet thick; and if we consider besides the mighty pressure of the superincumbent rocks which store them in the smallest compass, we cannot possibly doubt that one such stratum must have required thousands of years for its formation. Our wonder increases when we reflect that in many *coal-measures* (the series of beds intimately associated with the seams of coal) no less than a hundred thick and thin seams of coal alternate with layers of sandstone and shale, so that the reckoning would swell to millions were we able to fathom the ages of their successive growth.

It may well be asked how such vast masses of vegetable origin, which required the sun's light for their formation, came thus to be incased in stone thousands of feet beneath the surface of the earth? More than one theory has been advanced to solve this difficult problem, which can hardly be explained in any other manner than by a general, slowly progressing subsidence of the humid lowlands, alternating with periods of rest. Fancy a wide delta land, similar to Egypt or the Netherlands, covered with luxuriant forests, whose spoils, accumulating where they fall, form in the course of centuries a thick stratum of vegetable matter. This land then sinks, suddenly or gradually, under water, many a fathom deep, and remains there perhaps for ten thousand years, till a vast deposit is formed of sandstone and shale, brought down from the highlands by the rivers that come rolling from the interior, the pressure of which, aided by water, converts the stratum of wood into coal. By this deposit the bottom is gradually filled up, and the bay again converted into marsh or meadow, upon which again vegetation flourishes for a thousand years, till the materials of a second bed of coal are collected. A third submergence takes place, rocky strata are again deposited, the water again shoals into land capable of bearing plants, a third period of forests commences, and continues till the mass of vegetable matter destined to form a new bed of coal is accumulated It is unnecessry to pursue the series any further; let it suffice to say that in this manner coal followed upon sand, or sand upon coal, till in the carboniferous basin of Nova Scotia, for instance, a vertical subsidence of three miles was gradually filled up by the waste swept down from the higher lands, or by the accumulation of vegetable matter.

Great as are these changes of level, they do not indicate any more considerable or violent perturbations than those which take place at the present day, either from earthquakes or from slow oscillations of the soil. Large areas in the Pacific and elsewhere are known to be actually subsiding at the rate of three or four feet in the century, and when

measured on the scale of geographical time, the depression
which sunk the first carboniferous forests of South Wales or
Nova Scotia to the depth of ten or twenty thousand feet
probably proceeded at the same slow rate. Adding to these
vast epochs of gradual subsidence the long periods of rest
which intervened between them, it is perhaps no exaggeration to affirm that several millions of years may have been
required for the formation of a coal-field such as that of
Saarbrücken or South Wales. The fossil remains inclosed
in the various layers of the carboniferous beds alone suffice
to prove the immensity of time required for their accumulation, for the species of ferns or lycopods imbedded in the
lower seams of a coal-field are found gradually to disappear

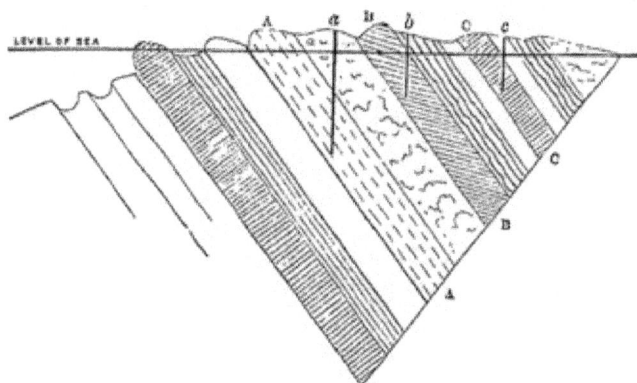

COAL BEDS RENDERED AVAILABLE BY ELEVATION.
a b c, shafts. A B C, coal-beds.

in the higher ones, while new species are continually appearing on rising in the series, until, finally, the plants of
the older seams have completely made way for newer forms.
Thus the coal formation has, during the vast ages of its
growth, changed more than once the aspect of its flora, and
the plants which flourished in its youth had long since disappeared from the earth when it approached its end.

Although all coal-fields must have originally been formed
in horizontal or slightly undulating situations, yet in many
cases they have undergone enormous derangements or dislocations from subsequent terrestrial changes; and to this
circumstance is mainly due there utility to man. Had they
been permitted to remain in their primitive *geological* position,

we probably never should have enjoyed the benefit of the
coal, because it would have been too deep for our reach.
We might have known it to be there, but it would have been
beyond our power to pierce a mile or two into the earth.
But, by a wonderful and merciful providence, the oscillations
to which the earth-rind is subject have frequently upheaved
them enormously out of their original positions; and the
elevated portions having often been denuded by water, large
patches of coal have thus been rendered available to man.

The various subterranean changes which have acted upon
the coal-fields during the course of unnumbered ages have
not only raised or sunk, but frequently dislocated, contorted,
ruptured, or broken them up in a most extraordinary
manner. In the coal-field near Mons, in Belgium, for instance, a vast lateral pressure has curved the strata again
and again, and even folded them four or five times into zig-
zag bendings, so that on sinking a shaft the same continuous
layer of coal is cut through several times.

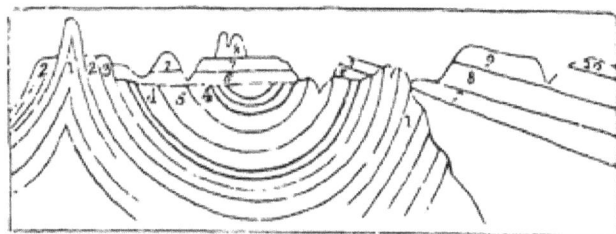

SECTION OF COAL-FIELD SOUTH OF MALMESBURY.

1. Old red sandstone.	5. Coarse sandstone.	9. Great oolite.
2. Mountain limestone.	6. New red sandstone.	10. Corn brash and
3. Millstone grit.	7. Lias.	forest marl.
4. Coal seams.	8. Inferior oolite.	

Frequently a concave form has been the result of these
terrestrial revolutions, and hence coal-fields are often called
coal-basins. Thus, in the coal-field south of Malmesbury, the
strata appear to dip from the surface, and rise again to it
after attaining a certain depth, so that a section of them
suggests the idea of a boat or basin.

Very commonly one portion of a continuous stratum or
series of strata has been broken away from the rest, and has
been displaced, either by elevation or depression, or shifting
on one side, for various distances. The amount of displacement is sometimes only a few inches, and at other times

several hundred fathoms, and the extent may be twenty yards or twenty miles.

We may easily conceive the difficulties which these disruptions frequently throw in the way of the miner, who, in following what he considers a valuable seam of coal, is suddenly stopped by coming in contact with a fault, a trouble, or a slip, as these phenomena are expressively called, and finds the coal shifted several yards above or below, or even completely lost. On the other hand, the miner, thus provokingly stopped in his labours, must not forget that it is perhaps owing to the very shifts he complains of that the outcrop of the coal has occurred at all in his neighbourhood, and that the coal is workable throughout a very large portion of the district in which he is interested.

A most important advantage is also derived from the existence of these numerous faults in coal strata—namely, that they intersect a large field of coal in all directions, and by the clayey contents which fill up the cracks accompanying minor faults, they become natural coffer-dams, which prevent the body of water accumulated in one part of the field from flowing into any opening which might be made in it in another part. A remarkable instance of the advantage arising from the presence of a great line of fault occurred in the year 1825 at Gosforth, near Newcastle, where a shaft was dug on the wet side of what is locally termed the Great Ninety Fathom Dyke, which there intersects the coal-field. The workings were immediately inundated with water, and it was found necessary to abandon them. Another shaft, however, was sunk on the other side of the dyke, only a few yards from the former; and in this they descended nearly 200 fathoms, or 1,200 feet, without any hindrance from the water.

The separation of a coal-field into small areas by dykes or faults is likewise very beneficial in case of fire in a coal-pit, for in this case the combustion is prevented from spreading widely, and destroying, as it otherwise would, the whole of the ignited seam.

'The natural disposition of coal in detached portions,' says the author of an excellent article in the *Edinburgh Review*,[*]

---

[*] Vol. cxi. p. 80.

'is not simply a phenomenon of geology, but it also bears upon national considerations. It is remarkable that this natural disposition is that which renders the fuel most accessible and most easily mined. Were the coal situated at its normal geological depth, that is, supposing the strata to be all horizontal and undisturbed or upheaved (sic), it would be far below human reach. Were it deposited continuously in one even superficial layer, it would have been too readily and therefore too quickly mined, and all the superior qualities would be wrought out, and only the inferior left; but as it now lies, it is broken up by geological disturbances into separate portions, each defined and limited in area, each sufficiently accessible to bring it within man's reach and labour, each manageable by mechanical arrangements, and each capable of gradual excavation without being subject to sudden exhaustion. Selfish plundering is partly prevented by natural barriers, and we are warned against reckless waste by the comparative thinness of coal-seams, as well as by the ever-augmenting difficulty of working them at increased depths. By the separation of seams one from another, and by varied intervals of waste sandstones and shales, such a measured rate of mining is necessitated as precludes us from entirely robbing posterity of the most valuable mineral fuel, while the fuel itself is preserved from those extended fractures and crumblings and falls which would certainly be the consequence of largely mining the best bituminous coal, were it aggregated into one vast mass. In fact, by an evident exercise of forethought and benevolence in the Great Author of all our blessings, our invaluable fuel has been stored up for us in deposits the most compendious, the most accessible, yet the least exhaustible, and has been locally distributed into the most convenient situations. Our coal-fields are, in fact, so many *bituminous banks*, in which there is abundance for an adequate currency, but against any sudden run upon them nature has interposed numerous checks, by locking up whole reserves of the precious fuel in the bank cellar, under the invincible protection of ponderous stone-beds. If we examine the nature of the mineral fuel thus provided for us in the bowels of the earth, we find a number of varieties greatly differing from each other in chemical composition and in combustible value. Thus the *anthracites* or *non-bituminous coals*, which contain from eighty-five to ninety-seven per cent. of pure carbon, are not easily ignited, and yield no flame and but little or no smoke.

The *bituminous coals*, on the contrary, contain a large proportion of volatile matter, amounting to as much as thirty, forty, or fifty per cent., and are consequently very inflammable. burning with a bright flame, considerable smoke, and a penetrating odour.

But as Nature in general does not love those sharp divisions to which theorists are so partial, thus also there is no fixed boundary between these two classes of mineral fuel; and we find an uninterrupted series of intermediate qualities between pure anthracite and the fattest coal.

It may be remarked that if coal were of one uniform chemical composition, its utility would be confined within narrower limits, as the bituminous, semi-bituminous, and anthracital varieties have each their distinguishing properties, which adapt each to special uses. Some kinds, from their richness in volatile bituminous matter, are excellent for the manufacture of illuminating gas, while, from their smaller proportion of carbon, they could hardly be used for the making of iron; and the anthracites, which yield little or no gas, are very serviceable for smelting or domestic purposes.

It appears from the researches of modern chemistry that the different varieties of coal are due to the progress of decomposition which wood and vegetable matter undergo when buried in the earth, exposed to moisture, and partially or entirely excluded from the air. Slowly evolving carbonic acid gas, and thus parting with a portion of their original oxygen, they become gradually converted into lignite or wood-coal, which contains a larger proportion of hydrogen than wood does. A continuance of decomposition changes this lignite into common or bituminous coal, chiefly by the discharge of carburetted hydrogen, or the gas by which we illuminate our streets and houses; and bituminous coal still continuing to evolve its volatile matter, not only after its being covered with strata thousands of feet in thickness, but even to the present day (as the fire-damp sufficiently proves), is thus ultimately transformed into anthracite, to which the various names of splint-coal, glance-coal, hard-coal, and culm, have been given.

When we consider the manner in which coal has been formed in swampy lowlands, or more particularly in river-deltas, which gradually subsided to a considerable depth beneath the level of the sea, we cannot wonder that, when compared with the whole extent of the globe, the area of the coal-fields is extremely limited, and confined to but a few favoured countries. In our times delta lands occupy but a small part of the continents and large islands, and there is no reason to suppose that they were more considerable during the carboniferous age, or at any other epoch. Besides, many of the ancient deltas, probably, never subsided at all, so that no coal could be formed on their site; and others, where coal strata were gradually piled up, may still be whelmed beneath the sea awaiting some future upheaval to become serviceable to future generations of man.

After the preliminary remarks on coal and the coal formation in general, I will now briefly describe the chief coal-producing countries of the globe. First on the list stands Great Britain, whose pre-eminence in industry and commerce is entirely founded on her vast deposits of coal. It is this invaluable mineral which sets those countless steam-engines in motion that perform the labours of a hundred millions of men; which spins and weaves the cotton of America, the silk of China, the wool of Australia, and the flax of Belgium into that amazing variety of tissues that serve to clothe almost all the nations of the globe; and which finally produces a greater quantity of the cheapest iron than the combined efforts of all the world. Thus our coals may well be called our black diamonds, and the comparison is indeed paying the latter too high a compliment, for larger masses of diamonds would be utterly worthless, while, by means of our coal, we are able to enjoy the produce of every zone.

Not only do our fifty-one coal-fields surpass in magnitude those which are disseminated over a far greater territory in Germany, Belgium, and France; but their local distribution and geological formation are as favourable as could possibly be wished. Furthest north we see the considerable deposits of Scotland extending from the coast of Fife to the valley of the Clyde. It is to them that Glasgow owes its half-a-million of inhabitants, and a wealth far surpassing that of all Scotland under the reign of 'bonnie' Queen Mary. In England, north of the Trent, along the Wear and Tyne, and even extending far beneath the sea, we have the coal-fields of Northumberland and Durham, with Cumberland and those of Yorkshire, Nottinghamshire, and

COAL BASIN OF CLACKMANNANSHIRE.
a, b. Coal seams. *c.* Limestone strata. *x, y.* Slips.

Derbyshire. After these comes the large field of Lancashire, or, as it is sometimes named, the Manchester Coal-field. Looking to the central districts, we see the coals of North and South Staffordshire and of Leicestershire. In the north-west we have the field of North Wales; in the more central west, the deposits of the Plain of Shrewsbury, Coalbrook Dale, and the Clee Hills; and in the south-west, the great coal-field of South Wales, and the minor ones of the Forest of Dean, of Somersetshire, and of Gloucestershire.

The inspection of a good geological map shows us at once how advantageously for commerce these several coal-stores are distributed. Every large coal-field in England and Scotland is hardly ever distant more than thirty miles from the next, so that from the Clyde to Somersetshire the whole interior of the country can easily, by means of canals and railroads, be provided with fuel. The east and west coasts of the land are nowhere above fifty miles from a coal-field; and even the most remote localities in the three kingdoms are able to provide themselves from distances within 150 miles.

But it is chiefly the neighbourhood of the sea which gives such an incomparable value to our most important coalfields; as, thanks to this advantageous situation, which none of the French, Belgian, and German coal-fields possess, Great Britain is enabled to provide not only her own coasttowus, but almost all the sea-ports in the world, with a cheaper fuel than can be produced in their own country. Even in Ostend, Belgian coal, rendered dear by canal transport, is unable to compete with that which is brought over sea from England; and Hamburg provides herself with fuel from Newcastle and Hartlepool, and not from the coal-fields of Saxony and Westphalia.

Coal is found in seventeen counties in Ireland, over an area of about 3,000 square miles. Yet, notwithstanding this great abundance of coal which the country possesses, and which is distributed throughout almost all parts of the island, from Limerick to Antrim, her capital and chief cities and ports have hitherto depended upon Great Britain for their supplies of mineral fuel, both for domestic and for manufacturing purposes. To those who are unacquainted with the actual circumstances, it appears scarcely credible that this fine country has made so little use of the coal which has been so bountifully bestowed upon her. Among other causes not political, which perpetuate this state of things, is the extraordinary facility and cheapness with which the ports of Ireland can be supplied from the great western coal-fields of Great Britain. The excellence, abundance, and cheapness of peat—which is not only the common fuel of the poor in the interior, and, indeed, of all classes in some districts, but is also brought in barges by the great canal, and consumed to a considerable amount in the capital itself--is another reason why the Irish coal-mines have, as yet, been so little worked.

When we consider the vast importance of coal, we cannot wonder at the paramount influence which it has exercised over the distribution of our population in modern times. While Salisbury, Winchester, and Canterbury—important towns of mediæval England—are reduced to atrophy from the distance and absence of coal-fields, Newcastle, Leeds, Manchester, Sheffield, Birmingham, Glasgow, and a host of other flourishing towns, may truly be said to be built on coal.

Where there are large coal-fields there is life and a prospect of almost unlimited prosperity, for they are sure to attract machinery and man. Take a geological map of a new and thinly-populated country; and if it be marked with coal-fields the spots where large cities will exist hereafter may be safely determined.

A more detailed examination of the chief coal-fields of England shows us the immensity of the mineral riches which are here still hoarded up for the benefit of future generations.

The superficial extent of the South Welsh coal-fields is about a thousand square miles. On its northern wing we find on an average twenty-one coal bands, forming an aggregate thickness of eighty and a half feet. In some parts of the south wing there are even as many as thirty-three bands of an aggregate thickness of one hundred and four feet of pure coal, so that the average depth of the field may be estimated at ninety-two feet.

Numerous transversal valleys intersect this magnificent coal-field, and afford the easiest means of working it in every depth. They also facilitate the transport of the coal on the canals and railroads which lead along their bottom or along their slopes to the harbours of the Bristol Channel. In the chapters on iron and copper, I have had occasion to describe the gigantic industries founded on these natural advantages, which have raised South Wales from being one of the most insignificant into one of the most important provinces of the empire.

All the coals of this basin are so rich in carbon as to yield from seventy to ninety per cent. of coke. They are either anthracites or only semi-bituminous, and are consequently not suitable for making gas, nor are they much liked for burning in open fire-places, as they do not emit a genial flame. On the other hand, the semi-bituminous Welsh coal is invaluable for burning in steam-engines, for it has been proved to generate one-quarter more steam than any other kind of English coal. One of its properties—by no means unimportant—is its non-liability to spontaneous combustion, which, it is well known, occasionally takes place with bituminous coals, and by which vessels have been lost at sea, and warehouses ignited on land. The value of the Welsh steam or slightly bituminous coal is enhanced by its quality of burning almost without smoke—a property hitherto slightly appreciated, but the importance of which in time of war is evident.

Steamers burning the fat bituminous coal can be tracked at sea at least seventy miles, before their hulls become visible, by the dense columns of black smoke pouring out of their pipes or chimneys, and trailing along the horizon. It is a complete tell-tale of their whereabouts, which is not the case with vessels burning Welsh or anthracital coal. From its compactness and density, the latter has likewise great advantages in point of stowage-space over the ordinary weak bituminous coals. For long voyages this concentration of power and economy of space may easily be appreciated. From its great superiority over other kinds, Welsh coal has been exclusively used for some time past by the French Government, and it is also employed in England by all the chief mail-packet companies.

Coal is not exported from the South Wales basin as much as from some other fields, owing to the enormous requirements of the large iron works, most of which consume as much as can be supplied by their collieries. Every week, however, sees increased supplies of Welsh coal thrown into the London market, and every year fresh collieries are opened to meet the demand. The total number of pits in South Wales in 1865 was four hundred and eighteen, which produced between seven and eight million tons annually. The New Navigation Pit at Mountain Ash may be cited as an example of the grand scale on which t e most important of these workings are conducted. The sh ft is eighteen feet in diameter inside the walling, and divided into four compartments, two of which are for the drawing of coal, one for sending the workmen up and down, and the fourth for the drainage. Notwithstanding the great depth of three hundred and seventy yards, a carriage containing two half-tons of coal can be wound up in one minute, and the whole colliery is estimated to supply more than one thousand tons a day. The mineral property extends over an area of seven miles long by three miles in width, covering from four thousand to five thousand acres of four-foot coal. From this one case the reader may form a slight estimate of the boundless resources of the whole basin.

The Great Central Coalfield, which ranges through South Yorkshire, Nottinghamshire, and Derbyshire, has a superficial extent of about one thousand square miles. In character it is closely allied to that of Newcastle, and is considered by some geologists to be a re-emergence of the same strata from beneath the covering of the magnesian limestone under which they suppose it to be concealed through the intervening space. It extends a little from the north-east of Leeds nearly to Derby, a distance of more than sixty-five miles. Its greatest width—twenty-three miles—is on the north, reaching nearly to Halifax on the west. On the south it extends towards the east to Nottingham, and is there about twelve miles wide. Though possessing some fine coal seams, it is of far inferior importance to the Manchester or South Lancashire Coal-field, from which it is separated by a lofty chain of hills. This highly valuable basin, which extends over about five hundred and fifty square miles, begins in the north-west of Derbyshire, and continues thence to the south-west part of Lancashire, forming an area somewhat in the shape of a crescent, having Manchester nearly in the centre.

Of the smaller coal-fields, the most important are the Whitehaven basin (one hundred and twenty square miles) and the Dudley or South Staffordshire area (ninety square

DUDLEY COAL-FIELD.
1. Limestone strata. 2. Coal.

miles) which is particularly valuable for the extensive iron works which it maintains. One portion of this coal-field is distinguished by the presence of one continuous bed of coal thirty feet thick, which for British coal is astonishing; and this mass extends seven miles in length and four in breadth. In this favoured district coal-seams five or six feet thick are called thin seams; in Newcastle they would be called thick seams.

Though not so extensive as the South Welsh area, the coal-fields of Newcastle and Durham are of far more ancient celebrity. Their produce is chiefly shipped on the Tyne, Wear, and Tees, small rivers hardly traceable on a map of the world, and yet far more important to commerce than the

mighty Orinoco or the thousand-armed Amazon. This magnificent coal area is bounded on the north by the river
Coquet, and extends southward nearly as far as Hartlepool,
a distance of about forty-eight miles. Its extreme breadth
is about twenty-four miles, and the whole superficial extent
may amount to about eight hundred square miles.

There are in all about fifty-seven different seams of coal in
the Great Northern Coal-field, varying in thickness from an
inch to five feet five inches and six feet, and comprising an
aggregate of about seventy-six feet of coal.

The pits are established chiefly with reference to one or
more of the three following seams. The most valuable is
called the *High Main Seam*, and is about six feet thick. The
next in value is the *Bensham Seam*, about three feet thick,
which is remarkable for its excellent quality as a domestic
coal, and for the enormous quantity of gas evolved from it in
the mine. The *Hutton or Low Main Seam*, averaging from
three feet six inches to five feet nine inches, is likewise of
very good quality, and is extensively worked. It must, however, be remarked that the same seam of coal is not generally
valuable in all places. Thus the High Main, which furnishes
excellent coal on the Tyne, becomes injured as it proceeds in
a south-easterly direction by being intermixed with a band
of coal of inferior quality, containing iron pyrites, &c.

The Newcastle Coal-field is generally worked at a great
depth, an expense of upwards of 50,000*l.* having in some
instances been incurred before the seam of coal was reached
which was to reward all this vast outlay and labour. The
most remarkable and enterprising work of this kind on
record is a sinking at Monk Wearmouth Colliery near
Sunderland. After piercing the superincumbent beds of
magnesian limestone and lower new red sandstone, the coal
strata were reached at a depth of three hundred and thirty
feet; but at the same time a spring was tapped which poured
water into the workings at the rate of three thousand gallons
per minute. This fearful influx was kept under by a steam-
engine of two hundred horse-power, and the work was made
sure by strong metal tubbing, and carried on successfully,
though not without extreme difficulty. On entering the
coal measures, however, a new and unexpected check was

experienced in the extra thickness of the uppermost coal
strata, for which no calculation had been made, and the
difficulties were increased when at the depth of one thousand
feet a fresh *feeder* or spring of water was tapped. Additional
expense and great loss of time were thus caused; but the proprietors persevered with real Anglo-Saxon pertinacity, un-
daunted by the apparently hopeless nature of the undertaking,
and by the fearful expenses incurred. Success crowned their
efforts, and finally, at a depth of 1,710 feet, the Hutton seam
of coal was reached, at a cost which, including the necessary
preliminary operations, could not have been less than 100,000*l.*

Another remarkable and costly piece of mining was that
of Gosforth Colliery, which lies about three miles north from
Newcastle, on the west bank of a romantic 'dean' or little
valley, through which the Ouseburn winds its way to the
Tyne. The sinking was commenced in 1825, and the coal
was won on Saturday, January 31, 1829. The High Main
coal was reached at twenty-five fathoms below the surface,
but near its first appearance the seam was thrown down in
an inclined direction by the great Ninety Fathom Dyke which
there intersects the coal-field. Hence it became necessary
to sink the shaft to the depth of 181 fathoms, in order to
come at the level of the lower range of coal. This having
been accomplished, a horizontal drift 700 yards long was
worked through the face of the dyke to the seam of coal a
little above its junction with the dyke. A great part of this
excavation had to be made through solid rock. To celebrate
the completion of this remarkable work, a grand subterranean ball was given at the very place of triumph, 1,100
feet below the surface of the earth. As the guests arrived
at the bottom of the shaft they went to the end of the drift
to the face of the coal, where each person hewed a piece of
coal as a memento of the day, and then returned to the ball-
room, which was brilliantly lit up, and where born-and-bred
ladies joined in a general dance with born-and-bred pitmen's
daughters. Between 200 and 300 persons were present, nearly
one-half of them belonging to the fair sex. It must be remembered that the pit had not been worked, so that no smoke and
dust exuded from its mouth or defiled the ball-room.

Some of the older coal-pits, where excavations have been

going on perhaps for a century or more, may be likened to large subterranean cities. The galleries of St. Hilda Colliery, near South Shields, are full seventy miles long; and Killingworth pits are said, on good authority, to have nearly one hundred and sixty miles of gallery excavation.

In some parts the operations of the collier have encroached upon the domains of the ocean. At the Howgill pits, for instance, west of Whitehaven, the excavations have been carried more than 1,000 yards under the sea, and about 600 feet below its bottom. But the most remarkable marine colliery which Great Britain has ever possessed was situated at Borrowstounness. The coal was found to continue under the bed of the sea in this place, and the colliers had the courage to work it half a mile from the shore, where there was an entry that went down into the submarine coal-pit. This was made into a kind of round quay, built so as to keep out the tide which flowed there twelve feet. Here the coals were laid, and a ship of that draught of water could lay her side to the quay and take in the coal. This wonderful pit, which belonged to the Earl of Kincardine's family, continued to be wrought many years to the great profit of the owners, and the astonishment of all that saw it; but at last an unexpected high tide drowned the whole at once, together with the labourers who were at work. 'While,' says the French geologist, Faujas de Saint Fond (who visited this remarkable mine about the end of the last century), 'the pitmen, by the dismal shine of their lamps, make the deep caverns resound with the blows of their pickaxes, ships driven by a fair wind sail over their heads, and the sailors, rejoicing at the beautiful weather, express their joy in songs: at another time a storm arises; the horizon is in flames, the thunder roars, the sea rages, the boldest tremble; then the pitmen, unconscious of the terrible scene, calmly pursue their labours, and think with pleasure of their homes, while the ship above is shattered to pieces and sinks; unfortunately, but too faithful a picture of the daily changes of human life.'

Nowhere in the world, perhaps, does human activity display a more restless energy than on the site of the Newcastle Coalfield, where, night and day, successive trains heaped with the black diamonds of England speed along far-stretching rail-

ways, and hurry down to river and ocean, until they are
unloaded and their contents shipped by machinery. Steam-
engines are unceasingly at work drawing coals and pumping
out water. Thousands of men are underneath our feet cutting down the coal by severe and peculiar labour, while
thousands above are receiving the loads and speeding them
forwards.

'Go where you will, there is a network of small railways,
leading from pit to pit in hopeless intricacy, but all having
a common terminus on the river's bank or the ocean's shore.
Go where you will, tall chimneys rise up before you, and here
and there a low line of black sheds, flanked by chimneys of
aspiring altitude, indicates that you are arriving at a colliery.
As you draw nearer, men and boys of blackest hue pass you
and peer at you with inquiring glances. Now trains of coal-
waggons rush by more frequently, noises of the most discordant character increase, and you know that you are at
the pit's mouth, when you behold two gigantic wooden arms
slanting upwards, upon which are mounted the pulleys and
wheels that carry the huge flat wire ropes of the shaft.
For a moment the wheels do not revolve - no load is ascending or descending—but the next minute they turn rapidly,
and up comes the load of coals or human beings to the surface. Perhaps the most impressive sight is a large colliery
fully engaged at nightwork, with burning crates of coal
suspended all around; and after this a view from some neighbouring eminence of all the far-flaming waste coal-heaps,
burning up the accumulation of waste and small coal
not worth carriage, ever added to the ever-consuming
mound, until the whole district appears like the active crater
of some enormous volcano.' *

The banks of the Tyne are in many places very high and
precipitous, and consequently render peculiar contrivances
necessary for the shipping of the coals. The means used
for this purpose are various: sometimes inclined tunnels,
through which the train of waggons is lowered in chains to
the water; sometimes slopes, along which the coals are
shovelled into the ship, or still better, the ingenious

* 'Edinburgh Review,' vol. cxi. p. 86.

mechanism of which William Howitt gives us the following description :—

'As you advance over the plain you see a whole train of waggons loaded with coal, careering by themselves without horse, without steam-engine, without man, except that there sits one behind, who, instead of endeavouring to propel these mad waggons on their way, seems labouring hopelessly by his weight to detain them.

'But what is your amazement when you come in sight of the river Tyne, and see these waggons still careering on to the very brink of the water!—to see a railway carried from the high bank, and supported on tall piles, horizontally,

SHIPPING COAL.

above the surface of the river, and to some distance into it, as if to allow those vagabond trains of waggons to run right off, and dash themselves down into the river!

'There they go, all mad together! Another moment, and they will shoot over the end of the lofty railway, and go headlong into the Tyne, helter-skelter. But behold! these creatures are not so mad as you imagine them. They are instinct with sense : they have a principle of self-preservation, as well as of speed, in them. See, as they draw near the river they pause, they stop! one by one they detach themselves; and as one devoted waggon runs on, like a victim given up for the salvation of the rest, to perform a wild summersault into the water below, what do we see?

It is caught! A pair of gigantic arms separate themselves from the end of the railway. They catch the waggon, they hold it suspended in the air, they let it softly and gently descend—and wither? Into the water? No; we see now that a ship already lies below the end of the railway. The waggon descends to it; a man standing there strikes a bolt, the bottom falls, and the coals which it contains are nicely deposited in the hold of the vessel. Up, again, soars the empty waggon in that pair of gigantic arms. It reaches the railway; it glides like a black swan into its native lake upon it; and away it goes, as of its own accord, to a distance, to await its brethren, who successively perform the same exploit, and then joining it, all scamper back again as hard as they can over the plain to the distant pit.'

The produce of the collieries situated further up the Tyne, where the river is no longer navigable by sea-going craft, is conveyed in a kind of oval vessels, called keels, to the port of Newcastle, or its out-stations, North and South Shields, where it is discharged into larger ships.

Newcastle may well be called the capital of King Coal. Once a town of military importance—as the old, grim-looking donjon-keep of Robert Curthose, the son of the Conqueror, still testifies—it entirely owes its modern importance to the treasures of coal adjacent to its walls. Its quays, black and sooty as the mineral on which its prosperity is founded, are lined with a dense row of counting-houses, and before them in the river still denser rows of colliers lie at anchor; while between both ebbs and flows a black-looking crowd—for all here wear the livery of the article to which all owe their bread. Some idea may be formed of the vast activity waving to and fro in this chief artery of the coal trade, from the fact that the annual arrivals in the Tyne are not less than 13,000 or 14,000, 10,000 of which are on account of the coal trade.

Sunderland, the great port of the river Wear, where annually more than 10,000 cargoes of coal are shipped to all parts of the world; Hartelepool, a town of modern date with magnificent docks; Stockton-on-Tees, and a number of minor places of shipment on the cost, likewise owe their prosperity te coal; so that probably no other article of trade

gives constant employment to so many vessels within so
confined a territory.

From Tynemouth Priory, a ruin romantically situated on
a bold promontory, the visitor frequently enjoys a magnificent marine picture; for when, after long-continued easterly
gales, the wind changes to a westerly breeze, many hundred
vessels—mostly colliers—put to sea together in a single tide,
and distribute themselves over the ocean with their prows
turned in almost every direction, some southward and coast-
wise, for English ports, for the Channel, and for the southern
countries of Europe; others, northward for Scotland and the
Norwegian costs; and others, again, due east, for Denmark
and the Baltic—all sinking deep in the water, weighed down
by that mineral fuel which is more valuable for England
than if it were replaced by the mines of Mexico or the diggings of Australia.

Yet a few years, and probably the dingy and crawling
craft which perform the chief part in this animated scene
will be abolished. Clipper screw-steamers are rapidly taking
their place, and the railroads daily transport a greater proportion of the seven or eight million tons of coal which
are annually devoured by our huge metropolis.

Before quitting the Northern coal districts, a few words
may be added on the swarthy population whose labours
bring their subterranean riches to the light. The chief
underground workmen are the hewers, who either remove
the coal with pickaxes, or sometimes blast it with powder.
To hew well is a work of skill as well as of strength, and
men must be early practised in it to earn high wages by
piece-work. In tolerably thick seams of coal, of five and six
feet and upwards, hewing is more a work of strength than
skill; but in the narrower seams skill predominates. In
these the arm is confined, the blow is shortened, the pick
is impeded. To gain space by adaptation of position, you
may see one hewer kneeling down on one or both knees,
another squatting, another stooping or bending double, and
occasionally one or more lying on their sides or on their
backs, picking and pegging away at the seam above them.
If the seam be hard as well as thin, and the man's position
confined, it is manifest that he cannot get his strength to

bear in full, or his full measure of coals. In such cases he
is bathed in perspiration, in a state of semi-nudity, enveloped
in floating and clinging coal-dust. If to this we add the
very faint light imparted by the Davy lamps, the constantly
thickening atmosphere, the exhalations from living beings,
exaggerated by heat, and not diminished by any free current
of air, and remember that eight hours is the usual day's
work of the hewer, we must surely confess that few men
have their strength more hardly tasked, or earn their bread
in a more laborious manner.

COAL HEWERS AT WORK.

To relieve this arduous toil, coal-cutting machines have
lately been devised, which are worked either by steam or by
compressed air, and will probably in time perform a great
part of the hewer's labour, as those already in employment
appear to be well adapted to the purpose for which they are
contrived, and further improvements in their construction
will no doubt be introduced. Coal-cutting machines, which
act either by picking or gouging, have been found to work
more economically than manual labour, while at the same
time much less coal is destroyed and reduced to slack. A
matter of still more importance is the diminished risk to
the persons and lives of the employed, who, when working in

a constrained position, and consequently unable to relieve themselves from the fall of a superincumbent mass of coal, are frequently crushed to death. The application of machinery to cutting coal gives another advantage of national importance, as, by enabling the working to be carried into the deeper seams of coal which lie at so high a temperature as to present serious or insurmountable difficulty to hand-work, it will render available to posterity new and hitherto inaccessible stores of coal.

The hewers may possibly fear to be thrown out of employment by its introduction on an extensive scale; but as it will relieve them from their most irksome drudgery, and allow them to reserve their strength for less injurious trial, they cannot be but thankful for the aid which it affords them.

They are usually paid according to the number of baskets or tubs they are able to fill. These are then conveyed by the putters through the smaller or lower galleries of the pit to the headways, where they are hoisted by the crow-men upon the rolleys or waggons for transporting the coals from the crow to the shaft. The roads along which the rolleys are driven are made sufficiently high for an ordinary horse by cutting away the roof or the floor. Some of them are two miles long, and are kept in repair by a rolley-wayman. Where tubs are used for the conveyance of coal the whole way, no crow is necessary, but a lad, termed a 'flatman,' who links the tubs together at the level or the flat.

Next to the hewers, the putters are the hardest labourers in the pit; and in some places their labour is even harder, for it is no easy matter to push corfs or tubs, weighing from six to ten hundredweight, along galleries which are often but three or four feet high, where the heat not seldom averages about 78° Fahr., and, in consequence of the increased pressure of the air, water boils at 220°. The term 'putter' includes the specific distinctions of the 'headsman,' 'half-marrow,' and the 'foal.' Where full tubs or baskets are to be pushed along the rails from the hewers to the crow and the rolley-drivers, the headsmen take the chief part; a half-marrow goes at each end of the train alternately with another half-marrow; while a foal always precedes the train. At the bottom of the shaft the 'onsetters' are stationed, who attach the tubs to the ropes which hoist them to the surface. Besides these various classes of workmen, we find the 'shifters,' who keep the galleries in repair, and the little 'trapper-boys,' whose duty it is to open the ventilating-doors whenever they hear the drivers or trains of coal-waggons coming on one side or the other. Their task—though humble, tedious, and requiring the least amount of intelligence—is of great importance, as the numerous doors which they guard must remain open only long enough for the passage of the trains, and must then be closed again immediately, or the current of air needed to ventilate the mine would be diverted in its course. It is hardly possible to imagine a more joyless childhood than that of these little fellows, condemned to sit in solitary gloom during the greater part of the day, and only comforted by the sudden shout or song of a team-driver, approaching with his train of waggons, and demanding the opening of the door.

Besides the workers underground, a number of labourers or artisans are constantly employed above pit from the 'banksmen,' whose duty it is to see all things living and lifeless up and down the shaft, to the 'staithmen,' who attend to the staith or shipping-place of coals. Many find constant occupation in the raff yard, where old waggons, ironwork, and woodwork, are duly hospitalled and refitted for fresh duty.

The daily work of the mine is conducted according to the strictest discipline. The 'resident viewer' is supreme, and has subordinate viewers, overseers, and wastemen, lamp-keepers, and other officers, who have each their departments, and discharge their duties assiduously.

Thus a first-rate northern colliery establishment—where a total of more than five hundred persons are variously employed —resembles a little community in itself. Men of all educations, arts, grades, and duties, and males of almost all ages from ten years, are here ; men, too, of all appearances—from the gentlemanly viewer to the doubtful wasteman, and from the underground workers-in-chief—the hewers—to the humble trapper-boys.

The peculiar nature of his underground occupations, which condemns the pitman, while working, to a position of great restraint, and taxes the limbs and muscles in a very unequal manner, naturally influences his outward appearance, so that he can be easily distinguished from every other operative.

His stature is diminutive, his figure disproportioned and misshapen; his legs being much bowed, and his chest protruding like that of a pigeon. His arms are long and oddly suspended. His countenance is not less strange than his figure, his cheeks being generally hollow, his brow overhanging, his cheekbones high, his forehead low and retreating, his complexion pallid. Many of these bodily peculiarities or malformations are probably hereditary. Pitmen have always lived in communities; they have associated only among themselves, and have thus acquired peculiar habits and ideas. They almost invariably intermarry, and it is not uncommon in their marriages to commingle the blood of the same family. They have thus transmitted natural and accidental defects through a long series of generations, and may now be regarded in the light of a distinct race of beings. In spite of the general march of intelligence, their education is still very imperfect, and they are just emerging from the greatest possible moral and intellectual darkness—an improvement due mainly to the Wesleyan Methodists. The untiring labours of this religious sect not only imparted to the colliery population in the North of England a higher tone of moral feeling, but in their efforts to instil religious principles into their minds, afforded them, through Sunday-schools, a slight amount of education and an imperfect capability of reading. These first seeds of improvement will, it is hoped, gradually ripen into fruit, and oppose a strong barrier to the prominent vices of colliers—gambling and intemperance.

A lack of mental and personal openness and boldness, a great inclination to injury and theft, the grossest superstition, and a want of the commonest economy and forethought, are likewise faults which are said to be very common among them. Deception is so much a practice with them, that they deceive when no earthly advantage can be obtained from their dishonesty.

On the other hand, the proofs of filial affection which

they exhibit, and the noble feelings and heroism which they display when explosions or accidents take place, prove that the groundwork of their character is good, and merely requires the influence of a better education to remove a great part of the blemishes which ignorance has engrafted upon an originally wholesome stock. Under every disadvantage, several eminent men have sprung from their class. Thomas Bewick, the celebrated wood-engraver, was early immured in pits; the late celebrated mathematician, Dr. Hutton, was originally a hewer of coal; Professor Hann, of King's College, in London, was a boy working underground in a northern colliery; and George Stephenson, the illustrious engineer—whose wonderful inventions have revolutionised the world, and who, after the lapse of many ages, will still be reckoned among England's most illustrious sons—began life as a trapper.

Though the use of coal was already known to the ancient Britons, yet the first public notice of the mineral is mentioned by Hume to have been in the time of Henry III., who, in the year 1272, granted a license to dig coals to the town of Newcastle. Somewhat later, in 1291, the abbot and monastery of Dunfermline in Scotland obtained a similar grant. The first coal is said to have been brought to London about the year 1305, where it was used only by smiths, dyers, and soap-boilers. The smoke, which was supposed to be injurious to health, caused great annoyance to the wealthier inhabitants of the city, so that in 1316 its use was prohibited by a decree of Edward I. This ordinance seems, however, to have been but little attended to; for a few years later inspectors were named to levy fines in case of its non-observance; and if these proved ineffectual, to demolish the fireplaces arranged for the burning of coal. The complaints against this fuel continued several centuries, for as late as 1661 King Charles II. was prayed to remedy the nuisance by banishing from town manufacturers who require large quantities of coal.

Yet, in spite of all the interdictions, complaints, and prejudices arrayed against it, coal continued to grow in use; for as early as 1615 Newcastle gave employment to about four hundred vessels, one-half of which number supply the demands of London. French ships even then fetched coals in that port, and the Hanse towns conveyed them to Flanders.

About the middle of the seventeenth century the coal trade, notwithstanding an increase of price, required nine hundred vessels; and fifty years later half a million of tons were exported from Newcastle, requiring fourteen hundred vessels for their carriage. During the eighteenth century the northern coal trade constantly increased with the steady growth of London, which in 1770, although not possessing one-sixth of its present population, already consumed seven hundred thousand tons; and it would have been impossible for the colliers to satisfy the constantly growing demand if the invention of the steam-engine had not lent its powerful aid to raise larger quantities of coal from a greater depth, and to drain many works which otherwise would have been deluged with water.

The other English coal-fields began to be worked at a much later period than that of Newcastle, but rapidly grew in importance with the vast increase of our manufactories.

The extraction of coal is indeed constantly increasing at a truly gigantic rate. Thus, in 1845 our whole annual production was rated at thirty-five millions of tons; in 1859 it amounted, according to trustworthy returns, to sixty-eight millions; in 1865 it had advanced to ninety-six million tons; and in 1873 the 3,527 collieries of the United Kingdom produced no less than 127,016,747 tons of coal; enough to encompass the earth with a girdle of coal three feet wide and about seven high.

The question of the duration of our coal-fields is evidently one of great national interest. It has of late excited the attention both of statesmen and philosophers, but unfortunately it is more easily put than answered. While some authorities give us the cheering assurance that we have enough to last us for the next two thousand years at least, others limit our supply to three or four centuries, or assign even a couple of hundred years as the period when our decendants will have to seek their coals in the mines of other countries. The quantity of fuel left in the Newcastle basin—the most anciently worked of our coal-fields—was estimated by Mr. Hall, in 1854, at 5,121,888,956 tons. Dividing this total by 20,000,000 of tons as the present annual consumption, the future supplies of this famous coal-field would thus be limited to about two hundred and fifty years—a very short period in the history of a nation. The immense consumption of coal in the iron furnaces and foundries of Staffordshire will probably lead to an exhaustion of that coal-field even before Northumberland and Durham, for its area is scarcely more than one-half of the area of the Northern Coal-field. It has, indeed, one very thick seam of coal of from thirty to forty feet, but this will not alone compensate the difference. The coal-fields of Yorkshire, Lancashire, and Derbyshire, situated amongst the numerous iron-works and manufactories, as well as large populations, justify a similar prophecy; but, on the other hand, better prospects are held out by the great coal-field of South Wales.

After deducting the coal practically unattainable from its depth, sixty thousand millions of tons may be considered a liberal estimate of the available mass. At the present rate of extraction (ten millions of tons) this would give a supply for the next six thousand years; but supposing the other sources to fail, the extraction of coal from the South Wales basin would of course be increased to such an extent as to limit its duration to six or seven centuries. It may be remarked that the largest estimates of future coal supply are based on the assumption that mines may be worked at a depth of four or five thousand feet; but this is very problematical. Mechanical skill may indeed pierce shafts to this depth, or even deeper; but the increase of temperature which is raised by one degree for about every successive seventy feet, along with the increasing density of the air, must ever oppose insuperable obstacles to human labour at such a distance below the level of the sea. To the natural heat and density arising from depth must be added the corruptions arising from human perspiration, which are constantly on the increase during working hours in working places. 'We speak,' says the author of an excellent article in the *Quarterly Review*,[*] 'from some brief personal experience of what these things are at a depth of nearly eighteen hundred feet, where the actual temperature varied from eighty-five to eighty-six and a half degrees. Such experience is necessary to qualify any man to judge of the vertical limit of human labour, and we hesitate to fix it at more than two thousand five hundred feet, and should fix it at that depth only for the hardiest of hewers and haulers of coal.'

The pressure of superincumbent strata, which renders the upholding of the roof, even at fourteen or fifteen hundred feet, a problem of ceaseless anxiety and expense, must also be taken into account. At depths much exceeding two thousand feet, it is very doubtful if the roofs could be securely upheld except at such an outlay as would considerably raise the cost of extraction, while the coal itself would be more and more dense, and therefore more and more difficult to dislodge. For these various reasons, all the strata of coal situated below the depth of two thousand five hundred feet, or at the very utmost three thousand feet, may be considered as practically unworkable; and thus sober-minded calculators, on comparing the available solid contents of our coal-fields with the rate of extraction, have come to the conclusion that a thousand years is the maximum of the probable future supply of England and Wales. Adding to this the Scotch and Irish coal, which are not included in the estimate, and swelling our account with lignite and peat, we have at any rate sufficient materials for keeping our fires burning for a good time to come, and may safely leave all desponding views on the subject to distant generations.

Next to England, no European country has so rapidly increased its coal production as the German empire; where, thanks to the railroads, the consumption of mineral fuel is yearly extending over a wider range, and gradually supplanting in many localities the use of wood. Official accounts inform us that in 1873, 727,861,000 cwt. of black coal, and 200,000,000 cwt. of lignite—together, about 45 million tons—were produced; a mass considerably greater than the joint production of France and Belgium, and equal to about one-third of the production of Great Britain. The chief coal-fields are those of Upper Silesia, of the Ruhr, of the Saar, of Waldenburg (in Lower Silesia), of Dresden and Zwickau (in Saxony), of Aachen (Aix-la-chapelle), Ibbenbüren, and Minden, which not only supply the greater part of Germany, but also yield a considerable exportation to France, Switzerland, and Holland.

The German ports on the North Sea and the Baltic still largely consume British coal, which, however, has been entirely driven from the Rhine; and Berlin, which in 1860 burnt 202,970 tons of English coal, consumed little more than one-half that quantity (123,401 tons) in 1865, in spite of a considerable increase of population; while at the same time the consumption of coal from Upper Silesia increased from 61,700 to 823,712 tons.

The small but thriving kingdom of Belgium, where the collieries of Liège, Namur, and Hainaut give rise to a commercial activity unequalled on the Continent, occupies the third rank among the coal countries of Europe, its production in 1863 having amounted to 10,500,000 tons. The provinces of Namur and Liège consume almost all the coal they produce, while Mons and Charleroi, in Hainaut, export more than three millions of tons to France.

This country, which, in 1868, produced 12,800,000 tons of coal, requires at least 25,000,000 for its consumption, and imports the difference from Belgium, England, and Germany. The chief coal-basins are situated in the departments of the Loire, du Nord, Saône et Loire, and Gard, which furnish about seven-eights of the whole production. Austria, whose principal coal mines are situated in Bohemia, produces about 3,500,000 tons. Spain possesses magnificent coal-fields in the Asturias and Santanders, but as yet they have been but little worked.

Beside the coal-basins of the mother country, Britain is richly provided with coals in many of her colonies. In New South Wales and Tasmania, in Labuan and Farther India, in Hindostan and New Zealand, in British Columbia and Honduras, valuable basins or seams of coal have been discovered; and a magnificent coal-field, far surpassing in magnitude those of the British Islands, extends from

Newfoundland, by Cape Breton, Prince Edward Island, and Nova Scotia, across a large portion of New Brunswick. Thus far it has been but little worked, in countries but thinly peopled, and covered for the most part with boundless forests ; but as from its general proximity to the sea it offers every advantage for mining operations, a brilliant future may safely be predicted for the lands it underlies.

The coal-fields of the United States are of still more ample proportions, as they surpass in extent all the known coal-basins of the world besides. Beyond the Alleghany Mountains we find the magnificent Appalachian Coal-field, traversing eight of the principal States in the American Union, from the northern frontiers of Pensylvania to Alabama, and covering a space of about sixty-five thousand square miles.

Of scarcely inferior extent are the vast coal-fields of Indiana, Illinois, and Kentucky, which nearly equal in magnitude the whole of England ; and another smaller but highly important coal region is situated between the lakes Erie, Huron, and Michigan, not to mention the minor coal-basins scattered here and there from Texas to Missouri, and from New York to Maine.

As yet, the Americans have not derived full benefit from their extraordinary coal deposits ; but the possession of so vast an accumulation of *power* allows us to predict a future of almost boundless enterprise and production for that wonderful country.

While in most of our coal-seams deep shafts have to be sunk to obtain the coal, and steam power has to be constantly employed to prevent its submersion, the Appalachian Coalfield is intersected by three great navigable rivers, the Monongahela, the Alleghany, and the Ohio, all of which lay open on their banks the level seams of coal. At Brownhill, on the first of these rivers, the main seam of bituminous coal, ten feet thick, breaks out in the steep cliff at the water's edge. Horizontal galleries may be driven everywhere at very slight expense, and so worked as to drain themselves, while the cars laden with coal, and attached to each other, glide down on a railway so as to deliver their burden into barges moored to the river's bank. The same seam may be followed the whole way to Pittsburg, fifty miles distant. Being nearly horizontal, it crops out as the river descends, at an continually increasing, but never at an inconvenient, height above the Monougahela. Besides this main seam, another layer of workable coal, six feet thick, breaks out on the slope of thé hills at a greater height. Here almost every proprietor can open a coal-pit on his own land, and the stratification being very regular, he may calculate with precision the depth at which coal may be won.

One of the most remarkable collieries in the world is that of Mauch Chunk (or the Bear Mountain) in Pennsylvania, where an enormous bed of anthracitic coal, nearly sixty feet thick, and probably caused by the doubling back of a twenty-eight feet seam upon itself, is quarried in the open air; the overlying sandstone, forty feet thick, having been removed bodily from the top of the hill, which, to use the miners' expression, has been 'scalped.'

Summary of the Coal Produce of the United Kingdom, 1873.

	TONS.		TONS.
North Durham and Northumberland	12,204,340	Staffordshire, South, and Worcestershire	9,463,559
Cumberland	1,747,064	Lancashire, North and East	9,560,000
South Durham	17,436,045	Lancashire, West	7,500,000
Westmoreland	1,972	Gloucestershire	1,858,740
Yorkshire	15,311,778	Somersetshire	
Derbyshire		Monmouthshire	4,500,000
Nottinghamshire	11,568,000	North Wales	2,450,000
Warwickshire		South Wales	9,841,523
Leicestershire		Scotland, East	10,142,039
Cheshire	1,150,500	Scotland, West	6,715,733
Shropshire	1,570,000	Ireland	103,435
Staffordshire, North	3,892,019		

Total of the United Kingdom, 127,016,747 tons.

Of this vast quantity 16,718,532 tons were used in the manufacture of iron, and 12,617,566 tons, of a declared value of 13,188,511*l.*, exported to foreign countries.

In the same year London consumed 2,665,680 tons brought by sea, 11,195 by canal, and 5,147,413 by railway, forming a total of 7,824,288 tons.—*Hunt's Mineral Statistics.*

CHAPTER XXXIII.

BITUMINOUS SUBSTANCES.

Formation of Petroleum—Enormous Production of the Pennsylvanian Wells—
Asphalte used by the Ancients—Asphalte Pavements—The Pitch Lake of
Trinidad—Jet—Its Manufacture in Whitby.

THE class of bituminous minerals exhibits a long series of
inflammable substances, which are supposed to be derived
from the decomposition of organic matter in the rocks containing them. Some (Petroleum Rock-naptha) issue in a
fluid state from the earth, while others pass by insensible
gradations from petroleum into pittasphalte or maltha (viscid
bitumen), and the latter as insensibly into the solid form of
asphalte. Certain bitumens, again, differ but slightly in composition from bituminous coals, so that it is, in reality, very
difficult to draw a decided line between them. Hence it is
highly probable that in petroleum we see the product of a
primeval vegetation which, under the influence of chemical
change and heat, has partly assumed a liquid form, and
oozing from the deep-seated strata in which it was confined
by terrestrial revolution, now permeates the superficial rocks,
or exists collected in subterranean cavities, whence it issues
in jets and fountains whenever an outlet is made by boring.

Petroleum springs have been known for many ages in
Burmah, where there are about one hundred wells from one
hundred and eighty to three hundred and six feet deep, each
lined with horizontal tubes, but not all now worked; at Baku,
in the neighbourhood of the holy fires, already mentioned;
near the village of Amiano, in Parma, whence enough was formerly obtained to light the streets of Genoa; at Zante, one of
the Ionian islands, which has furnished oil for more than two
thousand years, its petroleum spring having been mentioned

by Herodotus; at Agrigentum, in Sicily, which, according to
Pliny, furnishes a mineral oil that was collected and used for
burning in lamps; on the banks of the Kuban, and many
other localities; but it is only since the discovery of the immense sources of supply in the north-eastern States of America and in Canada that petroleum has become not only an
article of the greatest commercial importance, but a blessing
to millions in all parts of the world. It gladdens the long
winter evenings of the Icelandic peasant, and enlivens the hut
of the Australian settler; it has found its way into the remotest glens of the Alps, and to the distant sea-ports of
China. No wonder that its economical and cheerful light
has caused its consumption to increase with a rapidity almost
without a precedent in the annals of commerce.

In 1861 the exportation of 250 barrels of petroleum to
Europe was mentioned in the American papers as an 'important transaction,' and in 1870 no less than 141,208,550
gallons were exported from the ports of Philadelphia and
New York.

The petroleum production of Pennsylvania amounted in
1870 to 5,650,000 barrels, and that of West Virginia and
Ohio to 511,000.

When we reflect that this amazing mass of liquid bitumen
must necessarily be increased from year to year to meet a
constantly increasing demand, it might almost be feared
that, in spite of the prodigality of nature, its subterranean
reservoirs must one day be exhausted.

Asphalte, a mineral pitch of a deep black colour and a
conchoidal brilliant fracture, is frequently found swimming
on the surface of the Lake Asphaltites, or Dead Sea, in
Judæa, which receives its name from the circumstance. It
also occurs in considerable quantities in Lower Mesopotamia,
particularly at Hit, whence it was exported to Egypt, where
it was used for embalming as early as the reign of Thothmes
III., about 1,400 years B.C. The ancients also frequently
employed it combined with lime in their buildings. Not
only were the brick palaces of Nineveh and Babylon
cemented with this material, but some of the old Roman
castles in this country are found to hold bitumen as the
cement by which their stones are secured.

'It is a remarkable fact,' says the late Dr. Ure, 'that the substance thus employed in the earliest constructions upon record, should for so many thousand years have fallen well-nigh into disuse among civilised nations; for there is certainly no class of minerals so well fitted as the bituminous, by their plasticity, fusibility, tenacity, adhesiveness to surfaces, impenetrability by water, and unchangeableness in the atmosphere, to enter into the composition of terraces, foot pavements, roofs, and every kind of hydraulic work. Bitumen, combined with calcarious earth, forms a compact semi-elastic solid, which is not liable to suffer injury by the greatest alternations of frost and thaw, which often disintegrate in a few years the hardest stone; nor can it be ground to dust and worn away by the attrition of the feet of men and animals, as sandstone, flags, and even blocks of granite are. An asphalte pavement rightly tempered in tenacity, solidity, and elasticity, seems to be incapable of suffering abrasion in the most crowded thoroughfares—a fact exemplified of late in a few places in London, but much more extensively and for a much longer time in Paris.' Many of the asphalte pavements made in England have, indeed, proved failures; but as the proper proportions of the respective ingredients may not have been maintained, further trials are advisable. At present, although bitumen is employed, and with seeming advantage, as a cement between paving-stones, its use in the formation of foot pavement has been confined within narrow limits.

In Europe the most extensive mine of asphaltic rock is undoubtedly that of the Val de Travers in the canton of Neufchâtel; but the most remarkable deposit of bitumen in the world is the celebrated Great Pitch Lake in the island of Trinidad. With regard to its formation, Sir Charles Lyell remarks that the Orinoco, which discharges its vast volume of water right opposite to the island, has for ages been rolling down great quantities of woody and vegetable bodies into the surrounding sea, where, by the influence of currents and eddies, they may be arrested and accumulated in particular places. The frequent occurrence of earthquakes and other indications of volcanic action in these parts lends countenance to the opinion that these vegetable substances may have undergone, by the agency of subterranean fire, those transformations or chemical changes which produce petroleum; and this may, by the same causes, be forced up to the surface, where, by exposure to the air, it becomes inspissated, and forms the different varieties of pure and earthy pitch or asphaltum so abundant in the island. To my friend Mr. C. M. Collins, Chief Engineer R.N., then an officer in H.M.S. 'Scourge,' who visited the Pitch Lake early in 1849, I am indebted for the following interesting particulars about this grand scene of subterranean activity. The Pitch Lake is three miles in circumference; the bitumen is solid and cold near the shore, but gradually increases in temperature and softness towards the centre, where it may be seen boiling up through the cracks at the surface. The lake is a level plain, with small pools of fresh water in every direction, and around which grow trees and various tropical plants, as shown in the engraving, which gives one an excellent idea of the place.

The best pine-apples in the West Indies, called black pines, flourish on the border of the lake, together with the poisonous Manchineel tree, the mahogany, and tufts of graceful palms.

In 1849 many experiments were tried, under the superintendence of the celebrated Earl of Dundonald, in order to ascertain if the bitumen could be used as fuel on board steamships, but without success. Eighty tons of bitumen were taken by the 'Scourge' and landed at Bermuda Dock-yard, where many other experiments were tried, still under the earl's superintendence, till at length one of the convict hulks, the 'Medway', was lit up with gas, which was so successfully done that the town of Halifax, Nova Scotia, was afterwards for a time lit up with it. Though of a superior kind, its use was, however, discontinued, as coal could be procured from Cape Breton and Pictou, N.S., at a cheaper rate. In spite of these unsuccessful efforts to utilize the pitch, there can be no doubt that ere long a material of such valuable properties, and furnished so abundantly by nature, must become an important article of commerce. The wonder is, that it has been so long neglected.

Though Jet is frequently considered to be wood in a high state of bituminisation, yet the fact that we find this beautiful substance surrounding fossils, and casing adventitious masses of stone, seems to show that a liquid, or, at all events, a plastic condition, must at one time have prevailed in its formation. This opinion is further strengthened by the circumstance that petroleum strongly impregnates the rock in which it is found, giving out a strong odour when it is exposed to the air.

Jet occurs chiefly in the neighbourhood of Whitby in Yorkshire, the estates of Lord Mulgrave being especially productive. The jet miner attentively searches the slaty rocks, and finding the jet spread out, often in extreme thinness, between their laminations, follows it with great care, and is frequently rewarded by its thickening out to two or three inches.

The art of working jet is of very ancient date in this country, for the Romans certainly employed it for ornamental purposes, and probably found it in use among the Britons whom they conquered. Lionel Charlton, in the 'History of Whitby,' says that in one of the Roman tumuli he found the earring of a lady having the form of a heart, with a hole in the upper end for suspension from the ear. There exists no doubt that, when the Abbey of Whitby was the seat of learning and the resort of pilgrims, jet rosaries and crosses were common. The manufacture was carried on till the time of Elizabeth, when it seems to have ceased, and was not resumed till the year 1800, when Robert Jefferson, a painter, and John Carter, made beads and crosses with files and knives. A stranger coming to Whitby, and seeing them working in this rude way, advised them to try to turn it; they followed his advice, and found it answer. Several more then joined them, and from that time the trade has been gradually increasing; so that at present the total annual value of the mourning ornaments made at Whitby and Scarborough amounts to no less than 125,000*l.* About 250 men and boys are employed in searching for jet, and between 600 and 700 are engaged in its manufacture.

CHAPTER XXXIV.

SALT.

Geological Position of Rock Salt—Mines of Northwich—Their immense Excavations—Droitwich and Stoke—Wieliczka—Berchtesgaden and Reichenhall—Admirable Machinery—Stassfart—Processes employed in the Manufacture of Salt—Origin of Rock-salt Deposits.

COMMON salt is so necessary to man, and of such vast importance to the manufacturer and agriculturist, that the processes by which it is obtained are justly reckoned among the chief branches of industry.

In many of the warmer countries of the globe it is procured simply by the evaporation of sea-water in shallow lagoons; in others, it gushes forth in briny springs, or occurs in inland lakes, pools, and marshes, or is extracted in the solid form of rock-salt from the bosom of the earth.

The geological position of rock-salt is very variable; it is found in all sedimentary formations, and is generally inter-stratified with gypsum, and associated with beds of clay. In England its chief deposits occur in the new red sandstone in the region around Northwich, in Cheshire. They consist of two beds which are not less than one hundred feet thick, and are supposed to constitute large insulated masses about a mile and a half long and nearly 1,300 yards broad. The uppermost bed occurs at seventy-five feet beneath the surface, and is separated from the lower mass by layers of indurated clay, thirty-one and a half feet thick, with veins of rock-salt running between them. Hitherto only the lower bed has been worked, for the upper deposits are of inferior purity. These valuable mines were accidentally discovered in 1670, during an unsuccessful sinking for coal; and as ever since that time they have furnished a constantly increasing quantity of salt, amounting during the last few years to more than 1,250,0000 tons, the vastness of the excavations may easily be imagined. To support the roof, which is about twenty feet above the floor, and extends in some cases over several acres, huge pillars not less than fifteen feet thick have been left standing at irregular intervals, thus forming immense rows of galleries, which even when illuminated by thousands of lights are lost in a dim and endless perspective, which may well remind the spectator of the fabled Hall of Eblis.

As the salt is detached from the rock by blasting, the grandeur of the scene is not a little heightened by the frequent explosions re-echoing through the spacious vaults, and booming like thunder from some dark and distant gallery. For the transport of the salt underground, the roomy passages are traversed in every direction by tram-roads, on which waggons drawn by horses easily convey it from the place of extraction to the bottom of the shaft. The chief part of the Cheshire salt (both white and rock) is sent down the river Weaver to Liverpool. As it is rarely found of sufficient purity for immediate use, it is first dissolved in water, and afterwards reduced to a crystalline state by evaporating the solution. The necessary coals are mostly brought by canal from the neighbourhood of St. Helens, and salt taken as a return freight; so that, as in a clock-work where one wheel catches into another, nothing is wanting to render its manufacture as economical as possible.

Among the mineral treasures which nature has so prodigally bestowed upon Great Britain, the salt-mines of Cheshire hold a conspicuous rank, as they not only provide chiefly for our own vast consumption, but also for that of many other countries. In 1873 they produced 1,485,000 tons, of which 841,226 tons (value 789,185l.) were exported chiefly to British India (222,995 tons), the United States (242,444 tons), and Russia (84,528 tons).

Next in importance to the Cheshire mines are the brine-pits of Droitwich and Stoke, in Worcestershire, the former of which are said to have been worked in the time of the Romans, and now chiefly supply the London market. In some parts large areas are gradually subsiding, so that in the town itself whole streets have had to be removed ; in others the ground has suddenly sunk and formed considerable crateriform hollows, not seldom above a hundred feet deep. Great fears are entertained of similar catastrophes on a still more extensive scale. All these terrestrial revolutions are no doubt owing to the action of springs which are constantly washing away and weakening the pillars sustaining the roofs of ancient excavations.* In 1873 the produce amounted to 279,750 tons, of which about 60,000 were exported to foreign countries.

At Droitwitch the borings are only 175 feet deep ; and so abundant is the supply of brine that, if the pumps cease working, it speedily rises to within nine feet of the surface, and if left unremoved soon overflows.

The most renowned salt-mines on the continent of Europe are undoubtedly those of Wieliczka, a small town of about 6,000 inhabitants, situated to the south of Cracow, in a fruitful valley on the northern borders of the Carpathian mountains. 'After descending 210 feet,' says Mr. Bayard Taylor, an American traveller who visited them a few years ago, ' we saw the first veins of rock-salt in a bed of clay and crumbled sandstone. Thirty feet more, and we were in a world of salt. Level galleries branched off from the foot of the stair-case ; overhead a ceiling of solid salt, under foot a floor of salt, and on either side grey walls of salt, sparkling here and there with minute crystals. Lights glimmered ahead, and on turning a corner we came upon a gang of workmen, some hacking away at the solid floor, others trundling wheel-barrows full of the precious cubes. Here was the chapel of St. Anthony—the oldest in the mines— a Byzantine excavation, supported by columns, with altar, crucifix, and life-size statues of saints, apparently in black marble, but all as salt as Lot's wife, as I discovered by putting my tongue to the nose of John the Baptist. The humid air of this upper story of the mines has damaged some of the saints. Francis, especially, is running away like a dip candle, and all of his

* It may be remarked that similar catastrophes not seldom happen when the miner is at work below the foundations of a town. Essen, in the neighbourhood of Mr. Krupp's gigantic establishment, is giving way; and at Iserlohn, the Westphalian Birmingham, many houses have become uninhabitable from this cause.

head is gone except his chin. The limbs of Joseph are
dropping off as if he had the Norwegian leprosy, and Lawrence has deeper scars than his gridiron could have made,
running up and down his back. A Bengal light, burnt at
the altar, brought into sudden life this strange temple, which
presently vanished into utter darkness, as if it had never been.

'I cannot follow, step by step, our journey of two hours
through the labyrinths of this wonderful mine. It is a be-
wildering maze of galleries, grand halls, staircases, and
vaulted chambers, where one soon loses all sense of distance
or direction, and drifts along blindly in the wake of his conductor. Everything was solid salt except where great piers
of hewn logs had been built up to support some threatening
roof, or vast chasms, left in quarrying, had been bridged
across. As we descended to lower regions, the air became
more dry and agreeable, and the saline walls more pure and
brilliant. One hall, 108 feet in length, resembled a Grecian
theatre, the traces of blocks taken out in regular layers
representing the seats for the spectators. Out of this single
hall 1,000,000 cwt. of salt had been taken, or enough to
supply the 40,000,000 inhabitants of Austria for one year.

'Two obelisks of salt commemorated the visit of Francis I.
and his Empress in another spacious, irregular vault, through
which we passed by means of a wooden bridge, resting on
piers of the crystalline rock. After we had descended to the
bottom of this chamber, a boy ran along the bridge above
with a burning Bengal-light, throwing flashes of blue lustre
on the obelisks, on the scarred walls, vast arches, the entrances to deeper halls, and the far roof, fretted with the
picks of the workmen. The effect was truly magical. Presently we entered another and loftier chamber, yawning
downwards like the mouth of hell, with cavernous tunnels
opening out of the further end. In these tunnels the workmen, half naked, with torches in their hands, wild cries,
fireworks, and the firing of guns (which here so reverberates
in the imprisoned air that one can feel every wave of sound),
give a rough representation of the infernal regions, for the
benefit of the crowned heads who visit the mines. A little
further we struck upon a lake four fathoms deep, upon which
we embarked in a heavy, square boat, and entered a gloomy
tunnel, over the entrance of which was inscribed (in salt

letters) " Good luck to you ! " Midway in the tunnel, the halls
at either ends were suddenly illuminated, and a crash, as of
a hundred cannon bellowing through the hollow vaults,
shook the air and water in such wise that our boat had not
ceased trembling when we landed in the further hall. A
tablet inscribed " Heartily welcome ! " saluted us on landing.
Finally, at the depth of 450 feet, our journey ceased, although
we were but half way to the bottom. The remainder is a
wilderness of shafts, galleries, and smaller chambers, the
extent of which we could only conjecture. We then returned
through scores of tortuous passages to some vaults, where a lot
of gnomes, naked to the hips, were busy with pick, mallet, and
wedge, blocking out and separating the solid pavement. The
process is quite primitive, scarcely differing from that of the
ancient Egyptians in quarrying granite. The blocks are
first marked out on the surface by a series of grooves; one
side is then deepened to the required thickness, and wedges
being inserted under the block, it is soon split off.

'The number of workmen employed in the mines is 1,500,
all of whom belong to the "upper crust"—that is, they live
on the outside of the world. They are divided into gangs,
and relieve each other every six hours. Each gang quarries
out, on an average, a little more than 1,000 cwt. of salt in
that space of time, making the annual yield 1,500,000 cwt.!

'The men we saw were fine, muscular, healthy-looking
fellows; and the officer, in answer to my questions, stated
that their sanitary condition was quite equal to that of field
labourers. He explicitly denied the ridiculous story of men
having been born in these mines, and having gone through
life without ever mounting to the upper world.'

As far as explored, the salt-bed occupies a space of 9,000
feet in length and 4,000 in width, and consists of five successive stages or *stockwerke*, separated from each other by
intervening strata of from 100 to 150 feet in thickness, and
reaching to a depth of 1,500 feet. Notwithstanding the
immense amount of salt already quarried from this wonderful
deposit, which, according to authentic records, has been
worked ever since the twelfth century, and perhaps even much
earlier, it is estimated that, at the present rate of exploitation,
the known supply cannot be exhausted under 300 years.

It is a remarkable circumstance that sources of sweet
water are found in the mines, in close proximity to the salt.—
a circumstance which is owing to the latter not forming a
continuous stratum, but being imbedded in large nests or
insular masses in the tertiary clay of the mountain, so that
in several places the water filtering from the top is able to
gush forth in the subterranean galleries without any saline
admixture.

In the summer of 1868 a serious accident happened to the
mines of Wieliczka, which at one time was supposed to
threaten their total destruction. In the hope of discovering
valuable potash salts, such as occur at Strassfurt, in the
vicinity of the rock-salt, a boring was imprudently attempted
at a great depth, through a contiguous aquiferous stratum ;
and the consequence was that, instead of meeting with the
expected result, a powerful spring was tapped, which, pouring forth an immense volume of water, filled the lower
galleries. The inhabitants of the village of Wieliczka, which
is situated above the mines, were terrified. They not only
feared for the ruin of the mine, which afforded them their
chief means of subsistence, but dreaded also the falling-in of
their houses, in consequence of the melting of the salt-pillars
which upheld the flooded galleries.

Fortunately their fears proved to be exaggerated, as the
inundation, which remained confined to the lower galleries, is
now being rapidly subdued by means of powerful steam-
pumps, and measures have been taken for blocking up the
spring. Even supposing the water to have continued pouring in with undiminished force, and without any effort being
made to drain it off, the excavations are so vast that it would
have taken many years to fill them.

Besides the mines of Wieliczka, Austria possesses many
other considerable deposits of rock-salt in Gallicia (Bochnia),
Hungary, and Transylvania. In Salzburg (Ischl, Halstadt
Hall), Tyrol, and the neighbouring mines of Berchtesgaden,
in Bavaria, the salt does not occur in large solid masses, fit
to be at once extracted from the bosom of the earth, but
imbues masses of gypsum and anhydrite, which become
quite light and porous when the salt has been removed by
water. As it would be too expensive to remove this compound, an ingenious method has been contrived for intro-

ducing water into the mines from above, and drawing it off again through an adit or lower gallery as soon as it is saturated with salt.

The brine thus obtained at Berchtesgaden is then conveyed, by means of pipes or conduits, to Reichenhall and Rosenhain where the necessary fuel for its evaporation is near at hand. The distance from Berchtesgaden to Rosenhain is no less than thirty leagues; and between both places many high mountains, steep rocks, and narrow gorges intervene. The works which lead the brine over this distance, and through all these natural obstacles, may therefore justly be considered as a masterpiece of mechanical skill. As the pipes must often ascend mountains, the highest elevation above the pit being 1,218 feet, and then again slant down into ravines, as in many parts rocks had to be levelled and forests cut down for the purpose of laying them, and as they are·subject to frequent damage in a severe climate, it may easily be imagined that the greatest engineering power was required for the execution of so grand a work.

Hydraulic machines serve to raise the brine over the mountains, and the water-power of the rivulets descending from the heights is used for forcing it upwards. The contrivance is so admirable that the small machine at Berchtesgaden raises 270 cwt. of brine to a height of 311 feet by means of an equal weight of water descending from a height of 375 feet. In some parts the tubes of this colossal duct run along the high road, in others tunnels have been pierced to shorten the distance.

Thus the Alpine rock-salt requires much ingenuity to be rendered productive, while in other parts we find rock-salt cropping out on the surface of the earth, so as to be very easily worked. The soil of many extensive wastes in Asia and Africa is covered and impregnated with salt, which has never been inclosed by superimposed deposits. Near Lake Oroomiah, in Persia, it forms hills and extensive plains, and it abounds in the neighbourhood of the Caspian Sea.

In the valley of Cardona in the Pyrenees two thick masses of rock-salt, apparently united at their bases, make their appearance on one of the slopes of the hill. One of the beds, or rather masses, has been worked, and measures about 130

yards by 250, but its depth has not been determined. It
consists of salt in a laminated condition, and with confused
crystallisation. That part which is exposed is composed of
eight beds, nearly horizontal, and having a total thickness of
fifteen feet, but the beds are separated from one another by
red and variegated marls and gypsum. The second mass
not worked appears to be unstratified, but in other respects
resembles the former; and this portion, where it has been
exposed to the action of the weather, is steeply scarped and
bristles with needle-like points, so that its appearance has
been compared to that of a glacier.

The rock-salt deposit of Ilezk, about fifty miles to the
south of Orenburg, in Russia, is still more remarkable. On
the sides of a crater-like pit, which has been dug into the
mass where it most nearly approaches the surface, the rock-
salt is seen standing in perpendicular walls. On the west
side of the gap a convenient staircase leads to the bottom.
The salt is hewn in large square blocks, which are afterwards
sawn into smaller pieces of eighty-five pounds. The regular
annual produce is fixed by Government at 700,000 pud,
or about 10,000 tons, of which part is furnished *gratuitously*
to the neighbouring Kirghise hordes, who no doubt are
made to pay dearly for their salt in some other manner.

The labours in the mine are only carried on during the
summer, and begin by pumping out the water which has
settled at the bottom of the pit from the melting snow;
while the transport to the next river by means of sledges
takes place during the winter. Were this mine situated in
a less barbarous or more accessible country, it might easily
rival, or even surpass, the produce of Cheshire. To the
south of the pit, where the regular mining operations, such
as they are, take place, a great number of old pits or holes
may be seen, in which the Cossacks, Baschkirs, and Kirghise
used to provide themselves with salt before Government
undertook the regular working of the mine in 1754. These
pits, some of which are sixty feet square, and from six to
eight feet deep, are generally full of a saturated brine of a
brownish colour, and are made use of for bathing by the
Kirghise, who justly consider them as an excellent remedy
for many diseases.

Before 1856 all the salt produced in Prussia was obtained from brine-springs, but since that time enormous beds of rock-salt have been discovered in various parts of the kingdom. Those of Stassfurt, near Madgeburg, which extend over a surface of many square leagues, in a bed more than a thousand feet thick, produced 2,256,000 cwt. in 1866, and would alone suffice for the supply of the whole of Germany. Besides rock-salt, these mines contain also an inexhaustible quantity of highly valuable potash salts, which are largely used as artificial manures, and provide the materials for the numerous chemical manufactories which within a few years have converted an obscure hamlet into a flourishing seat of industry. At Speerenberg, about twenty miles to the south of Berlin, the earth-borer, though driven to a depth of 4,000 feet, has not yet reached the bottom of an enormous deposit of rock-salt, and at Segeberg, in the province of Sleswig-Holstein, a shaft is now being sunk for the purpose of working a rich bed of rock-salt recently discovered. Railroads are constructing for the purpose of connecting this new source of wealth, which is destined to furnish salt to a great part of North Germany and Denmark, with the ports of Hamburg, Lübeck and Kiel, and thus in a few years a hitherto unknown village will rank amongst the most thriving communities of Prussia.

The method of preparing the rock-salt, and the processes employed in manufacturing salt from brine-springs, are nearly the same in all salt-works. The first process is to obtain a proper strength of brine, either by saturating fresh water with the salt that has been brought up from the mine, or pumping up the salt water from springs that have become saturated by passing through saliferous beds. The brine obtained in a clear state is put into evaporating pans and brought as quickly as possible to a boiling heat (in the case of strong brine 226° F.), when a skin is formed on the surface, consisting chiefly of impurities. This is taken off, and either thrown away or used for agricultural purposes, and the first crystals which form are likewise raked away and thrown aside as of little value. The heat is then kept up to the boiling point for about eight hours, during which time evaporation goes on steadily, the liquid gradually diminishing, and the salt being deposited; it is then raked out, put into moulds, and placed in a drying stove, to render it perfectly dry and ready for sale.

When salt is to be prepared from the weak brines which are of common occurrence in France and Germany, the brine is concentrated by natural evaporation previous to the more costly application of artificial heat. Having been first raised by pumps, it is then allowed to trickle in a continuous stream down the surface of bundles of thorns exposed to the sun and wind, and built up in regular walls between parallel wooden frames. These evaporating works (*Gradirwerke*, or graduation-houses) are frequently of an immense extent. At Salza, near Schönebeck, for instance, the graduation-house is 5,817 feet long, the thorn walls are from 33 to 52 feet high in different parts, and present a total surface of 25,000 feet. According to the weakness of the brine, it must be the more frequently pumped up and made to flow down repeatedly over the thorns in different compartments of the building. An immense quantity of fuel is saved by this economical mode of evaporation.

The origin of rock-salt deposits is one of the most interesting geological questions. According to some authorities, they were the result of igneous agency, while others are of opinion that in every case they have been deposited from solution in water. Their usual occurrence in lenticular or irregularly-shaped beds, having a great horizontal extension, favours the aqueous theory, for masses protruded upwards, or sublimated by volcanic power, are generally found to occupy vertical fissures. To account for their formation we must suppose a sea such as the Mediterranean cut off by an elevation of the land at its mouth from its previous communication with the ocean, and gradually losing more water by evaporation than it receives by rain and rivers. As thus the amount of salt which it holds dissolved increases, deposits of rock-salt will ultimately form at the bottom of its deepest parts, and subsequent changes in the earth's surface may then either conceal them under superincumbent strata, as at Northwich, or leave them exposed, as in many of the African or Asiatic wastes.

CHAPTER XXXV.

SULPHUR.

Sulphur Mines of Sicily — Conflagration of a Sulphur Mine — The Solfataras of Krisuvick — Iwogasima in Japan — Solfatara of Puzzuoli — Crater of Teneriffe — Alaghez — Büdöshegy in Transylvania — Sulphur from the Throat of Popocatepetl — Sulphurous Springs — Pyrites — Mines of San Domingos in Portugal — The Baron of Pommorão.

THOUGH in every volcanic region of the globe sulphurous exhalations arise from a great number of craters or solfataras, yet sulphur is but rarely found in sufficient quantities to remunerate the miner's toil. In this respect the island of Sicily is unrivalled, for no other country possesses such masses of this valuable, and in many cases indispensable, mineral.

The numerous sulphur-pits of Sicily, which occur in crevices or hollows over a space of 150 geographical miles, are situated chiefly in the southern part of the island, in the districts between the sea-border of the province of Girgenti and the mountains of Etna, Mannaro, Castro Giovanni, and Catolica. They are no doubt the produce of a vast volcanic action which took place about the beginning of the tertiary period, when the sulphurous fumes, rising through countless clefts or fumaroles from the mysterious furnaces of the deep, condensed in the chalk and clay grounds of the superficial strata.

In former times, as long as the chief use of sulphur was confined to the fabrication of gunpowder, its production was comparatively insignificant; but since the manufacture of sulphuric acid has become a branch of industry of continually increasing importance, sulphur, the ingredient necessary to its formation, has considerably risen in value, and now constitutes the chief article of Sicilian exportation.

Girgenti, the most important town on the south coast of the island—though its dirty miserable streets and its 15,000 inhabitants form a melancholy contrast to the wealth and luxury of ancient Agrigentum, within whose lofty walls a population of 800,000 souls is said to have existed—owes to the increase of the sulphur trade the slight dawn of prosperity which has enlivened it during these latter years.

All the sulphur-pits in the south-west of the island send their produce to the port of Girgenti, and on every road one meets with long files of mules and asses loaded with sacks of sulphur.

The grape disease, against which this mineral is everywhere used in France and Italy as the only successful remedy, has given a new impulse to the sulphur trade by raising the price to about three times its former value. The merchants of Girgenti did not neglect this opportunity for making their fortunes; for as soon as the grape disease became a national calamity for the chief wine-producing countries, they bought up large tracts of sulphur-grounds, and thus acquired considerable wealth.

A visit paid by Dr. Häckel * in 1859 to the sulphur pits near Girgenti proves that mining operations in Sicily are still in the primitive condition described by all former travellers. Not even our commonest improvements are known; the pickaxe and the spade are almost the only implements employed, and with these the earth is excavated in the most slovenly manner, wherever a vein promises to be productive. The materials thus loosened from the rock are carried out of the mine in baskets and thrown into large heaps, from which the sulphur is extracted in the following wasteful manner. The conical mounds are covered with a mantle of moist clay, in which some openings are left for the emission of the smoke, and set fire to at the bottom. The melted sulphur collects in grooves or channels, and flows into square forms, where it congeals into a solid mass. This method, which is said to have been first introduced by the Saracens, causes, of course, a great loss of sulphur; but distilling ovens heated by coal have been found too expensive to answer. Dr. Häckel

---

* 'Zeitschrift für allgemeine Erdkunde,' No. 81, Juni 1860.

traversed one of the longest excavations, which was sometimes so narrow that he could only with difficulty pass, and then expanded into high vaults whose roof was ornamented with beautiful crystals of celestine and gypsum. The workmen were completely naked, on account of the oppressive heat which reigns in these pits; and their dark brown skins, sprinkled with light yellow sulphur-dust, gave them a very strange and savage appearance. Most of the inhabitants of Girgenti are at present employed in the sulphur mines, and comparatively few are engaged in cultivating the beautiful gardens and fields that extend from the foot of the town to the sea, and occupy the site where once the ancient city of Agrigentum rose from the shore to the terraced hills which are still crowned with the ruins of her colossal temples.

On an average the sulphur-ores of Sicily yield about sixteen per cent. of brimstone, and the quantity annually produced has increased from 94,985 tons in 1851 to 184,173 tons in 1866. Besides Girgenti, the chief ports from which sulphur is exported are those of Licata, Terranova, Siculiana, Palermo, Messina, and Catania.

One of the most remarkable events in the history of the Sicilian sulphur mines occurred during the last century in the solfatara of Sommatino. This celebrated pit, which is situated on the precipitous right bank of the Salso Valley, took fire in 1787, through the negligence of the workmen, and as may easily be imagined from the inflammable nature of the materials, the conflagration caused the complete abandonment of the pit. After two years, however, during which the fire raged incessantly, the mountain suddenly burst asunder on its south-eastern flank, and a stream of melted sulphur, gushing forth from the cleft, precipitated itself into the neighbouring river. This phenomenon, which was evidently caused by Nature having performed on a vast scale an operation similar to that by which the sulphur is usually extracted from the ore, produced a mass of the purest brimstone, amounting to more than 40,000 tons, so that the owners of the pit, who had given up their property as totally lost, became enriched by the very circumstance which had seemed to menace them with utter ruin.

Next to the sulphur mines of Sicily, those of Teruel and

Lorca in Spain, which in 1862 furnished 12,639 tons, are the most considerable in Europe. The mines of Perticara di Talamella in Italy, annually yield about 4,000 tons; and the Austrian sulphur-pits of Swoszowice near Cracovia and of Radoboy in Croatia, produced 1,867 tons in 1865.

In all these mines the sulphur has been deposited or condensed in times long past, undoubtedly in the same manner as it is formed in the solfataras of the present day, where the decomposition of the volcanic gases on reaching the atmosphere causes the precipitation of sulphur.

Most of these still active solfataras are unproductive in a commercial sense, either from their inaccessible position in the crater basins of enormous volcanoes, or from their situation in remote deserts, or from the small quantities of the mineral forming on their surface.

The solfataras of Krisuvick in Iceland, for instance, are separated from the nearest ports by such rugged lava-fields as to render the cost of transport an almost insurmountable obstacle to their being worked with profit. But though undeserving of the mercantile speculator's attention, these northern sulphur-pits rank among the most striking natural wonders of Iceland.*

The remote solfataras of Japan afford a more abundant supply. 'The sulphur,' says Kämpfer, in his history of that singular country, 'is the produce of a small island which, from the great quantity it affords of this substance, is called "Iwogasima," or the Sulphur Island. It is not above a hundred years since the natives first ventured to explore that desert spot, which, from the smoke rising from its surface, was previously supposed to be the abode of demons. At length a bold adventurer obtained leave to visit the dreaded island. He chose fifty resolute men to accompany him on his hazardous expedition, and on landing found, instead of the fiends he expected to encounter, a volcanic soil, covered in many parts with thick deposits of sulphur, and emitting dense volumes of smoke from countless fumaroles. Ever since that time the island yields a considerable revenue to the Prince of Satzuma.'

One of the most celebrated solfataras is that of Puzzuoli,

near Naples. It may be considered as a nearly extinguished crater, and appears, by the accounts of Strabo and others, to have been before the Christian era in very much its present state, giving vent continually to aqueous vapours, together with sulphureous and muriatic acid gases, like those evolved by Vesuvius. This remarkable spot has attracted the attention of naturalists and poets since the remotest antiquity, and Homer mentions it in his immortal narrative of the peregrinations of Ulysses. The process for the separation of the sulphur, which is condensed in considerable quantities amongst the gravel collected in the circle which forms the interior of the crater, is conducted in the following manner. The mixture of sulphur and gravel is dug up and submitted to distillation, to extract the sulphur; the gravel is then returned to its original place, and in the course of about thirty years is again so rich in sulphur as to serve for the same process once more.

'The crater of the Peak of Teneriffe,' says Leopold von Buch, ' is now but an immense solfatara. The sulphureous vapours which escape from every part of the vast cauldron decompose the rock, convert it into white clay, and cover it in many places with beautiful crystals of sulphur. By this constant chemical action the soil towards the centre of the crater has been rendered so soft that in many places great caution is necessary to avoid sinking into the yielding mass, which has a temperature higher than that of boiling water.'

A remarkable sulphur formation occurs on the rocks surrounding the crater of the volcano Alaghez, situated in Northern Armenia. The sulphur is precipitated in thick crusts on their walls, and as the summit of the crater is inaccessible, the people of the neighbourhood, in order to collect the sulphur, fire at it with musket-balls, and pick up the fragments thus detached.

Close beneath the summit of the Patuka, in Java, is a circular lake about fifteen hundred feet in diameter. The borders are covered with a rich vegetation; the water is clear and colourless, but appears yellowish from the reflection of the sulphur which covers the whole bottom of the lake. In 1818 Reinwardt found in this piece of water an islet completely composed of sulphur.

After the eruption of the Tashem Idjem, another Javanese volcano, in 1796, such quantities of sulphur were formed that several hundred shiploads could be gathered and exported as the produce of this single volcanic paroxysm.

The mountain Büdöshegy in Transylvania, exhibits the remarkable phenomenon of sulphur-caves. On entering one of these vast subterranean crevices, incrustations of sulphur are seen to cover the lower part of the walls; but respiration is still easy and free. On advancing a few steps the air acquires a sharp acidulous taste, and the feet begin to feel a warmth which gradually increases to an intolerable heat. On advancing still further the lights are extinguished. A speedy retreat is necessary, and imprudent visitors have been known to pay for their curiosity with their lives.

In the island of Milo there are likewise numerous caverns, the walls of which are incrusted with sulphur and alum. A visit to these grottoes, which annually yield about five hundred tons of pure sulphur, is not without danger from the suffocating fumes that issue from their crevices.

Some of the Arabian volcanoes also produce considerable quantities of sulphur, such as the Dufan, which is called Djebel-el-Kebril, or 'Sulphur Mountain,' by the Arabs, and is mentioned in the writings of Herodotus.

Though the craters of volcanoes are generally almost inaccessible, yet history mentions a curious instance where the most extraordinary exertions were made for collecting sulphur above the regions of perpetual snow. During the wondrous campaign which ended in the overthrow of the empire of Montezuma, Cortez, being in want of powder, sent a party under Francisco Montaño, a cavalier of determined resolution, to gather sulphur from the smoking throat of Popocatepetl, which rises, with its silvery sheet of everlasting snow, to the height of 17,852 feet above the level of the sea. After traversing the lower region—which was clothed with a dense forest, so thickly matted that in some places it was scarcely possible to penetrate it—the track of the Spaniards opened on a black surface of glazed volcanic sand and of lava, the broken fragments of which, arrested in its boiling process in a thousand fantastic forms, opposed continual impediments to their advance. They now came to the limits of perpetual snow, where new difficulties presented themselves, as the treacherous ice gave an imperfect footing, and a false step might precipitate them into the frozen chasms that yawned around. To increase their distress respiration in these aërial regions became so difficult that every effort was attended with sharp pains in the head and limbs. At length they reached the edge of the crater, which presented an irregular ellipse at its mouth more than a league in circumference. A lurid flame burned gloomily at the bottom, sending up a sulphurous steam, which, cooling as it rose, was precipitated on the walls of the cavity. The party cast lots, and it fell on Montaño himself to descend in a basket into this hideous abyss, into which he was lowered by his companions to the depth of four hundred feet! This was repeated several times, till the adventurous cavalier had collected a sufficient quantity of sulphur for the wants of the army.

Many mineral springs owe their medicinal properties to the hydrosulphuric acid which they contain, and whose decomposition frequently gives rise to the formation of sulphur. When the large marble slab which covers the imperial source at Aix-la-Chapelle is removed at the end of every twenty years, about two hundred pounds of sulphur are collected from the walls above the spring.

Combinations of sulphur with metals, particularly with iron and copper (pyrites), occur in much more considerable masses and in a far greater number of localities than the pure uncombined mineral. Formerly the sulphides of iron ($52\frac{1}{2}$ per cent. sulphur, $47\frac{1}{2}$ per cent. iron) and copper served only for the fabrication of vitriol and alum; but since the progress of chemical science has allowed them to be profitably used for the manufacture of sulphuric acid, they have acquired a far greater importance. Our own mines, which are situated chiefly in the county of Wicklow, produced in 1873 58,924 tons of iron pyrites or mundic, besides which an additional quantity of 520,347 tons of coppery pyrites was imported from Spain, Portugal, Norway, and other countries.

In Southern Spain the mines of cupriferous pyrites—particularly in the province of Huelva, and in the Sierra de Tharsis, which on account of the copper they contain were already worked in times anterior to the Roman occupation, and give proof of their ancient importance by the vast dimensions of their excavations—sent us no less than 246,692 tons in 1873. In Portugal the mines of San Domingos, in the province of Alentejo, likewise afford a remarkable example of the mining industry of the Romans, in the ancient adit which' served for draining the works. They merely used the ores that were richest in copper, and rejected the poorer qualities, which form the immense mounds of scoriæ round the mouth of the excavations. After having been abandoned for many centuries the mines of San Domingos are once more diligently worked. Their newly-acquired importance is due chiefly to the enterprise of Mr. James Mason, now Baron of Pommorão. A railroad nine miles long unites the mine with the left bank of the Guadiana, which has been rendered navigable for larger vessels for a length of ten. miles. Moreover the port of Pommorão has been excavated in the steep bank of the river for the convenient shipping of the pyrites. Before 1858 a solitary hermitage was the only dwelling at San Domingos, which is now a thriving village of five hundred houses, with a handsome church and a railroad station. The number of the workmen employed in 1866 amounted to two thousand, and large works were being erected for the separation of the copper. In 1859 the produce of these mines amounted to 7,887 tons, and in 1873 they exported 199,559 tons to England. Such are the wonderful changes which can be brought about when the right man finds the right place for the employment of his energies. France produced in 1866 about 100,000 tons of pyrites, Prussia 38,248 tons, and Belgium 28,956 tons, so that the total production of Europe now probably amounts to more than a million tons. In Canada the ore of a vast deposit of pyrites is exported to the United States, where it serves for the fabrication of sulphuric acid. Thus a substance scarcely noticed twenty years ago has become an important article of commerce in both hemispheres.

Various Modes of its Collection on the Prussian Coast—What is Amber? The
extinct Amber Tree—Insects of the Miocene Period inclosed in Amber—Formidable Spiders—Ancient and Modern Trade in Amber.

AMBER is a resinous substance, the produce of extinct
forests that now lie buried in the earth or under the
bottom of the sea.

It is found abundantly on the Prussian coast of the Baltic,
where it is collected in many ways. After stormy weather
it is frequently cast ashore by the surf, or remains floating
on the water. The amber-fishers, clothed in leather dresses,
then wade into the sea, and secure the amber with bag-nets,
hung at the ends of long poles. They conclude that much
amber has been detached from its bed when they discover
many pieces of lignite floating about. In some parts the
faces of the precipitous cliffs along the shore are explored in
boats, and masses of loose earth or rock, supposed to contain
the object of search, are detached with long poles having iron
hooks at their ends. It is also dredged for on an extensive
scale at the bottom of the Frische and Curische Haffs, and
further inland large quantities are dug up out of the earth.
That which is washed ashore generally consists of small
pieces, more or less damaged, while the specimens obtained
by digging or dredging are frequently of large size and of a
tuberous form, so that, though inferior in quantity to the
former, their value is probably ten times greater.

Digging for amber is a favourite pursuit of the peasantry;
and though in many cases it proves unsuccessful, yet sometimes it is highly remunerative. Near the village Kowall, a
few miles from Dantzig, avenues of trees were planted a few

years back along the high road. On digging one of the holes destined for their reception, a rich amber nest was found. Favourable signs induced the landowner to persevere in digging, and at length, at a depth of about thirty feet, such rich deposits of amber were found as enabled him to pay off all the mortgages on his estate.

The territories where amber is found extend over Pomerania and East and West Prussia, as far as Lithuania and Poland, but chiefly in the former provinces, where it is found almost uniformly in separate nodules in the sand, clay, or fragments of lignite of the upper tertiary and alluvial formations. It also occurs in the beds of streams, and in the sand-banks of rivers. How far its seat may extend under the Baltic is of course unknown. Amber is likewise met with on the coast of Denmark and Sweden, in Gallicia and Moravia, near Christiania in Norway, and in Switzerland, near Basle. It is occasionally found in the gravel-pits near London,. and specimens have been dug up in Hyde Park. At Aldborough, after a raking tide, it is thrown on the beach in considerable quantities, along with masses of jet.

On the Sicilian coast, near the mouth of the River Giaretta, many pieces of a peculiar blue tinge are collected and sent to Catania to be cut and polished.

Single pieces and even large deposits of amber are said to have been discovered on the coasts of the Caspian Sea, in Siberia, Kamtschatka, and China, in North America and Madagascar. These accounts, however, require confirmation, as several other fossil or non-fossil resinous substances so strongly resemble amber as to have deceived even well-informed naturalists.

What is this substance, and how has it been produced? There is now no longer any doubt that, like other vegetable resins, it has been secreted by trees which have long since disappeared from the surface of the earth, but once formed extensive forests on the islands or shores of the vast sea which at that time covered the plains of Northern Europe as far as the foot of the Ouralian chain.

How those islands disappeared, and how those primeval forests came to be buried under land and sea, becomes apparent from the changes that have taken place in the

South Baltic lands since the last two thousand years in consequence of partial upheavings and subsidences. According to the oldest Prussian chronicler, Peter of Duesburg, whose narrative begins with the year 1226, the waves of the Baltic at that time reached as far as the present town of Kulm, and a century later vessels sailed as far as Thorn. The present delta of the Vistula was a shallow morass, dotted here and there with a flat island, and continued in that state until towards the end of the thirteenth century, when dykes, raised by the industry of man, prevented the constantly recurring inundations of the river, and converted gloomy swamps into fertile meadows.

In other parts we find the sea incroaching upon the land. Since the times of the Teutonic Order a whole province between Pillau and Balga has been submerged by the floods of the Frische Haff; and the first Christian church in Prussia, originally built five miles from the sea, now stands close to the shore. Dense fir-forests rose in gloomy monotony, but a thousand years back, where now the Baltic rolls its waters; and where at that time ships lay at anchor, we now find hillocks of sand. After such changes in comparatively so short a time, we cannot wonder that the islands of the amber period should have been replaced by other lands and another sea.

We are indebted for the first accurate observations on the nature of the amber-tree to Professor Goeppert, who proved, by the microscopical examination of the cells of fossil pieces of wood that were veined or streaked with amber, and thus evidently had secreted the resin, that they proceeded from several coniferæ, belonging to the extinct genus Pinites.

In many of our pines and firs we frequently find between the annual rings crevices or interstices filled with resinous matter, but far less abundantly than in the amber-trees. The only existing coniferous plant that can in any way be compared to them is the *Dammara australis*, of New Zealand, at the base of whose trunk masses of resin, weighing twenty or thirty pounds, are frequently found. In Brazil Von Martius often saw similar lumps at the foot of the copal-tree, which dropping from the trunk had collected in considerable masses between the roots, thus showing that the large rounded or globular pieces of amber must have been formed in the same manner, while the thin and flat straight or cupuliform pieces were moulded upon the rind, or between the annual rings of the tree.

When we consider the abundant secretion of the amber-trees, and the numberless ages during which they may have flourished, we cannot wonder that, since the oldest historic times, every violent storm which stirs up the ancient forest-grounds at the bottom of the Baltic casts the valuable fossil ashore, and that in all probability future generations will still be able to collect it in undiminished quantities.

Interesting in itself, amber acquires a still greater scientific importance through the remains of extinct plants and animals which are found imbedded in its substance, as in a transparent shrine. As, in the present day, many a luckless fly is caught in the recently secreted resins of the coniferæ and hymeneæ, thus also amber, while still in a semi-fluid state, became the tomb of numerous insects and spiders. So wonderful is their preservation that they seem to have lived but yesterday, and yet how many millenniums may since have passed away; for although the amber formation belongs to the miocene period, and is consequently of modern date when compared with the forests of the coal-formation, we still are separated from it by a vast series of ages.

INSECTS OR VEGETABLE SUBSTANCES ENCLOSED IN AMBER.

The extinct organic world which is thus beautifully revealed to us so greatly resembled the present vegetable and animal creation that, on a superficial examination of the fossil remains contained in a rich amber collection, one would hardly suppose them to be anything very uncommon. The unlearned observer who connects the idea of a past world with grotesque or gigantic forms, shakes his head with an incredulous smile, and thinks that he has often seen

similar flowers blooming in the fields, or met with similar
insects in the forest. Even the naturalist is uncertain, until
a closer inspection teaches him that each of these so wonderfully preserved plants or animals possesses some distinct
characteristics which widely distinguish it from the analogous
forms of the present day.

Of the plants of the coal period we find not a single one
existing in the Amber Forest; the vegetation is much more
complicated and various in its aspect; and the numerous
coniferæ indicate a climate similar to that of the present
northern regions. But arboreal growth was by no means
confined to the coniferæ, for evergreen oaks and poplars
flourished along with them.

Heath plants, chiefly belonging to genera similar to
Andromeda, *Kalmia*, *Rhododendron*, *Ledum*, and *Vaccinium*,
as testified by numerous leaves, formed the underwood of
these forests—a vegetation similar in character to that of the
Alleghany Mountains.

The deep shade of these primitive woods prevented the
evaporation of water, the ground remained damp and swampy,
and the mouldering leaves produced a thick layer of humus;
on which flourished, no doubt, a dense cryptomagous vegetation, as well-preserved ferns, mosses, lichens, confervæ, and
small mushrooms (partly growing as parasites on dead insects), sufficiently testify.

But the vegetable remains of that ancient period are far
surpassed in variety and number by the embalmed relics of
the animal creation. Among 2,000 specimens of insects
collected by Dr. Berendt, to whom we owe the best mono-
graph on amber, this naturalist found more than 800
different species.

Flies, phryganeæ, and other neuroptera, crustacea, mille-
pedes and spiders, blattidæ in every phase of development
beetles, bees, and a large variety of ants, show a great similarity to the insect life of the present day, and justify the conclusion that the contemporaneous animal forms, whose size
or peculiar habitat prevented their being embalmed in amber,
were comparatively no less abundant.

Whether man already existed at the time of the amber-
formation is a question which, of course, could only be

thoroughly settled by the discovery of some specimen of human workmanship *imbedded in* the fossilised resin. At all events, the amber-rings of rude workmanship which have been found at a considerable depth below the surface of the earth, along with rough pieces, sufficiently prove that man must have been a very old inhabitant of the Baltic regions, · for those remarkable specimens of his unskilled industry have evidently preceded the catastrophe which buried the rough amber under the earth, and must have been exposed for the same lapse of time to the influences of the soil, as they are all found covered with the same dull and damaged crust.

That birds enlivened the amber-forest might well have been supposed, as there was no want of fruits and mealy seeds for their subsistence; but their existence is proved beyond all doubt by a feather which Dr. Berendt discovered in a piece of pale yellow transparent amber. To what bird may this remarkable relic of the past have belonged, and when may the wing to which it was attached while living have cleaved the air?

No fish or reptile has ever yet been found in amber, however frequently fraud may have attempted to imbed them in a resinous case for the deception of ignorance. It is, indeed, hardly conceivable that the finny and agile inhabitants of the waters could ever have allowed themselves to be caught in the resins of a terrestrial forest, though some small and less active reptiles may occasionally have been entrapped.

Of all the insects and spiders, and the more rare crustaceans inclosed in amber, not a single specimen belongs to a species of the present time; but though the species have disappeared, almost all the animals of those primitive woods, as far as they are known, belong to genera of the present time; so that upon the whole, the proportion of the still flourishing genera to such as are extinct is as eight to one.

It is remarkable that, along with many specimens similar to the present indigenous types, some are found with a tropical character, whose representatives are at present existing in the Brazilian forests, while others are completely without any analogous forms in the present creation; as, for instance, those strange Archnidans, the Archæi, which, armed with toothed mandibles longer than the head, and provided with strong raptorial claws, must have been most formidable enemies to the contemporaneous insects.

Amber was held in high estimation by the nations of antiquity, and reckoned among the gems on account of its rarity and value. Ornaments made of this substance have been found among the vestiges of the lacustrine dwellings of Switzerland, and afford a convincing proof that even in pre-historic times it was an article of commerce. Many centuries before the Christian era the Phœnician navigators purchased amber from the German tribes on the coast of the North Sea; these, in their turn, having obtained it, probably by barter, from the Baltic lands. Thus from hand to hand the beauti-ful fossil resin found its way to the courts of the Indian princes on the Ganges, and of the Persian kings in Susa and Persepolis. According to Barth,* the search for the Amber Land was most probably the aim of the journey which the celebrated traveller Pytheas of Massilia undertook 330 years before Christ, in the times of Alexander the Great.

Plato and Aristotle, Herodotus and Æschylus, have de-scribed and lauded in prose and verse the wonderful properties of amber, which was not only highly valued for its beauty, its aromatic smell, and its electro-magnetic power, but also for the medicinal virtues ascribed to it by a credulous age.

Under Nero the wealthy Roman senators and knights lavished immense sums on decorating the seats and tables, the doors and columns, of their state rooms with amber, ivory, and tortoiseshell; and even at a later period, under Theodosius the Great, when the declining empire was already verging to its fall, large quantities still continued to be imported from Germany.

Though no longer so highly prized as by the ancients, amber still continues to be a source of considerable profit to the Baltic provinces. Almost all the amber collected throughout the land finds its way to the seaports of Königs-berg and Dantzig, where it is sorted according to its size and quality. Good round pieces of a shape fit to be worked

* Urgeschichte Deutschlands.'

into ornaments, and weighing about half an ounce, are worth from nine to ten dollars per pound; a good piece of a pound weight fetches as much as fifty dollars; and first-rate specimens of a still more considerable size, and faultless in form and colour, are worth at least one hundred dollars, or even more, per pound. A mass weighing thirteen pounds has been found, the value of which at Constantinople was said to be no less than 30,000 dollars. Smaller pieces, from the size of a bean to that of a pea, such as are fit for the beads of necklaces or rosaries, are valued at from two to four shillings per pound, and the grit or amber rubbish which is used for varnishing, fumigating, or the manufacture of oil and acid of amber, is worth no more than from three to eighteen pence. It is much to be regretted that amber, when melted or dissolved, is incapable of coalescing into larger masses with the retention of all its former qualities, as then its value would be considerably greater. Large amber vases would then ornament the apartments of the wealthy, and the corpses of the illustrious dead might repose in transparent shrines, and their features be preserved from decay for many ages.

The trade in rough amber is almost exculsively in the hands of the Jews, who purchase it from the amber-fishers, or are interested in the diggings which are made on most of the littoral estates. Through the agency of the smaller collectors, it is then concentrated in the hands of the rich traders, who sell or export it in larger assortments.

The best qualities only of translucent, milky, or semi-opaque amber, find a ready sale in the Oriental market, where they are almost exclusively used for making the mouth-pieces to pipes, and these form an essential constituent of the Turkish tschibouque; for there is a current belief among the Eastern nations that amber is incapable of transmitting infection.

Every Turkish pasha sets his pride on a rich collection of pipes, as it is the hospitable custom of the Orient to offer a cup of coffee and a hookah or tschibouque to a stranger; and this fashion is of no small importance to the amber-dealers of the Baltic. A somewhat inferior quality is sent by way of Copenhagen and London to China, Japan, and to the East and West Indies.

Russia also consumes a considerable quantity of amber, which is very elegantly turned or manufactured in St. Petersburg and Polangen, and thence finds its way over the whole empire. Here, as among the Turks, only the translucent and perfectly opaque white qualities are esteemed; the latter being chiefly employed for the manufacture of the calculating tables which are commonly used by the Russian merchants. Necklaces of transparent amber are in great request among the peasantry of Hanover and Brunswick, where strings of pale-coloured crystalline beads weighing from half a pound to a pound are worth from fifty to sixty dollars.

Amber of a deeper colour and of a rounded form is chiefly exported to Spain, France, and Italy.

Thus each country chooses according to its taste among the abundant amber-masses which extinct forests furnish to the inhabitants of the Baltic coast-lands, and which trade, through a hundred known and unknown channels, scatters over the whole surface of the globe.

CHAPTER XXXVII.

MISCELLANEOUS MINERAL SUBSTANCES USED IN THE INDUSTRIAL ARTS.

Alum—Alum Mines of Tolfa—Borax—The Suffioni in the Florentine Lagoons—China Clay: how formed—Its Manufacture in Cornwall—Plumbago—Emery—Tripolite.

ALUM, a double salt, consisting of sulphate of alumina (the peculiar earth of clay) and sulphate of potash, or sulphate of alumina and sulphate of ammonia, was known to the ancients, who used it in medicine, as it is now used, and also as a mordant in dyeing, as at the present day. Their alum was chiefly a natural production, which was best and most abundantly obtained in Egypt. In later times Phocis, Lesbos, and other places, supplied the Turks with alum for their magnificent Turkey red; and the Genoese merchants imported large quantities from the Levant into Western Europe for the dyers of red cloth. In 1459 Bartholemew Perdix, a Genoese merchant who had been in Syria, observed a stone suitable for alum in the Island of Ischia, and burning it with a good result, was the first who introduced the manufacture into Europe. About the same time John di Castro learnt the method at Constantinople, and manufactured alum at Tolfa. This discovery of the mineral near Civita Vecchia was considered so important by John di Castro that he announced it to the Pope as a great victory over the Turks, who annually took from the Christians 300,000 pieces of gold for their dyed wool. The manufacture of alum was then made a monopoly of the Papacy; and instead of buying it as before from the East, it was considered a Christian duty to obtain it only from the States of the Church. With the progress of Protestantism, however, the manufacture began to spread to other countries. Hesse-Cassel began to make alum in 1554, and in 1600 alum-slate was found near Whitby in Yorkshire.

Alum-stone is a rare mineral, which contains all the elements of the salt, but mixed with other matters from which it must be freed. For this purpose it is first calcined, then exposed to the weather, in heaps from two to three feet high, which are continually kept moist by sprinkling them with water. As the water combines with the alum, the stones crumble down and fall eventually into a pasty mass, which must be lixiviated with warm water, and allowed to settle in a large cistern. The clear liquor on the surface, being drawn off, is to be evaporated and then crystallised. A second crystallisation finishes the process, and furnishes a marketable article. Thus the Roman alum is made; but its production being far from enough for the supply of the world, the greater portion of the alum found in British commerce is made from alum-slate and analogous minerals, which contain only the elements of two of the constituents, namely, clay and sulphur, and to which, therefore, the alkaline ingredient must be added.

Borax, or Borate of Soda, is a substance extensively used in the glazes of porcelain, and recently in the making of the most brilliant crystal when combined with oxide of zinc. Formally its chief supply was obtained under the name of Tincal, from Thibet, where the crude product was dug in masses from the edges and shallow parts of a salt lake; and, in the course of a short time, the holes thus made were again filled. Crude borax is also found in China, Persia, the Island of Ceylon, and in South America; but at present by far the largest quantity used in commerce is derived from the lagoons of Tuscany, where vapours charged with a minute quantity of boracic acid rise from volcanic vents.

Before the discovery of this acid, in the time of the Grand Duke Leopold I., by the chemist Hoeser, the fetid odour developed by the accompanying sulphuretted hydrogen gas, and the disruptions of the ground occasioned by the appearance of new suffioni or vents of vapour, had made the superstitious natives regard them as a diabolical scourge, which they vainly sought to remove by priestly exorcisms; but since

1818 the skill, or rather the industrial genius, of Count Larderel, originally a simple wandering merchant, has rendered these once abhorred fugitive vapours a source of prosperity to the country, and were they to cease, all the saints of the calendar would be invoked for their return.

By a most ingenious contrivance the waters which have been impregnated by the volcanic streams are concentrated in leaden reservoirs by the heat of the vapours themselves. The liquid, after having filled the first compartment, is diffused very gradually into the second, then into the third, and successively to the last, where it reaches such a state of concentration that it deposits the crystallised acid, which the workmen immediately remove by means of wooden scrapers. As this mode of gradual concentration requires very few hands and no artificial heat—a circumstance of great importance in a country without fuel—it may almost be said that the acid is obtained without expense. The produce of the lagoons amounts to more than 2,000 tons annually.

China-clay, or kaolin, a substance largely used for the fabrication of porcelain and in paper-making, is quarried from amidst the granitic masses of Dartmoor and Cornwall. It results from the decomposition of felspar, the chief constituent of granite, and is thus, to a certain point provided by nature. As it is, however, mixed with many grosser particles, it requires repeated washings in a series of small pits or tanks, until finally the water, still holding in suspension the finer and purer china-clay, is admitted into larger ponds. Here the clay is gradually deposited and the clear water on the surface is from time to time discharged by plug-holes on one side of the pond. This process is continued until, by repeated accumulations, the ponds are filled. The clay is then removed to large pans about a foot and a half in depth, where it remains exposed to the air until it is nearly dry—a tedious operation in our damp climate, as during the winter at least eight months are necessary, whilst during the summer less than half the time suffices.

When the clay is in a fit state it is cut into oblong masses and carried to the drying-house, a shed the sides of which are open wooden frames constructed in the usual way for keeping off the rain, but admitting the free passage of the air. The clay thus prepared is next scraped perfectly clean, and is then packed up into casks and carried to one of the adjacent ports, to be shipped for the potteries or the paper manufactories. It is of a beautiful and uniform whiteness, and is perfectly smooth and soft to the touch.

China-clay is an important article of commerce. The production of Cornwall and Devonshire amounted in 1873 to 180,197 tons, and it is reckoned that more than 300,000*l.* is annually spent in Cornwall in obtaining and preparing it.

As in our times no branch of industry remains unimproved, a machine has recently been invented which greatly accelerates the drying. The lumps of china-clay are placed in the compartments of the drying machine, and the whole is then rotated with great velocity. Thus the water is thrown off by the operation of the centrifugal force, and two tons of clay can be dried in five minutes.

The remarkable contrast between the extreme hardness and softness of the same primitive rock is a curiosity of geology well worth notice. 'A stone,' says the author of 'Cornwall and its Mines,' is primevally fused and cooled into as hard a substance as nature affords; so hard that it paves our streets in long enduring slabs and blocks. Tens of thousands of feet passing daily over it make no impression upon it. Years of travel only see it smooth and shining. The men who pass over it pass away, but it is still durable and unsoftened. Halls and mansions, clubs and palaces, are built of it, and it endures unmarked by Time's devouring tooth. Is it not adamantine? But lo! what is that elegant cup from which you are sipping your tea? It is of Worcestershire china, fine and almost pellucid. Well, it is made out of the soft ruins of that very granite whose endurance is proverbial. How remarkable that the best types of firmness and fragility should be found in the same stone! What the immense billows of the Atlantic are now beating upon, at the Land's End, without effect, that very substance has been found in such a friable natural condition as to be finally moulded into that vase which stands, elegant and admired, upon your mantlepiece, and which one puff of wind, or one whirl of a lady's silk dress, would dash down into innumerable and unmendable fragments.'

Valuable kaolins are likewise found in China and Japan, in Saxony and France; but the production is nowhere so extensive as in Cornwall.

The best quality of Plumbago, graphite, or black-lead, from which the finest lead-pencils were made, was formerly abundantly supplied by the mine of Borrowdale in Cumberland. At one time as much as 100,000l. was realised in a year, the plumbago selling at forty-five shillings per pound. Its high value, however, proved a source of loss to the proprietors, for robbery found here a most profitable field. Even the guard stationed over it by the owners was of little avail against determined thieves, for about a century ago a body of miners broke into the mines by main force, and held possession of them for a considerable time. The treasure was then protected by a building consisting of four rooms upon the ground floor, and immediately under one of them is the opening, secured by a trap-door, through which alone workmen could enter the interior of the mountain. The mine has now, however, not been worked for many years. Its 'nests' of plumbago seem to be exhausted, and it is no longer an object capable of exciting greed, or worthy of anxious protection. At present the mines of Batougal, near Irkutsk in East Siberia, furnish large quantities of graphite of the very best quality; and the plumbago from Ceylon is likewise highly esteemed. Inferior qualities—employed for counteracting friction between rubbing surfaces of wood or metal, for making crucibles and portable furnaces, for giving a gloss to the surface of cast iron, &c. &c.—are supplied by Lower Austria, Bohemia, Bavaria, Ticonderago in the State of New York, New Jersey, Arendal in Norway, Finland, &c. &c.

The name black-lead applied to plumbago, from its resembling lead in its external appearance, is very inappropriate, as not a particle of that metal enters into its composition. Plumbago might indeed more justly be called black-diamond, for, consisting, in its pure state, of carbon, it is consequently but another form of the chief of precious stones.

The beautiful oriental ruby has likewise a humble cousin among the useful minerals, as it is identical in composition with Emery, which, since the remotest antiquity, has been employed for the grinding of metals and glass. Emery was formerly almost exclusively furnished by the island of Naxos; but since 1847 new mines have been discovered in some other isles of the Grecian Archipelago, and in the neighbourhood of Smyrna, from which place it is now imported in large quantities. In 1850 the total production of the East amounted to 1,500 tons. In 1864 a considerable mine of emery was discovered at Chester (Hampden County, Massachusetts), and—as might be expected in a country where none of the gifts of nature are allowed to be wasted—already provides for the wants of all the manufactories of the United States.

Tripolite, a mineral related to the precious opal, as it consists almost entirely of silica, is likewise used for polishing stones, metals, and glasses. Its composition is truly remarkable, as it is actually formed of the exuviæ, or rather the flinty envelopes of diatoms, which belong to the minutest forms of vegetable life. They are recognised with such distinctness in the microscope that the analogies between them and living species may be readily traced, and in many cases there are no appreciable differences between the living and the petrified. As every cubic inch of tripolite contains millions of these exuviæ, and the stone not seldom occurs in deposits often many miles in area, imagination is at a loss to conceive the innumerable multitudes of organised atoms whose flinty remains have been piled up in these masses of hard rock.

CHAPTER XXXVIII.

CELEBRATED QUARRIES.

Carrara—The Pentelikon—The Parian Quarries—Rosso antico and Verde antico—
The Porphyry of Elfdal—The Gypsum of Montmartre—The Alabaster of Volterra—The Slate Quarries of Wales—'Princesses' and 'Duchesses'—'Ladies'
and 'Fat Ladies '—St. Peter's Mount near Maestricht—Egyptian Quarries—
Haggar Silsilis—The Latomiæ of Syracuse —A Triumph of Poetry.

BESIDES metals, and the various minerals mentioned in
the previous chapter, the solid earth-rind furnishes an
inexhaustible supply of marbles, slates and stones, for
building or paving; and their extraction occupies a vast
number of industrious hands.

In a popular work on geology, published some years ago,
Mr. Burat informs us that about 70,000 persons were employed in the 18,000 more or less important quarries at that
time worked in France, and that the produce of their labour
amounted to a value of more than 2,000,000*l.*

There can be no doubt that the quarries of England or
Germany are at least equally productive, and thus a very
moderate estimate leads us to the conclusion that the
quarries of Europe—from those which furnish the costliest
marble to those which yield the commonest building-stone—
employ at least half a million of workmen, and produce an
annual value of no less than 12,000,000*l.*—a sum which is
probably doubled or trebled before the heavy materials can
be placed in the hands of the consumer. A land ribbed with
stone, like England, has therefore a considerable advantage
over a flat alluvial plain like Holland, as it possesses in its
rocky foundations a source of wealth which nature has denied
to the latter.

Though several other stones, such as granite and porphyry,

are susceptible of a fine polished surface, and serve for the decoration of palaces and churches, yet marble or pure compact limestone is chiefly used for ornamental purposes, both on account of its beautifully variegated tints, and its inferior hardness, which allows it to be more easily worked. Our Derbyshire and Devonshire quarries supply a great variety of richly coloured marbles; but the best material for the sculptor is supplied by the limestone mountains of Carrara, which furnish a homogeneous marble of the purest white, with a fine granular texture, resembling that of loaf sugar. These far-famed quarries, which were worked by the ancients, having been opened in the time of Julius Cæsar, are situated between Spezzia and Lucca, in the Alpe Apuana, a small mountain-group no less remarkable for its bold and sharp outlines than for its almost total isolation from the monotonous chain of the Apennines, from which it is separated by a wide semi-circular plain. Where the Alpe Apuana faces the sea, it is chiefly formed of magnesian and glimmer slate, in which large masses of limestone are imbedded; but the more inland part of the group belongs entirely to the limestone formation, and abounds in romantic scenery and noble peaks towering to a height of six thousand feet above the level of the sea. Towards its north-western extremity rises Monte Sacro (5,200 feet in height), the famous marble mountain on whose slopes are scattered the quarries to which the small town of Carrara owes its ancient and world-wide celebrity.

The quarries themselves by no means afford an imposing sight, as they are mostly small, and very badly worked; but it is interesting to watch the transport of the huge blocks of superb material from the various glens in which the quarries are situated, while the numerous water-mills for cutting or polishing the marble enliven the whole neighbourhood.

In the town of Carrara numerous sculptors are constantly employed in rough-hewing the marble into various forms, such as capitals, friezes, busts, &c., &c., in order to diminish the cost of transport, or to discover faults in the stone before it is shipped. There are also shops where marble trinkets or ornaments are exposed for sale; but Florence, Leghorn, and

Genoa are the chief dépôts of ready-made marble articles, such as vases, urns, sculptured chimney-pieces, and copies of renowned statues. Different kinds of fruit are also executed in marble, and with the aid of colour made to imitate nature so closely as to deceive the eye.

In Carrara the inferior qualities of marble are used for building and paving, as it is here the cheapest material. The window and door-frames, the flooring and the chimney-slabs, in even the meanest houses, are made of marble, and form a striking contrast to the squalid poverty of the remainder of the furniture.

The quarries which furnished the material for the finest works of the Grecian chisel partake of the interest which attaches to every vestige of ancient art.

About eight miles to the north of Athens rises the Pentelikon, or mount Penteles, from whose flanks was excavated the marble that served for the construction of the Parthenon, of the Temple of the Olympian Zeus, and of the other matchless edifices of the Athenian Acropolis. No other quarries in the world can boast of their material having undergone a more beautiful transformation, for never has marble been more highly ennobled than by the genius of Phidias! The ancient roads ascending from the foot of the mountain to the quarries still show the traces of the sledges on which were transported the huge blocks of more than twenty tons in weight that now lie scattered among the ruins of the Acropolis.

On the summit of the Pentelikon the Athenians had placed a statue of Pallas Athene, that the goddess might overlook the land devoted to her worship. Here, from a height of 3,500 feet, she looked down upon the plain of Marathon, and many other spots of everlasting renown; but the outlines of the prospect are monotonous and naked, and require for their embellishment the beautifying remembrances of the past.

Mount Marpena, in the island of Paros, furnished the most renowned statuary marble of ancient times. It was called Lychnites, because its quarries were worked by lamp-light, in deep mine-like excavations; and the difficulty and cost of its extraction show how highly it was prized. It has a yellowish white colour and a texture composed of fine shining scales lying in all directions. The celebrated Arundelian marbles at Oxford consist of Parian marble, as does also the Medicean Venus. More than twenty centuries have elapsed since the Parian quarries were abandoned in consequence of the decay of Grecian art; but in our enterprising days a company has been formed (1857) for working the beautiful marble which has been recently discovered near St. Minas, not far from the site of the ancient quarries, and is said to be superior not only to that of Carrara but even to the renowned Lychnites of the ancients.

The quarries which in olden times furnished the beautifully coloured marbles called Rosso antico and Verde antico had likewise for many centuries been abandoned and totally forgotten. In 1846 they were rediscovered on the island of Tino and in the Maina by Professor Siegel, who soon after undertook to work them, and has furnished, among others, a large number of the beautiful columns of Rosso antico which decorate the interior of the court of the basilica of St. Paul's in Rome. The hammer of the quarryman once more resounds in the wilds of the Taygetos, and the lawless robber of the Maina already feels the beneficial influence of industry.

On the slopes of the mountain which bound the impetuous Oesterdal Elbe in Sweden, in a wild and desolate country, where the poverty of the people is so great that they frequently grind the bark of fir-trees to mix it with their bread, are situated the finest quarries of porphyry which Europe possesses. This beautiful stone, which attracts the eye even in its unpolished state, consists of a red-brown or blood-red mass, in which numerous small flesh-coloured felspar pieces are embedded. After having been rough-hewn on the spot, the blocks are transported to the neighbouring works of Eldfdal, where they are cut and polished into slabs, vases, chimney-pieces, and other articles fit for the decoration of palaces. The contrast is most striking when, after having traversed the barren neighbourhood, and still deeply impressed with the sight of poverty and distress on his road, the traveller suddenly finds himself before a group of handsome buildings which at once bear witness to the activity within.

Besides the red porphyry of Elfdal, that of the Altai Mountains in Asia deserves to be noticed. It consists of a brown-red mass with snow-white crystals, and is capable of a very fine polish. The quarries are situated on the face of a high rock on the left bank of the Kurgun, one of the wildest mountain streams of the Altai, about one hundred miles from the town of Kolywansk, where it is cut and polished.

Gypsum, or sulpate of lime, and the peculiar form of that mineral called Alabaster, are substances of considerable importance in the arts. Rendered more valuable by a slight admixture of carbonate of lime, the gypsum of Montmartre, near Paris, has long been celebrated for its excellence as a cement or stucco. It is found resting on a limestone of marine origin, and in some places appears immediately beneath the vegetable soil, so that it can be readily and conveniently worked without having recourse to subterranean excavation. These quarries furnish the whole of northern France with the well-known plaster of Paris, and the value of their annual produce amounts to not less than 100,000*l*.

When sulphate of lime or gypsum assumes the opaque, consistent, and semi-transparent form of alabaster, it is worked like marble. The pure white and harder varieties are usually employed for the sculpture of statues and busts; while the softer kinds are cut into vases, boxes, lamps, and other ornamental objects. The alabaster quarries in the neighbourhood of the ancient Etruscan town of Volterra are the most famous in Europe, and have afforded employment for many centuries to her industrious population. Volterra exports her beautiful produce to all parts of the world, even as far as the interior of China. Beggary is here unknown (a rare case in Italy), for even women and children are all employed in cutting, sawing, rasping, or filing alabaster. In the remotest antiquity, when the city was still called Volathri or Volaterræ, this industry was practised within her walls, and a collection of sepulchral urns and other works of Etruscan art contained in the town-hall bears testimony to her ancient skill. Now, however, art seems to have degenerated into mere manufacturing ability; the statues and other objects are almost always repetitions of the same models, and but very rarely some speculative person introduces a novelty, for the purpose of obtaining a somewhat higher price for his wares.

Great Britain possesses apparently inexhaustible quantities of alabaster in the red marl formation in the neighbourhood of Derby, where it has been worked for many centuries. The great bulk of it is used for making plaster of Paris, and as a manure, or as the basis of many kinds of cements. For these common industrial purposes it is worked by mining underground, and the stone is blasted by gunpowder; but this shakes it so much as to render it unfit for works of ornament, to procure blocks for which it is necessary to have an open quarry. By removing the superincumbent marl, and laying bare a large surface of the rock, the alabaster, being very irregular in form, and jutting out in several parts, can be sawn out in blocks of a considerable size and comparatively sound. This stone, when preserved from the action of water, which soon decomposes it, is extremely durable, as may be seen in churches all over this country, where monumental effigies many centuries old are still as perfect as when they proceeded from the sculptor's chisel. The Derbyshire alabaster, commonly called Derbyshire spar, gives employment to a good many hands in forming it into useful and ornamental articles. Another kind of alabaster also found in Derbyshire is crystallised in long needle-like silky fibres, which, being susceptible of a high polish and quite lustrous, is used for making necklaces, bracelets, brooches, and other small articles.

Besides her inexhaustible coal, iron, and lead mines, Wales possesses in her slate quarries a great source of mineral wealth. For this article, which many would suppose to be but of secondary importance, is here found in such abundance and perfection as to command a ready market all over the world. Thus, in North Wales the face of the mountains is everywhere dotted or scarred with slate quarries, of which by far the most important and largest are those of Llandegui, six miles from Bangor, in which more than three thousand persons are employed. This circumstance alone will give an idea of their extent, but still more their having their own harbour, Port Penrhyn, which holds vessels of from 300 to 400 tons, and whence slates are sent not only to all parts of Great Britain, but even to North America. The cost of the inclined planes and railroads which serve to transport the slates from the quarries to the port is said to have amounted to 170,000l. The masses of slate are either detached from the rock by blasting, or by wedges and crowbars. They are then shaped on the spot into the various forms for which they are destined to be used.

Though the quarries of Llandegai are unrivalled, yet there are others which give employment to workmen whose numbers range from a thousand to fifteen hundred. In those that export their produce by way of Carnarvon, which owes its prosperity almost entirely to this branch of industry, about 2,300 men were employed in the year 1842, and since then their number has very much increased. Walking on the handsome new quay of the town, the visitor everywhere sees slates ready for shipment in the numerous small vessels which crowd the picturesque harbour. They are heaped up and arranged or sorted in large regular piles according to their dimensions or quality, the 'ladies' and 'fat ladies' apart, as also the 'countesses,' 'marchionesses,' 'princesses,' 'duchesses,' and 'queens,' for by these aristocratic names the slate merchants distinguish the various sizes of the humble article they deal in.

Kohl, the celebrated traveller, remarks that, whenever a new branch of industry springs up in England, a number of active hands and inventive heads set to work to extend its application. This was also the case with the Welsh slates, when, about half a century ago, they first began to attract the notice of the commercial world. A polish was soon invented in London which gave them the appearance of the finest black marble. As they are easily worked with the aid of a turning lathe, they can thus be used for the manufacture of many useful and ornamental articles, which have all the lustre of ebony, and may be obtained at a much cheaper rate.

The quarries of St. Peter's Mount, near Maestricht in the Netherlands, are probably the most extensive in Europe. For the white tufaceous limestone of the mountain has been used from time immemorial both for building and manuring, and the enormous extent of its caverns is perhaps even owing in a greater measure to the agriculturist than to the architect. And yet St. Peter's Mount, situated near a town of moderate extent, would hardly have been excavated beyond the limits of a common quarry if the broad Meuse, flowing at its foot, had not opened towards Holland an almost unlimited market for its produce. Sure of being able to dispose of all the materials they can possibly extract from the earth, the quarrymen of the neighbourhood are thus yearly adding new passages to the labyrinths at which so many generations of their forefathers have toiled from age to age.

In spring and summer they are mostly occupied with agricultural pursuits, and then but rarely leave the light of day to burrow in the entrails of the earth by the dismal 'sheen' of smoky lamps; but as soon as the approach of winter puts a stop to the labours of the field, they descend into the cavities of the mountain, and begin to excavate the vast mounds of stone or grit which are to be shipped in the following spring.

Accompanied by a guide, the stranger enters these amazing quarries, that extend for miles into the interior of the mountain, and is soon lost in wonder at their endless passages, perpetually crossing each other to the right and left, and ending in utter darkness. The dismal grandeur of these dark regions is increased by their awful silence, for but rarely a drop of water falls from the vaults into a small pool below, and even the voice of man dies away without awakening an echo.

But in order to make one feel the full impression of night and silence, the guide, after penetrating to some distance into the interior of the cave, extinguishes his torch. A strange sensation of awe then creeps over the boldest heart, and by an almost irresistible instinct the stranger seeks the nearest wall, as if to convince himself that he has at least the sense of touch to depend upon where the eye vainly seeks for the least ray of light, and the ear as vainly listens for a sound. Then also he feels how dreadful must have been the agonies of the despairing wretches who, having lost their way in these dark labyrinths, prayed and wept and shouted in vain, until their last groans died away unheard.

It has not seldom happened that persons have hopelessly strayed about in these vast caverns, and there slowly perished, while but a few fathoms above the labourer was driving his plough forward or the reaper singing his evening song. In several parts of the caverns an inscription on the walls points out where and under what circumstances a corpse has been found, or a few traces of black chalk give a rude portrait of the victim. Here we read the short story of a workman who, losing his way, roamed about till the last glimmer from his torch died out in his burnt fingers; there that of another whose lamp by some chance was overturned, and who, plunged in sudden darkness, was no longer able to find his way out of some remote passage.

The French geologist, Faujas de Saint-Fond, who, in the year 1798 minutely explored the Mount, relates that one day the torches that were carried before him discovered at some distance a dark object stretched out upon the floor, which on a closer inspection proved to be a corpse. It was a shrivelled mummy, completely dressed, the hat lying close to the head, the shoes separated from the feet, and a rosary in its hand. From the dress Faujas conjectured it to be the body of a workman, and perhaps more than half a century might have passed since the poor man died in the quarry which had probably given him bread for many a year. The dry air, and the total absence of insects in these subterranean vaults, explained the mummy-like preservation of the corpse.

In the year 1640 four Franciscan monks resolved, for the greater glory of God and of their tutelary saint, to construct a chapel in a deserted part of the caverns. Taking with them a thick pack of thread, they attached one end of it to a spot where the trodden path ceased, and penetrating deeper and deeper into the unknown vaults as long as their thread lasted, finally came to a larger cavity or hall, which had probably not been visited for centuries, but which in consequence of their misfortune has since become one of the show-places of the quarry. At the entrance one of them drew with a piece of coal a still existing sketch of his convent on the wall, and wrote underneath the date of the discovery which was to be so fatal to him and his companions. What must

have been their dismay when, wishing to retrace their steps,
they discovered that, by some accident, the thread which
alone could lead them out of the labyrinth had been
severed, and that they were left without guide or compass
in the inextricable maze of the caverns.   The prior,
alarmed at their prolonged absence, and knowing that
they had visited the quarries, ordered them to be sought
for.   But such is the vast extent of the excavations that
it was only after seven days that their corpses were
found, lying closely together, their faces downwards and
their hands folded, as if their last moments had been spent
in prayer.

The caverns of St. Peter's Mount, generally devoted only
to the pacific labours of the quarryman or to unbroken
silence, have once been the scene of a bloody fight; and as
the quiet waters of the Todtensee, on the heights of the
Grimsel, have witnessed a sanguinary battle between the
French and the Russians, so these subterranean vaults have
once resounded with the din of fire-arms and the shouts
of embittered enemies.

During the siege above mentioned, which brought
Maestricht into the power of the French Republic, some
sharpshooters of the besieging army took up their position
in the quarries.   The Austrians, who occupied Fort St.
Pierre, on the back of the mountain, and had already made
several successful sorties, formed the plan of penetrating
into the caverns from the interior of the fort, in the hope of
surprising the enemy who occupied their entrances.   But as
the torches which lighted their silent march betrayed their
intentions, the Frenchmen cautiously and slowly advanced
upon them, surprised them with a sudden volley of musketry,
and drove back into the depths of the caverns all those who
were not made prisoners or killed.

To the geologist the quarries of St. Peter's are particularly
interesting, as the calcareous tuff of which they are composed
is extremely rich in valuable petrifactions.   Here, among
others, was found, in 1770, the famous skull of the Mososaurus
Hoffmanni, a giant lizard twenty feet long, which, before
the discovery of the still more colossal Ichthyosauri in
England and Bavaria, was considered as the most remark-

able fossil known, and now forms one of the chief ornaments of the Museum of the Jardin des Plantes in Paris.

There are no subterranean animals peculiar to the caverns of St. Peter's, such as have been found in large natural caves; but foxes and martens not seldom find here a secure retreat, and many a bat hibernates in their warm recesses.

Near the small village El Massara, which, like all the hamlets on the Nile, stands in a grove of date-palms, are situated the quarries which furnished materials for the temples and pyramids of Memphis. After having first traversed a wide plain of sand, the visitor reaches the foot of the Mokattam mountain. Here the ground is everywhere covered with enormous heaps of rubbish from the quarries, which look more like an attempt to cut the whole mountain into blocks and carry it away than simple excavations for building materials. At first sight the traveller might indeed almost fancy that all the eighteen thousand towns of ancient Egypt must have been dug out of these stupendous excavations. Here he may see the mountain cut through from top to bottom, and open spaces as large as the squares of a great capital on a level with the plain, while there the rock had been hollowed out into enormous halls, their roofs reposing on titanic pilasters. In one spot the rocky wall has been cut rectangularly, after which it runs parallel to the river for a distance of about fifteen hundred feet, and then again projects at a right angle towards the plain, so that in the space between the wings there would be room enough for a small town, yet all the stone which once filled this vast cavity has been sawn from the rock and carried over the Nile. Here and there enormous masses of stone, like those which in winter roll from the high Alps into the valleys, have fallen from the overhanging rock. Many bear evident traces that the hand of man originally severed them from the mountain; but their size was such as to baffle even the perseverance and mechanical skill of the Egyptians. No power could remove them from the spot; and thus they remained to be worn away by time, which as yet, however, has hardly left a trace of its passage upon their chiselled sides.

From the mounds of rubbish that lie before the openings of the quarries there is a magnificient view of the long line of pyramids which mark, to the west, the extreme limits of the fertile land. Here, indeed, is a place for meditation, for nowhere do stones preach a more impressive sermon!

The quarries of Haggar Silsilis, in a wild mountainous country between Assuan and Edfu, are on a still grander scale. Passages as broad as streets, running on both sides between walls fifty or sixty feet high, now stretching straight forward, now curving, extend from the eastern bank of the river into the heart of the mountain, where spaces have been hollowed out large enough to embrace the Roman Colosseum! To the north numberless Cyclopean caverns have been hewn out, and enormous colonnades stretch along the foot of the mountain. The rough-hewn irregular roof rests upon immense square, or many-sided, pillars, frequently eighty or a hundred feet in circumference. Enormous blocks, already completely separated from the rock, rest on smaller ones, ready to be transported, while in other parts the labours of the quarrymen were suddenly arrested by the invasion of the foreign conqueror, who put a stop to the dominion of the priests. The Bedouin, astonished at these vast works, so alien to his roving habits, exclaimed at their aspect, with a wondering mien, ' Wallah! if these unbelievers had lived until now, they would have carried away the whole mountain and levelled it with the ground!

Syracuse, the proud city that vied with Rome itself, and the remembrance of whose magnificence and glory, both in arts and arms, will live as long as classical literature, is now reduced to a heap of rubbish, for her remains deserve not the name of a city. But though even her ruins have mostly disappeared, '*etiam periere ruinæ*,' yet the vast quarries which furnished the materials for her palaces and temples still bear witness to her ancient grandeur.

The Latomiæ of the Capuchins, thus named from a convent of that order situated on the rock above, now form a noble subterranean garden, one of the most romantic and beautiful spots imaginable. Most of it is about one hundred feet below the level of the earth, and of an incredible extent. The whole is hewn out of a rock as hard as marble, composed of a concretion of shells, gravel, and other marine bodies. The bottom of this immense quarry, from which probably the greatest part of Syracuse was built, is now covered with an exceedingly rich soil; and as no wind from any point of the compass can touch it, it is filled with a great variety of the finest shrubs and fruit-trees, which bear with prodigal luxuriance, and are never blasted. The oranges, citrons, pomegranates, and figs, are all of a remarkable size, and, frequently growing out of the hard rock, where there is no visible soil, exhibit a very uncommon and pleasing appearance. This quarry is the same that Cicero eloquently describes as the vast and magnificent work of the kings and tyrants of ancient Syracuse, hewn out of the rock to a prodigious depth. It also served as a prison for the Athenian soldiers that were made captives during the Peloponnesian war, and even now these vast excavations might be used for a similar purpose, as their high and overhanging walls forbid the possibility of escape, and ten men would be able to guard ten thousand without danger. Yet the genius of Euripides liberated many of his countrymen from this deep pit of bondage; for, happening to sing a chorus of one of his immortal tragedies, it moved the tyrant to restore them to liberty, in honour of their illustrious fellow-citizen. Never has a more grateful offering been awarded to a poet's genius, and never has the magical power of the Muses celebrated a nobler triumph.

CHAPTER XXXIX.

PRECIOUS STONES.

Diamonds—Diamond-cutting—Rose Diamonds—Brilliants—The Diamond District
in Brazil—Diamond Lavas—The great Russian Diamond—The Regent—The
Koh-i-Noor—Its History—The Star of the South—Diamonds used for Industrial
Purposes—The Oriental Ruby and Sapphire—The Spinel—The Chrysoberyl—
The Emerald—The Beryl—The Zircon—The Topaz—The Oriental Turquoise—
The Garnet—Lapis Lazuli—The Noble Opal—Inferior Precious Stones—The
Agate-cutters of Oberstein—Rock Crystal—The Rock-crystal Grotto of the
Galenstock.

IN former ages superstition ascribed a strange mysterious
power to precious stones. Gems of conspicuous size or
lustre were supposed to confer health and prosperity on their
owners, to preserve them in the midst of the most appalling
dangers, or even to give them a command over the world of
spirits.

The crucible of modern chemistry has, indeed, effectually
dispelled these illusions of a poetic fancy; but the precious
stones have lost nothing in value by their nature being
better known. They are still the favourite and most costly
ornaments of wealth and beauty, and they still deservedly
rank among the wonders of creation. For surely no fabled
talismanic virtues can be more worthy of admiration than
that natural power which in the secret laboratories of the
subterranean world has caused their atoms to unite in lustrous crystals, and imparted to such vulgar materials as
carbon, clay, or sand, the gorgeous reflections of the rainbow
or the glorious colours of the setting sun.

The diamond, it is almost unnecessary to say, is the chief
of precious stones, none other equalling it in brilliancy and
refractive energy. Although generally colourless, like pure
rock-crystal, yet it is also found of every variety of tint, from
a roseate hue to crimson red, or from a pale yellow to dark

green and blue, or even black. Colourless diamonds are in general most highly esteemed, but coloured stones are sometimes of an exquisite beauty, and of corresponding value. Blue is an exceedingly rare colour, and one of this shade, the celebrated Hope diamond, which weighs forty-four and a half carats,* and unites the charming colour of the sapphire with the prismatic fire of the diamond, is valued at 25,000*l.*

As the rough stones are rarely found with an even or transparent surface, the assistance of art is required to develop their full beauty. The diamond, being by far the hardest of all substances, can only be cut and polished by itself. Hence the lapidaries begin their operations by rubbing several diamonds against each other while rough, after having first glued them to the ends of two wooden blocks thick enough to be held in the hand. It is the powder thus rubbed off the stones, and received in a little box for the purpose, that serves to grind and polish them.

The process of diamond-cutting is effected by a horizontal iron plate of about ten inches' diameter, called a *schyf* or mill, which revolves from two thousand to three thousand times per minute, and is sprinkled over with diamond dust mixed with oil of olives. The diamond is fixed in a ball of pewter at the end of an arm resting upon the table on which the plate revolves; the other end, at which the ball containing the diamond is fixed, is pressed upon the wheel by movable weights at the discretion of the workman.

The method of cutting and polishing diamonds was unknown in Europe before the fifteenth century, but appears to have been practised long before in India, though in a rude manner. The original facetting of the Koh-i-Noor was the work of an unknown and prehistoric age.

The diamonds which were employed as ornaments before that period—as for instance the four large stones which enrich the clasp of the imperial mantle of Charlemagne, as now preserved in Paris—remained in their rough and uncut state.

The invention is ascribed to Louis von Berguen, a native of Bruges, then the great emporium of Western trade and luxury, who in the year 1476 cut the fine diamond of Charles the Bold; and ever since that time Antwerp and Amsterdam

---

* The carat is equal to 3⅕ grains Troy weight.

have maintained the first rank in the practice of an art which might be supposed to have a more appropriate seat in London or Paris, the centres of modern wealth and fashion.

Diamonds are generally cut either as rose diamonds or as brilliants. The rose diamond is flat beneath, while the upper face rises into a dome, and is cut into facets. The brilliant, which is always three times as thick as the rose diamond, is likewise cut into facets, but so as to form two pyramids rising from a common central base or *girdle*. Each pyramid is truncated at the top by a section parallel to the girdle, which cuts off $\frac{5}{18}$ of the whole height from the upper one, and $\frac{1}{18}$ from the lower one. The superior and larger plane thus produced is called the *table*; and the inferior and smaller one is called the *collet*. Although the rose diamond projects bright beams of light in more extensive proportion often than the brilliant, yet the latter shows an incomparably greater play, from the difference of its cutting. In executing this there are formed thirty-two facets of different figures, and inclined at different angles all round the table on the upper side of the stone, while on the under side twenty-four other facets are made round the small table. It is essential that the facets of the top and the bottom shall correspond together in sufficiently exact proportions to multiply the reflections and refractions, so as to produce the gorgeous display of prismatic colours which renders the brilliant so pre-eminently beautiful.

From the hardness of the diamond, its cutting is a very tedious and expensive operation, requiring more time in the proportion of fifty to one than the cutting of the sapphire, which comes next to it in hardness.

Experiment has determined that the diamond consists of pure carbon, so that the same substance which in its common black state is utterly worthless in very small quantities, becomes the most costly of precious stones, when it makes its appearance in the crystalline form. Already Newton, by observing the extraordinary refractive power of the diamond, had been led to place it among combustibles; but Cosmo III., Grand Duke of Tuscany, was the first who proved the truth of this bold conjecture by actual observation. He exposed diamonds to the heat of the powerful burning-glass of Tschirnhausen, and saw them vanish in a few moments into air. The formation of the diamond in nature is one of the many problems which ' our philosophy ' has not yet enabled us to solve. Time is an element which enters largely into nature's works; she occupies a thousand or even thousands of years to produce a result, while man in his experiments is confined to a few years at most.

The most anciently renowned diamond districts are situated in the Indian peninsula, in the kingdoms of Golconda and Visapour, extending from Cape Comorin to Bengal, at the foot of a chain of mountains called the Orixa, which appear to belong to the trap-rock formation. Tavernier describes them as giving employment to thousands of workmen, but they seem now to be all but exhausted.

We are but little acquainted with the diamond mines of Landak in Borneo, though Ida Pfeiffer, on her second voyage round the world, obtained permission to visit them—a favour but rarely accorded to strangers by the suspicious potentate to whom they belong. So much is certain, that very few stones from this quarter find their way to the civilised world, which at present draws its chief supplies from the mines of Serro do Frio and Sincora in Brazil, and from the newly-discovered diamond-fields of the Cape.

When diamonds were first found in the Serro do Frio, about the beginning of the last century, the real value of the glittering crystals was so little known that they were made use of as card-marks by the planters of the neighbourhood. An inspector of mines, who had been some time in India, was the first who discovered their true nature. Wisely keeping his secret to himself, he collected a large quantity of them, and escaped with his treasure to Europe. In 1729 the governor of Brazil, Don Lourenço de Almeida, sent some of the transparent stones of the Serro to the court of Lisbon, with the remark that he supposed them to be diamonds, and thus the attention of Government was at length attracted to their value. By a decree of the 8th of February, 1730, the diamond district was placed under the rule of an Intendant, armed with the most arbitrary powers. Not only all strangers were carefully excluded from its limits, but not even a Portuguese or a Brazilian was allowed to tread its forbidden ground without a special permission; its population was limited to a scanty number, nor durst the foundation of a new house be laid unless in the presence of magistrates and mining inspectors. A system of secret delations was introduced worthy of the worst times of the Inquisition, and many an innocent person was banished, imprisoned, or transported to Africa, without even knowing his accuser, or the trespass laid to his charge. In one word despotism seemed to have exhausted all her inventive powers for the purpose of securing to the Crown the monopoly of the costliest gem on earth.

But in spite of every precaution, it was impossible to put down the contraband trade in diamonds. The audacity of the smugglers increased with the obstacles placed in their way, so that a far more considerable quantity of diamonds were secretly sold and exported than ever came into the hands of Government. Traversing the deep forests on almost inaccessible mountain paths, the bold *free-traders* met, at some place of appointment, the negroes who had been able to secrete some of the precious stones, and paid them a trifle for diamonds which beyond the limits of the district were worth at least twenty times the price given. Sometimes even the smugglers searched for diamonds themselves in the unfrequented wilderness. While some were washing the sands, others kept watch upon an eminence, and gave notice of the approach of the soldiers, who were constantly patrolling the district.

The heaviest penalties could not prevent the inhabitants of the Serra from defrauding the Crown, and Herr von Tschudi ('Travels in South America in 1857–1861') was told many amusing instances of their smuggling contrivances. One of them had concealed a diamond of twenty-five carats in the handle of his riding-whip, for which purpose he had practised for many weeks the art of plaiting the thin leather straps which covered it, and another had secreted his precious stones in a kettle with a double bottom.

When the Brazils became an independent country, the monopoly of the diamond trade was abandoned by the new Government, and any speculator was allowed to search for diamonds on payment of a slight duty. The precious stones are found chiefly in alluvial deposits (*Cascalho virgem*), in the beds of torrents, or along low river-banks, and frequently large quantities of overlying rubbish (*Cascalho bravo*) have to be removed before the diamond-bed can be reached. The mining labours are generally performed by slaves, though some of the poorer miners, or Faiscadores, have no other assistance but that of their own families. The work varies with the seasons. During the dry period of the year the cascalho is removed from the beds of the dessicated brooks, and dams are raised or canals dug for the purpose of turning off the stream into another channel, while the wet season is made use of for washing the sands. While this operation is going on, an overseer seated on a high chair keeps a sharp look-out upon the negroes; but in spite of all his attention, and of the severe punishments that await them in case of discovery, they know how to secrete many a diamond, by rapidly throwing it into their mouth, and concealing it under their tongue or swallowing it. In the Portuguese times an Intendant, complaining to the overseers of the frequency of theft, accused them of negligence, but was told that no vigilance in the world could prevent it. To convince himself of the fact, he ordered a negro who enjoyed the reputation of being a most expert hand at secreting diamonds to be brought before him, and placing a small stone in a heap of sand, promised him his liberty in case he should succeed in appropriating the stone without being detected. The negro began to wash the sand according to the usual method, while the Intendant was observing him all the time with the eyes of a lynx. After a few minutes he asked the slave whether he had found the stone. 'If the word of a white man can be trusted,' answered the black, 'I am from this moment free;' and taking the diamond out of his mouth, he handed it to the Intendant.

The negroes employed in the diamond washings are generally hired by the miners at so many milrees a week. Although their labour is very severe, they generally prefer it to any other, as on Sundays and Feast-days they are allowed to search on their own account (of course in places not previously occupied), and have, moreover, an opportunity of stealing diamonds.

The profits thus lawfully or unlawfully made they generally spend in drinking, a slave but very rarely saving money for the purpose of purchasing his freedom. In the Portuguese times, while the mines were still worked on account of Government, a negro who was fortunate enough to find a stone weighing an oitava (17½ carats) was at once rewarded with his liberty; at present only small rewards are given. Formerly most diamonds were found in the district of Tejuco, the capital of which received in 1831 the significant name of Diamantina; but in 1844 new mines were discovered in the Serro do Sincora, in the province of Bahia, whose richness eclipses that of the most brilliant times of Tejuco.

The total produce of Brazil is estimated at about 300,000 carats, annually worth on the spot from 300 to 500 milrees the oitava (17½ carats). The miners rarely make a fortune, as their expenses are very great; the chief profits of the diamond trade fall to the share of the merchants, who purchase the stones in the mining districts and then sort and export them. The price of diamonds is subject to considerable fluctuations, which, proceeding from the markets of London, Paris, and Amsterdam, are most sensibly felt in the diamond districts, for the great European houses in whose hands the trade of the rough stones is concentrated, and who dispose of considerable capital, are able to wait for better times, while the small Brazilian trader or miner is soon obliged, for want of money, to sell his stones at any price. After the breaking out of the Crimean war, diamonds were very much depreciated at Diamantina. They were offered for sale at absurdly low prices, and even then a purchaser was rarely to be found. The market improved very slowly; but when the war was at an end, the prices once more rose to a height which had never been known before. The commercial crisis in North America and Europe at the end of 1857, and in the beginning of 1858, caused a new reaction, the effects of which Tschudi was able to note during his stay at Diamantina. 'Good ware' (*fazenda regular e boa*), consisting of stones averaging a vintem * in

---

* In the Brazilian diamond trade, the oitava (17½ carats) is considered as the unity of weight. It is subdivided into 4 quartas or 32 vintems; the vintem is equal to 2¹⁴⁄₁₀₀ grains. Stones of half a vintem still pass as good ware (*fazenda ainda boa*), when well-shaped and colourless. Middling ware (*fazenda mediana*) consists of from 64 to 100 stones to the oitava, while all below that weight is sold as refuse. •

weight, which a few months before had been paid with 500 milrees the oitava, were now offered for 300. A stone of six vintems was sold in March 1858 for 170 milrees; six months before it would have been worth 240 or 260.

While the price of the smaller stones of about a vintem or less is regulated by the exporters in Rio, conjointly with the European houses, fancy prices are asked at Diamantina for larger stones of several carats. A fine diamond of an oitava sells for about three contos of rees (360*l.*), and one of two oitavas, or thirty five carats, is often sold on the spot for ten to twelve contos (1200*l.*–1440*l.*).

When cut and polished a brilliant of the first or purest water in England, weighing one carat, is valued at 12*l.*, a rose diamond of the same weight at 8*l.*, while the value of all those of a larger size is calculated by multiplying the square of the weight in carats by twelve or eight, except for those exceeding twenty carats, the price of which increases at a much more rapid rate. The enormous value ascribed to large diamonds is, however, merely fanciful, for they are worth neither more nor less than what purchasers may be inclined or able to give for them. According to the above valuation, stones weighing 100 carats would be worth at least ($100 \times 100 = 10,000 \times 12$) 120,000*l.*; a sum which has probably never yet been paid for a diamond of that weight. A very trifling spot or flaw of any kind lowers exceedingly the commercial value of a diamond. The number of large diamonds is very small. Among ten thousand stones, the Brazilian mines furnish but one that weighs ten carats; diamonds above twenty carats are very rare, and in all Europe there are but five diamonds of more than one hundred carats.

The largest of these is the magnificent gem of the first water, without fault or blemish, which sparkles in the Imperial sceptre of Russia. It weighs 194¾ carats; its largest diameter is one inch three lines and a half; its height ten lines. It is of East Indian origin, and once figured with another similar stone in the throne of Nadir Shah. When this tyrant was murdered, it was stolen, and ultimately came into the possession of an American merchant named Schafrass, who purchased it with several other valuable stones of an Afghan chieftain in Bagdad, for the round sum of 50,000 piasters. In 1772 the Empress Catherine II. bought the diamond of Schafrass, who had meanwhile settled in Amsterdam, for 450,000 silver roubles and a patent of nobility.

Of somewhat smaller size, but of unparalleled beauty, is the magnificent diamond called the Pitt or Regent, which Mr. Pitt, grandfather of the famous Lord Chatham, into whose possession it had come while governor of Madras, sold to the Duke of Orleans, Regent of France, for 135,000*l.*, having himself paid 12,500*l.* for it to Tamohund, the most famous native dealer in India. It originally weighed 410 carats, but has been reduced to 136¾ by cutting it into a brilliant—an operation which is said to have lasted two years. It is now the first among the jewels belonging to the French Government; Napoleon used to wear it in the hilt of his sword.

During the five years the stone remained in his possession, Governor Pitt is said to have lived in such constant dread of having it stolen, that he never made known beforehand the day of his coming to town, nor slept two nights following in the same house. If this story be true, great indeed must have been his relief when he parted with his gem, which, though small in weight, was to him a true millstone in anxiety.

The diamond of the Grand Duke of Tuscany, which weighs 139½ carats, and is consequently a trifle larger than the Regent, once belonged to Charles the Bold of Burgundy, and at the battle of Nancy fell into the hands of an ignorant trooper, who plucked the gem from the helmet of the unfortunate duke and sold it for a crown. At a later period it came into the possession of the Court of Tuscany, and is now the first crown jewel of the Emperor of Austria.

The most celebrated diamond in the world is undoubtedly the Koh-i-Noor, or 'mountain of light,' which, according to Hindu legend, was worn by one of the heroes of the Great War which took place about four thousand years ago, and which forms the subject of the epic poem the Maha-Bhârata. After numberless vicissitudes and peregrinations, we find it in the possession of the Grand Moguls, and in 1739 in that of Nadir Shah, who, on his occupation of Delhi, compelled

Mohammed Shah, the great-grandson of Aurungzeb, to give up to him all the valuables contained in the imperial treasury. According to the family and popular tradition, Mohammed Shah was imprudent enough to wear the Koh-i-Noor in front of his turban at his interview with his conqueror, who being

'the mildest-mannered man
With all true brooding of a gentleman,'

insisted on exchanging turbans in proof of his regard; and is said to have bestowed upon the diamond thus politely annexed the name of Koh-i-Noor. After the fall of his dynasty, the stone became the property of Ahmed Shah, the founder of the Abdali dynasty of Kabul, and, when Mr. Elphinstone was at Peshawur, was worn by his successor Shah Shuja, on his arm. When Shah Shuja was driven from Kabul, he became the nominal guest and actual prisoner of Runjeet Singh, who, following the good example of Nadir Shah, *gently* persuaded his protégé to part with his diamond for the revenues of three villages, not one rupee of which he ever realised. 'By what do you estimate its value?' asked the Sikh Maharajah of his victim, as the surrendered Koh-i-Noor lay on the arm of his new master. 'By its good luck,' said Shah Shuja, 'for it hath ever been his who hath conquered his enemies.'

Subsequent events fully proved the truth of this remark, for when the Punjab was annexed by the British Government, it was stipulated among other conditions that the Koh-i-Noor should be presented to the Queen of England. But in spite of its promising name, the 'Mountain of Light' was but of inferior lustre, for the Orientals are mere bunglers in the art of diamond-cutting, and lay greater weight upon the size than upon the brilliancy of a jewel. Hence it was resolved to have it recut by the most skilful Amsterdam lapidaries, who came over to England for the purpose. The operation, which was performed with the assistance of a small steam engine, and cost no less than 5,000*l.*, was perfectly successful; and though the Koh-i-Noor, which formerly weighed $186\frac{1}{2}$ carats, has been reduced by its conversion into a brilliant of $102\frac{3}{16}$ carats, it has gained so much in lustre that it now fully deserves the name assigned to it by the hyperbolical phraseology of the East.

All these large diamonds originally came from India; but latterly they have been rivalled by a stone of Brazilian origin, originally weighing 254½ carats, but reduced by cutting to 125, which has received the poetical name of 'Estrella do Sul' or Star of the South. It is a singular fact that as yet this beautiful gem has brought good luck to none of its possessors. An old negro woman accidentally found it in a diamond mine at the Rio da Bagagem, in Minas Geraes, among a heap of pebbles that had been previously washed. She gave it to her master, who did not even reward her with her liberty, and superstition has traced all the ill-luck attached to the stone to that ungenerous act. The first proprietor of the diamond was a needy man, who for a trifling sum had been allowed by the proprietor of the mine to search for stones with the few slaves he possessed. The proprietor now claimed the diamond, alleging that it had not been found on the premises hired by the former; and a law suit, profitable of course to none but the lawyers, was the consequence. To be able to defend his cause, the possessor of the stone pawned it to the Brazilian Bank for about 8,000l., for which he had to pay fifteen per cent. commission and a high interest.

The law suit, as may naturally be supposed, lasted so long that the man died before it was decided, and but a very small sum of money remained to his widow. After passing through several hands, the stone was purchased at Rio for a million of francs, by a Dutch jeweller, who, to make up this considerable sum, was obliged to borrow money at the usual high interest of the place. He took the stone with him to Amsterdam, where it was cut, at an expense of about 4,000l. The great desideratum was now to find a purchaser for the magnificent jewel, which, however, did not prove of the first water. It was offered for sale to several crowned heads, and no princely bridal was allowed to pass without an attempt to dispose of the 'Star of the South;' but all efforts proved fruitless, for it seems that the monarchs of our days are of opinion that the exorbitant sums formerly paid for diamonds of an uncommon size may be invested in a

much more profitable manner. The unfortunate speculator
died of a broken heart. By the latest accounts the stone
is still in Paris, held in pawn by a commercial house, which
most probably will keep it, as the accumulated interest of
years must of course absorb its whole value.

The uncut stone belonging to the King of Portugal, and
weighing 1,680 carats, is now well known to be not, as was
supposed, the greatest diamond in the world, but a mere
white topaz. The quality of another stone of $138\frac{1}{2}$ carats,
found near the Rio Abata in 1791, and likewise in the Portuguese treasury, has not been determined. So much, however, is certain, that the Portuguese Crown possesses (or
possessed a few years ago) the richest collection of diamonds
in the world, the value of which was estimated at about
8,000,000l.(?) Of all the stones annually sent to Lisbon
by the General Intendant of the diamond districts, the king
selected the largest and finest for the royal treasury, and the
others were sold. When King Joao VI. returned in 1821
from Brazil to Portugal, he carried along with him almost
as many diamonds as Voltaire's Candide on his escape from
Eldorado. The jewels were deposited in sealed bags in the
vaults of the Lisbon Bank, where they remained forty years
as a dead capital. In 1863 it was at length very wisely
resolved, with the consent of the Cortes, to sell these rough
diamonds, and to invest the proceeds for the benefit of the
civil list.

Though diamonds are usually washed out from the soil,
yet they also generally occur in regions that afford a laminated granular quartz rock, called *Itacolumite*, which in thin
slabs is translucent and more or less flexible. This rock
occurs in the mines of Brazil and the Urals, and also in
Georgia and North Carolina, where a few diamonds have been
found. Before taking leave of the prince of gems, I will
not omit mentioning that it is not only the costliest of
all ornaments, but serves also for several more humble
though more useful purposes, as for cutting glass by the
glazier, and all kinds of hard stones by the lapidary.
Small drills are made either of imperfect diamonds, or
of fragments split off from good stones in their manufacture for jewelling. They are used for drilling small holes

in rubies and other hard stones, for piercing holes in china where rivets are to be inserted, or in any other vitreous substance, however hard. Diamonds have also been recently used for arming the end of the borer in a new rock-boring machine, for scooping out holes in the hardest rocks, such as granite and porphyry. The use of diamond dust within a few years has increased very materially with the increased demand for all articles wrought by it, such as cameos and intaglios.

Since the discovery of America, when prodigious masses of gold and silver were suddenly poured into Europe, no age has witnessed a greater development of subterranean wealth than the present. The gold fields of California and Australia, the petroleum wells of Pennsylvania, the Comstock silver lode, and the copper mines of Burra Burra, have successively astonished the world, and now diamonds have been found glittering among the arid sands of the deserts to the north of the Cape Colony. Almost every packet from that part of the world brings considerable consignments of South African diamonds to London, where they are regularly sold every month by public auction. In colour and brilliancy they fully equal the East Indian and Brazilian diamonds, and surpass them in average size and weight. The consequence of this rapid influx of large stones of the first water has been a considerable fall in price for diamonds of this character. Thus, at one of the last sales (February 28) a brilliant of 20 carats (cut from a 39-carat stone), which prior to the Cape discoveries would have found ready sale at 5,000l. or more, was knocked down unsold at 2,000 guineas.

The mineral substance that ranks next to the diamond, whether we estimate it by its hardness, the splendour of its colour, or its rareness, is that called by the mineralogists Corundum. It consists of pure crystallised alumina (the oxide of the now well-known metal aluminium), variously tinted by the addition of small quantities of iron or chromium. To this class belong the ruby, the sapphire, the topaz, the emerald, the amethyst, and other stones of gorgeous colour, distinguished by the epithet 'oriental' prefixed to the name.

The Oriental Ruby, or Red Sapphire, is the red stone *par excellence* of jewellery; and, from its fiery lustre, probably identical with the carbuncle of Pliny, and the anthrax or 'burning coal' of theophrastus. Its finest colour is a most rich and lovely crimson, known as the pigeon's blood tint; but its scarlet tints are also most beautiful. The red rays of the prism falling on a ruby produce a charming effect. Pegu is the land of rubies, and Australia now likewise furnishes stones of excellent quality. A perfect ruby above three and a half carats is more valuable than a diamond of the same weight. If it weigh one carat it is worth ten guineas; two carats, forty guineas; three carats, one hundred and fifty guineas; six carats, above one thousand guineas. The largest oriental ruby known to be in the world was brought from China to Prince Gagarin, Governor of Siberia. It came afterwards into the possession of Prince Menschikoff, and is now a jewel in the imperial crown of Russia.

The blue variety of the corundum is the Oriental Sapphire of the jeweller. There is one hue of it of a soft pure azure, distinguished from the commoner kinds by its retaining its fine blue even by candlelight, when an ordinary sapphire looks purple or black. Unlike the ruby, it occurs in specimens of a considerable size. A good blue stone of ten carats is valued at fifty guineas. If it weighs twenty carats its value is two hundred guineas, but under ten carats the price may be estimated by multiplying the square of its weight in carats into half a guinea; thus one of four carats would be worth eight guineas. A sapphire of a marble-blue colour, weighing six carats, was disposed of in Paris by public sale for 70*l.*; and another of an indigo blue, weighing 6 carats and 3 grains, brought 60*l.*, both of which sums much exceed what the preceding rule assigns, from which we may perceive how far fancy may go in such matters.

The Spinels, whose transparent and most precious forms consist essentially of alumina combined with magnesia, and tinted perhaps with iron, include two resplendent stones, the Spinel Ruby, a scarlet variety of considerable fire and of rich colour, and the Balais or Balass Ruby, thus called from one of the most celebrated localities of the spinel in former times, namely Beloochistan or Balastan. The latter is of a delicate and rarely deep rose colour, showing a blue tint when looked through, and a redder one when it is looked at. Both of

these minerals are termed rubies by the jewellers, and the
deep-tinted kinds are sometimes sold for the true stone.
In fact, nearly all the large and famous gems that pass under
the name of rubies belong to this species, as for instance the
ancient ruby in the crown of England, which was presented
to Edward the Black Prince by Don Pedro the Cruel, and
the enormous stone time-honoured in Indian tradition, that
came along with the Koh-i-Noor into Her Majesty's possession.   Such was the superstitious value attached to it by its
former proprietor, Runjeet Singh, that he would sooner have
lost a province than this stone.   When the weight of a good
spinel exceeds four carats, it is said to be valued at half the
price of a diamond of the same weight.

The Chrysoberyl, called also by the jewellers the Oriental
Chrysolite, is a stone of almost adamantine lustre and trans-
parence.   It is a compound of alumina and the rare oxide
glucina, a constituent of the beryl.   It has usually a peculiar,
sometimes a very delicate greenish yellow or primrose colour,
and is then one of the most beautiful of gems.   The finer
specimens are from Brazil.

The Emerald and the Beryl are one and the same mineral
—a silicate of alumina and glucina, which owes to a small
trace of iron its blue, pink, or yellow tints, or else to
a little chromium the transcendent green which characterises
it as the emerald.   The colour of this beautiful gem is so
pleasant to the eye that the ancients attributed to it the
power of strengthening and relieving the sight when fatigued
by previous exertions.   Both from its beauty and rareness
they held it in high estimation, and Pliny ranks the emerald
in value immediately after the diamond and the oriental
pearl.   In the Egyptian tombs real emeralds are sometimes
found as the ornaments of regal mummies, and they have
not seldom been discovered among the ruins of Rome, or at
Herculaneum and Pompeii.   Scythia, Bactria, and Egypt
were renowned among the ancients as the countries which
furnished the most beautiful emeralds.   At present these
precious stones are obtained chiefly from New Granada and
Siberia, in which latter country they occur of much larger
size, but of less beauty, and consequently far inferior value.
The first Siberian emeralds were discovered in the year 1831,

in the neighbourhood of Catharinenburg, by some peasants, while making tar, and other mines were opened in 1834 ten versts distant from the former. Here was found an enormous stone, fourteen inches long and twelve broad, and weighing 16¾ lbs. troy, and another superb specimen consisting of twenty crystals, from half an inch to five inches long, and as much as two inches thick, embedded in a matrix of mica-schist. Both these monstrous gems now rank among the chief ornaments of the Imperial Mineralogical Cabinet.

When the Spaniards conquered Peru, they found many beautiful emeralds in the possession of the natives. The largest of these stones, about the size of an ostrich egg, was adored as a god in one of the temples, and other emeralds of a smaller size placed around it were honoured as its children. In their blind fanaticism, the otherwise greedy Spaniards shivered the god and his family to pieces, but it is more than probable that the wisest of the band quietly picked up the fragments and afterwards disposed of them to advantage. The finest stones used to be found in the valley of Manta ; but the Indians kept the mines secret, to avoid being obliged to work in them, or perhaps out of hatred against their oppressors. At present, the American emeralds are chiefly obtained from the valley of Tunka, in the province of Santa Fé de Bogota, in New Granada.

The price of emeralds varies considerably, according to their purity, the beauty of their colour, their lustre, and their size. Before the discovery of America, they were uncommonly dear, all knowledge of the old mines having been lost, so that the emeralds still used as ornaments were all ancient. Afterwards their value decreased, when a greater quantity was imported from Peru ; but recently they have again risen in price, as America at present furnishes but few good stones. A splendid specimen in the possession of Mr. Hope, weighing six ounces, cost 500l. ; another fine American emerald belonging to the Duke of Devonshire is two inches long, and weighs above eight ounces, but, owing to flaws, it is but partially fit for jewellery.

The Beryl, which exhibits every gradation of tint, from a pale azure blue to a fine mountain green, and also occurs in a pale orange yellow variety, is found in great perfection at

Oduntschilou and Mursinsk in Siberia. A beautifully clear crystal, ten inches long, discovered in the latter locality in 1828, and forming part of the mineralogical museum at Petersburg, is valued at 8,000*l*. Formerly Brazil and Cangayum in the Deccan were in much repute as fields in which the beryl was found, and many a brilliant little stone has been furnished by the Mourne Mountains in Ireland.

The Zircon consists of the mixed oxides of silicon and of the rare element zirconium, and is one of the heaviest and most lustrous of gem-stones. Its colourless variety is the nearest match in brilliancy and refractive energy for the diamond, while the deep orange-tinted red zircon is that transcendent gem, the true hyacinth, which makes a very superb ring-stone.

The Topaz, a silicate of alumina and fluorine, is found chiefly in Siberia, Brazil, and Saxony, and is also met with in the granitic detritus of Cairngorm in Aberdeenshire. The colourless Brazilian variety (*Pingo d'agoa*, or waterdrop) surpasses rock-crystal in purity and refractive power, and being of the same weight as the diamond, is sometimes mistaken for it. The pale yellow topaz when heated in a crucible assumes a rose-red colour, and is then called by the jewellers ruby of Brazil. The Saxon topaz, on the other hand, becomes white when exposed to heat, and thus deprived of colour is sold for the diamond. In ancient times the topaz was highly esteemed; but, in spite of its beauty, it is not now considered of very great value, from its being too frequently found, and is sold in the rough for about forty shillings per pound.

When of a beautiful 'forget-me-not' blue, and above the size of a pea, the Oriental Turquoise, which in inferior specimens is but of little value, fetches a considerable price, so that fine stones of about half an inch in length are worth 15*l.* or 20*l.* The turquoise, which consists of phosphate of alumina coloured by oxide of copper, occurs chiefly in the mountainous range of Persia, whence it is brought by the merchants of Bochara to Moskau; but the Shah is said to retain for his own use all the larger and finely tinted specimens.

Major Macdonald gives the following account of a new field for the turquoise which he discovered in Arabia

Petræa. 'In the year 1849, during my travels in Arabia in search of antiquities, I was led to examine a very lofty range of mountains, composed of iron sandstone, many days' journey in the desert; and whilst descending a mountain of about six thousand feet high, by a deep and precipitous gorge, which in the winter time served to carry off the water, I found a bed.of gravel, where I perceived a great many small blue objects mixed with the other stones; on collecting them I found they were turquoises of the finest colour and quality. On continuing my researches through the entire range of mountains, I discovered many valuable deposits of the same stones, some quite pure in pebbles, and others in the matrix. The action of the weather gradually loosens them from the rock, and they are rolled into the ravines, and, in the winter season, mixed up by the torrents with beds of gravel, where they are found.'

The Occidental or Bone Turquoise, which has generally but one-fourth of the value of the oriental, is said to be fossil bones or teeth, coloured with oxide of copper.

The Garnet, a silicate of some base which may be lime, magnesia, oxide of iron, or chromium, is in its finer specimens one of the most beautiful coloured products of nature's laboratory. By jewellers the garnets are classed as Syrian, Bohemian, or Cingalese, rather from their relative value and fineness than with any reference to the country from which they are supposed to have been brought. Those most esteemed are called Syrian garnets, not because they come from Syria, but after Syrian, the capital of Pegu, which city was formerly the chief mart for the finest garnets. Their colour is violet purple, which in some rare instances vies with that of the finest oriental amethyst; but it may be distinguished from the latter by acquiring an orange tint by candlelight. The Bohemian garnet is generally of a dull poppy-red colour, with a very perceptible hyacinth orange tint, when held between the eye and the light. When the colour is a full crimson, it is called pyrope or fire-garnet, a stone of considerable value when perfect and of large size. Garnet is easily worked, and when facet-cut is nearly always (on account of the depth of its colour) formed into thin tables, which are sometimes concave or hollowed out on the under side. Cut stones of this latter kind, when skilfully set, with a bright silver foil, have often been sold as rubies.

Though Lapis Lazuli, a silicate of soda, lime, and alumina, with the sulphide of iron and sodium in minute quantities, is without transparency, and without much lustre, yet its beautiful azure-blue tints, often interspersed with yellow specks, and veins of iron pyrites, which, from their brilliant appearance in the comparatively dull blue stone, might easily be mistaken for gold, entitle it to be ranked among the semi-precious stones. The finest quality, which sells in the mass for 30l. per pound, is used for jewellery, and for making costly vases and ornamental furniture. Lapis lazuli was also the source from which the beautiful pigment ultramarine was obtained; but this colour is now prepared artificially at a very cheap rate. This beautiful mineral is found in crystalline limestone of a greyish colour on the banks of the Indus, and in granite in Persia, China, and Siberia.

In the long list of the crystalline or hyaline quartzes, consisting of silex or silica in various degrees of purity, there is but one variety, the Noble Opal, that ranks among precious stones of the first quality. In this beautiful gem, minute fissures are apparently striated with microscopic lines, which, diffracting the light, flash out rainbow tints of the purest and most brilliant hues. The Noble Opal, which is one of the favourite jewels of modern times, was no less highly esteemed by the ancients, 'for in this stone,' says Pliny, 'we admire the fire of the ruby, the brilliant purple of the amethyst, the lustrous green of the emerald, all shining together in a wonderful mixture.' For the sake of a magnificent opal, set in a ring, and valued at 20,000l. of our money, the senator Nonius was exiled by Mark Antony. He might have escaped banishment by presenting his opal to the covetous triumvir; but he preferred exile with his gem to staying in Rome without it.

The Precious Opal is so rare a stone that, with all our mining enterprise and geological research, we know of only two certain localities for it, namely, in Hungary and in Mexico, though some specimens are said to have been found in the province of Honduras and in the stormy Feroe Islands.

The opal mines of Hungary, situated at Czernewitza, in the county of Saros, belong to the Crown, and are at present farmed by Herr Goldschmidt of Vienna, for 10,000 florins annually. About 150 workmen are employed, and as good stones occur but rarely, and are of a corresponding value, it may easily be imagined that, what with the constant fear of being robbed, and that of not being able to cover his expenses, poor Herr Goldschmidt is no less to be pitied than Governor Pitt while in possession of his diamond.

The finest and largest opal in the world is in the Imperial Mineralogical Cabinet in Vienna. It has the most magnificent play of colours, chiefly green and red, weighs seventeen ounces, and is irregularly cut, so as not to diminish the mass. An Amsterdam jeweller is said to have offered half a million of florins for this unique gem. Unfortunately, its large size prevents its being used as an ornament, as for this purpose it would have to be cut to pieces, which would be an unpardonable piece of Vandalism.

Their relative beauty increases the value of opals so considerably that fine stones of a moderate bulk have, in modern times, been frequently sold at the price of diamonds of equal size. The so-called 'Mountain of Light,' an Hungarian opal in the Great Exhibition of 1851, weighed 526½ carats, and was estimated at 4,000l. The black opals, which allow the red fire of the ruby to flash out from the dark ground-colour of the stone, are also highly esteemed. Besides the commoner varieties of opal, such as semi-opal, opal jasper, wood opal, different kinds of quartz crystal, including amethyst, Cairngorm stone, and a long and beautiful array of jaspers and chalcedonies, such as agate, onyx, sard, plasma, and chrysoprase, may be placed in a list of stones of the second degree in point of value, if that value be estimated by rarity and price.

The cutting, or grinding and polishing of most of these stones, which are commonly comprised under the name of agates, is chiefly carried on in the small towns of Oberstein and Idar, situated in the picturesque valley through which the rapid Idar flows into the Nahe, a tributary of the Rhine. The sterile soil yeilds but a scanty produce, but the neighbouring hills abound in chalcedonies, which afford the people an ample imdemnity for the barrenness of their land. As early as the fourteenth century, the art of agate-cutting was introduced into this remote valley, from Italy, where it had long been practised; but for a long time the trade was conducted in a very rude manner. The workmen themselves undertook the sale of their ware, and wandered as pedlars to the neighbouring castles or towns, where they could hope to dispose of their agates to the best advantage. It was not before the middle of the last century that the industry of Oberstein made a considerable progress. Gold and silver-smiths settled in the small town to set the stones as they came out of the hands of the grinders, and gradually a more wealthy class of traders was formed, who undertook long voyages, and extended their operations to distant countries. The fairs of Frankfort and Leipzig were regularly attended by the merchants of Oberstein and Idar, and some of them even ventured as far as Smyrna or Archangel. The taste and the ability of the workmen improved as their market extended, but now the want of the raw material began to be felt. The neighbouring hills were no longer able to meet the demand, the stones continually rose in price, the better qualities could hardly be procured, and thus the agate manufactory was menaced with decline, when a fortunate circumstance gave it a new impulse.

Some of the inhabitants of Idar who in 1827 had emigrated to Brazil discovered in their new home an inexhaustible supply of stones. Enormous masses of chalcedony were found scattered as boulders near the banks of some rivers or disseminated in the plains, and could be sent as ballast at a trifling expense across the ocean. Thus almost all the rough material that Oberstein needs comes at present from Brazil or even India, and only the rarer varieties of agate-jaspis are at present collected in the neighbourhood of Idar. In possession of the best materials, supplied by a number of localities, and comprising all imaginable varieties of chalcedony—carnelion, plasma, heliotrope, jaspis, rock-crystal, amethyst, topaz, lapis lazuli, malachite—and commanding a market which extends further and further over the globe, the prosperity of Oberstein and Idar steadily increases. One hundred and eighty-three water-mills with 724 large grinding-stones), situated along the romantic Idar, give employment to about 3,000 workmen, and the value of the manufactured stones amounts to at least 220,000*l.* annually.

No stones are so porous or so easily coloured by artificial means as the varieties of chalcedony. In ancient times the onyxes from the Nerbudda were 'baked in ovens,' and to this day, in the neighbourhood of Brooch, the nodules of onyx dug in the dry season from the beds of torrents are packed in earthern pots with dry goat's-dung, which is set on fire. By this baking process the grey or dark green iron hydrate which permeates their pores and gives them a dull colour is changed into the red oxyde, which imparts to the improved stones rich hues of orange and hyacinthine red, and the more ornamental of the mottled onyxes that come from Cambay are those thus artificially beautified.

The art of baking and colouring is now fully understood in Oberstein. Some agates consist of impermeable white bands or layers alternating with others of a grey or dull colour, and of a porous nature. When placed in honey and exposed to a moderate heat for eight or ten days, the saccharine matter penetrates into the microscopical pores. Then the stones are boiled in sulphuric acid, which, carbonising the honey, imparts a deep black colour to the porous layers which it had permeated, and by thus setting off the white layers to the best advantage, changes a previously almost worthless stone into a beautiful onyx or sardonyx. An Italian who came to Oberstein to buy rough agates for the cameo-cutters of Rome made the Germans acquainted with this method, which had long been practised by his countrymen. By other chemical processes, some of which are generally known, while others are kept a secret, rich yellow, or apple-green, or blue tints are imparted by the agate-dealers of Oberstein to the rough produce of nature. A description of all the varieties of quartz used for ornamental purposes would lead me too far; but a few words on rock-crystal may not be uninteresting.

This beautiful mineral occurs in many varieties, such as the violet, rock-crystal, or amethyst, the most beautiful specimens of which are procured from India, Ceylon, and Persia; the false topaz when yellow, the morion when black, the smoky quartz when smoke-brown. The limpid and colourless kinds are often called Bristol or Irish diamonds after the various localities in which they are found. Rock-crystal frequently occurs in the Alps, as is well known to every traveller in Switzerland. Small rock-crystals have hardly any value, but considerable prices are paid for very large specimens, which are accordingly much sought for by chamois-hunters and goatherds. About a century since a quartz cave was opened at Zinken, which afforded 1,000 cwt. of rock-crystal, and at that early period brought 300,000 dollars. One crystal weighed 800 pounds.

In 1867 a party of tourists, descending from the solitudes of the Galenstock, discovered, in a band of white quartz traversing a precipitous rock-wall about a hundred feet above the Tiefen Glacier, some dark spots which the guide, Peter Sulzer, of Guttannen, declared to be cavities in which undoubtedly rock-crystals would be found. The weather being unpropitious, no search was made at the time; but a few weeks after Sulzer and his son revisited the spot, and after having clambered up to the holes with great difficulty, found that they communicated with a dark cavity, from which the intrepid explorers extracted some pieces of black rock-crystal with the curved handles of their alpenstocks.

In the August of the following year the Sulzers, accompanied by a few friends from Guttannen, to whom they had imparted the secret, made a more decisive attempt to force their way into the cave, by widening the entrance with gunpowder. To clamber and maintain one's position on a nearly vertical rock on ledges only a few inches broad is at all times a matter of no small difficulty; but this difficulty is very much increased when at the same time the hammer and other implements for blasting are to be handled. The weather was also very bad, and every now and then a dreadful gust of wind threatened to hurl the hardy adventurers from the rock upon the glacier. Hail and rain stiffened their limbs; and thus they passed a miserable night closely huddled together on a narrow projection before the cavity. Wet to the skin, and their teeth chattering with cold, they resumed their labours on the following morning, and at length sufficiently widened the entrance to open a passage into a cave which was found to penetrate to a considerable depth into the mountain. The cave was filled nearly up to its roof with a mound consisting of pieces of granite and quartz mixed with chlorite sand; but here and there, imbedded in the rubbish, glistened the large planes of jet black morions which showed that their toil had not been fruitless. Originally the crystals had grown from the sides or the roof of the cave; and who can tell the ages that were required for their formation, or the mysterious circumstances that favoured their growth?—then at an equally unknown time the concussion of an earthquake, or maybe their own weight, had detached them from the rock to which they clung, and precipitated them upon the floor. Upon the whole more than a thousand large crystals were found in the cave, many of them weighing from fifty pounds to more than three cwt.

After the first explorers had collected about a ton, the whole able-bodied population of Guttannen, provided with hammers, spades, ropes, baskets and trucks, came forth to carry away the remainder. As the report had spread that the Canton of Uri, on whose territory the cave was situated, intended to stop their proceedings, they worked night and day with feverish haste, and in the space of a week had entirely stripped it of its treasures, which were partly conveyed to the new Furca Road, and partly transported over the glaciers to the Grimsel. One of the party fell into a crevice with a crystal of a hundred pounds upon his back, but extricated himself, though he was obliged to abandon his prize. Thus when the authorities from Uri made their appearance on the spot, nearly all the crystals had been removed out of their reach. Seven of the finest specimens, each rejoicing in an individual name, like the mammoth-trees of America, now form a magnificent group in the museum of Berne, to which they were sold for 8,000 francs. The 'king,' 32 inches high and 3 feet in circumference, weighs 255 pounds; the 'grandfather,' though of inferior height, makes up for this deficiency by a superior girth and weighs 276 pounds. Many other fine crystals were sold to various museums and private collections for six or seven francs per pound, so that Sulzer's discovery will long be gratefully remembered in the annals of the poor village of Guttannen.

INDEX.

ABE

ABEN ABOO, last Morisco chief of
   Granada, his end, 174
Abraham, his purchase of the field of
   Machpelah with silver money, 297
Abydos, rock-hewn cemeteries of, 205
Abyssinia, rock-churches of, 186
Aconcagua, height of the volcano of, 54
Adelsberg, cave of, vast dimensions of
   the, 135, 138
— entrance to the Cave of, 137
— stalagmital formations of, 140
— traversed by a river, 150
— fungi in the, 157
— subterranean animals found in the,
   162, 163
— insects in the, 163
Adit levels, drainage by, 269
Adullam, David's refuge in the cave of,
   169
Æolian caverns, 198–200
— those of Terni, 198
— fables respecting, 199
Africa, future services of Artesian wells
   to, 51
— cannibal caves of South, 234
Agates, 496
— of Oberstein, 496, 498
Aidepsos, antiquity of the hot baths of, 44
Ajunta, rock-temples of, 182, 183
Alabaster, origin of, 4
— of Montmartre, 468
— of Volterra, 468
— of England, 469
Alaghez, sulphur of the crater of the vol-
   cano of, 445
Allamaia, subterranean water-courses of, 150
Albano, Lake of, the crateriform hollow
   forming the, 132
Albert the Great, his discovery of arsenic,
   385
Alchemists, their search for gold, 371
Aloschga, fire temple of, 91
Aleutian Archipelago, formation of a new
   volcanic island in the, 60
Aleutian Mountains, volcanoes of the, 61
Alexander the Great, wealth of, 286, 498

AME

Aldborough, amber found on the coast at,
   450
Algeria, Artesian wells of, 51
Algiers, great part of, destroyed by the
   earthquake of 1755, 118
Aljaska, volcanoes of the peninsula of, 61
Almaden del Azogue, quicksilver mines
   of, 371–373
— mines of New Almaden in California,
   378
Alpujarras, destructions of the Moors of
   Granada in the caves of, 173, 174
Alston, situation of the town of, 366
Alston Moor, horses used in the mines of,
   262
— great drain of Nent Force Level, 270
— lead mines of, 365, 366
Altaï, copper mines of the, 326
— porphyry of the, 468
Alten Fjord, copper mines of, 324
Aluminium-bronze, 387
Aluminium, discovery and uses of, 387
Amber, modes of collecting, on the Prussian coast, 449
— diggings near Dantzig, 449, 450
— various places in which it is found, 450
— what is amber? 450
— the extinct amber-tree, 451
— insects inclosed in amber, 452–455
— ancient and modern trade in, 455–457
— constituents of, 458
— mines of Tolfa, 458
— manufacture of, 459
Amblyopsis spelæus, of the Mammoth
   Cave of Kentucky, 168
America, number of active volcanoes in
   Western and Central, 61
— copper mines of, 326
— ancient copper mines of, 327
— iron industry of, 362
— lead mines of, 367
— silver mines of, 300–314
— coal-fields of North, 424
— fossil monkeys of South, 24
— animals of the Pliocene period and of
   the present day, 24

AMM

Ammonites, number of species of the, 18
— characteristics of the, 18
Ammonites Henleyi, 9
Anaitis, golden statue of the goddess, 286
Anchorites, caves of, 178
Ancyloceras gigas, 19
Audernach, on the Rhine, glacial beer cellars of, 192
— entrance to the glacière of, 201
Andes, sea-shells found on the, 34
— fish disgorged from the volcanic caverns of the, 69
André, St., town and church of, buried by a landslip of Mount Granier, 127
Andreasberg, St., depth of one of the pits of, 247
Animals, impressions produced on, by an earthquake, 113
— subterranean, 159–168
— divine honours paid to them by the Egyptians, and converted into mummies, 205
— caves containing remains of extinct animals, 213
Anoplotheriums, size and characteristics of the, 23
Antæopolis, rock-hewn cemeteries of, 205
Anthony, St., of Egypt, his rock-cave, life, and death, 178, 179
Anthracites, or non-bituminous coal, 401, 402
— value of, for steam-engines, 405
Antimony, first mention of, 383
— uses of, 383
Antioch, earthquake of, in the reign of Trajan, 97
— its subsequent subversion by an earthquake, 97
Antiparos, Grotto of, 134
Antuco, eruption of the volcano of, in 1835, 79
Apalachian coal-field, its enormous extent, 424
Apollo, at Delphi, golden statue of, 285
Apteryx australis of New Zealand, 216
Aptornis, Professor Owen's resuscitation of the, 217
Aqueducts of the Romans, 41
— of the Turks, 41, 42
Aqueous rocks, countless ages of the formation of the, 1, 5
— incomplete knowledge of these sedimentary formations, 1
— aqueous strata disturbed by igneous formations, 4
Arabia, sulphur of, 446
Arcadia, consecrated caves to Artemis and Pan in, 167
Arcueil, artificial mushroom-beds at, 158
Arica, effects of an earthquake sea-wave at, 109

BAK

Argentiferous veins of the Chaosthal and the Veta madre, their length, 247
Armenia, hermits in, 179
Arnaud, St., Colonel, his massacre of the Arabs in the cave of Sholus, 176
Arracan, mud volcanoes of the coast of, 93
Arsenic, discovery of, 385
— supply of, 385
Artesian wells, subterranean heat shown by, 32
— theory of, 48
— of the inhabitants of the Sahara, mentioned by Olympiodorus, 48
— the well of Grenelle at Paris, 49
— Artesian well sunk in the London basin, 49
— various uses of Artesian wells, 50
— those of Algeria, 51
— future importance of, in Africa and Australia, 51, 52
Ashes thrown out by volcanic eruptions, 66, 67
Asia Minor, earthquakes of, in the reign of Tiberius, 97, 100
Asphalte, 426
— found swimming on the Dead Sea, 427
— uses of, 427
— pavements made of, 428
Assuan, rock-hewn cemeteries of, 205
Astrophyllites comosa, 392
Augustus, Emperor, and the sacrilegious soldier, story of, 286
Aurignac, sepulchral grotto of, 228, 229
Australia, future importance of Artesian wells to, 52
— stalactital caves of, 141
— ossiferous caves of, 216
— discovery of gold in, 289
— Sir Roderick Murchison's surmises, 289
— copper mines of, 329
Austria, coal-fields of, 423
— salt mines of, 433–436
Auvergne, carbonic acid gas springs of, 88
— mænæ, or crateriform hollows, in, 132
Avaricum (Bourges), Cæsar's siege of, 317
Avernus, Lake of, formed in an extinct crater of a volcano, 57
Aviculopecten subulnatus, fossils of, 15
Axmouth, landslip at, 128
— Sir C. Lyell's account of it, 128
Azores, earthquakes in the, 100
Azure Cave of Capri, beauty of the marine excavation called the, 143

BABYLON, golden image of Belus at 285
Baghlad, coins of, 287
Baghilt coal mine, in Wales, drowned, 273
Bahaud, Port, upheaval of the land at, 36
Baku, burning springs of, 91
— new mud volcano near, 95

BAL

Balearic Islands, troglodytes of, 234
Ballarat, gold mines of, 291
Baltic, changes on the shores of the, 451
Banca, tinstone of, 335
Bann Bridge, subsidence of the land at, 36
Barbary, earthquake of 1775 in, 118
Barigazzo, burning springs near, 90
Bath, thermal springs of, 43
Bats, clusters of, in caverns, 159
Baumann's Cave, in the Harz Mountains, 136
— fatal expedition, 136
Bean shot and feathered shot of copper-works, 321, 322
Bear, grisly, of the Rocky Mountains, 125
— bones of huge and formidable extinct species found in caverns, 123, 125
— remains of bears found in caves, 213
Beatus, St., his cave on the Lake of Thun, 181
— pilgrimages to his cave, 181
Beauheyl, or 'living streams' of tin, 337
Beaujonc, scenes of the inundation of the mine of, 274
Beckford, his remarks on the Grotto of Pausilippo and Virgil's tomb, 242, 243
Beetle, cavern, in the cave of Adelsberg, 163
— in the Mammoth Cave of Kentucky, 167
Belemnites of the Lias and Oolite, 19
— size and characteristics of the, 19, 20
Belemnite, restored, 19
Belgium, lead mines of, 367
— production of zinc in, 381
— coal-fields of, 423
Belus, image of, in the temple of Babylon, 285
Belzoni, his aptitude for his work, 203
Benedict, St., his cave near Subiaco, 180
Berchtesgaden, salt mines of, 436
Bergmann, his experiments with platinum, 382
Berguen, Louis von, discovers the art of cutting diamonds, 478
Beryl, the, 491, 492
Bethlehem, Church and Grotto of the Nativity at, 188
Bewick, Thomas, a coal-hewer in early life, 419
Bilma-el-Moluk, the royal tombs of Thebes, 202-204
Biscayana, Veta de la, silver mine of, 304
— its great wealth and subsequent abandonment, 304
Billiton, tinstone of, 335
Birds, cave-haunting, 160
Birmah, mud volcanoes of, 93
— rock temples, 184
Bismuth, first mention of, 383
— whence furnished, 383
Bituminous substances, 426
Black Country, iron furnaces of the, 351
Black lead. See Plumbago.

BRI

Blast furnaces for iron, 352
— benefits of the hot-, 353
Blasting in mines and its dangers, 258-260
Bleyberg-à-Montzen, lead mines of, 367
Blothrus spelæus, of the Cave of Adelsberg, 163
— its pursuit of the cavern-beetle, 163
Blowers in coal-mines, 279
Bogs, effects of bursting of, 130
Bohemia, ice-caves of, 197
— gold coins of, 287
— gold of, 288
— silver mines of, 299
— their produce, 300
— tin mines of, 336
— iron mines of, 358
— coal-fields of, 426
Bolivia, active volcanoes of, 61
Bolsena, Lake of, formed in the extinct crater of a volcano, 57
Bonifacio, in Corsica, caverns of, 144, 145
Borax, or borate of soda, former chief supply of, 459
— obtained as a crude substance in various places, 459
— the suffioni of the Florentine lagoons, 460
Boring for minerals, 249
— Williams's account of the emotions of the boring party, 249
— mode of operation, 250, 251
— prices in the North of England for boring, 250 note
— implements used for boring, 250
Burneo, diamond mines of, 480
Borrowstoness Colliery, 410
Bosio, Anthony, his discovery of the catacombs, 209
Boston, in America, smelting-houses of the Bay of, 328
Botallack mine, in Cornwall, 317-319
— the blind miner of, 319, 320
Bourbon, Isle of, volume of the lava stream of the eruption of 1787, 75
Bracciano, Lake of, formed in the extinct crater of a volcano, 57
Brachiopods of the Silurian seas, 12, 13
Bradstein, ice-cave of, 197
Brazil, ossiferous caves of, 216
— iron-ores of, 363
— lead mines of, 367
Bressay, islet of, its marine caverns, 142
Breton, Cape, rain-drops of the Carboniferous period preserved at, 29
Brienz, village of, twice burned by a landslip, and twice reconstructed, 127
Brilliants, 479, 484
Britannia metal, 335
— manufacture of, 383
Brittany, traces of depression of the land on the coast of, 37

BRI

Brixham, bone-caves of, 227
Bronze, an alloy of copper and tin, 332'
— implements of, found in Switzerland, 332
Brownhill. in North America, bituminous coal-field at, 424
Brûle, near St. Etienne, burning coal-mine at, 283
Brunswick, New, coal-fields of, 424
Buch, Leopold von, his observations as to the rise of the land of Sweden, 35
Büdöshegy, in Transylvania, sulphur caves of the mountain, 446
— visit to the caves of, 446
Bufador, or the water-spout of Pope Luna, 146
Buffalo, food of the, 26
Burgbrohl, carbonic acid gas spring of, and quantity it produces, 88
Burra-Burra copper mine, in Australia, 329
Busingen, destruction of the village of, 124
Bustamente, Don José, his draining gallery, 304

CADIZ, effects of the great earthquake of 1755 on, 117, 118
Cadmium, discovery and uses of, 386
Calabria, earthquake of 1783 in, 98
— conduct of the peasants in, 99
— movement of the sea during the earthquake, 107
— depth of the original shock of 1857, 111
Calamine, zinc produced from, 380
— worked in Prussia, Belgium, and England, 381
Calamites nodosus, 393
Caldera, copper mines of, 326
California, upheaval of the land at, 34
— discovery of gold in, 288
— immense flood of emigration into, 289
— gold washing at, 295
— copper mines of, 328
— iron discovered in, 362
— quicksilver of, 378
Callistus, catacomb of, discovery of the, 210
Calobozo, sounds accompanying earthquakes at, 103
Camborne, copper mines of, 317
Cambrian rocks, antiquity of the, 2, 3, 10
— fossils of the, 10, 11
Cambyses, his enormous wealth, 286
Campagna, different kinds of stone of the, 308
Canada, iron pyrites of, 448
Canary Islands, earthquakes of the, 100
— maare, or crateriform hollows, of the, 132

CAT

Cane, Grotto del, cruel experiments on dogs at, 89
Canstadt, in Wurtemburg, mills kept at work in winter by Artesian wells, 50
Capac Urcu, the volcanic cone of, blown to pieces, 67
Caraccas, town of, destroyed by an earthquake, 131
Carbonic acid gas springs, 88
— those of Germany, 88
Carboniferous period, fishes of the, 13
— vegetable and animal remains of the, 14, 18
— insects of the, 15
— rain-drops of the, preserved at Sydney, in Cape Breton, 29
— proof of the density of the atmosphere of the, 29
— plants of the, 391
Carburetted hydrogen, springs of, 90-93
Carclaze tin mine, 341
Cardiganshire, lead mines of, 366
Cardona, rock-salt of the valley of, 437
Cardrew mine, in Cornwall, drainage of, 270
Carguairazo, fish disgorged from the eruption of the volcano of, 69, 70
Carinthia, dollinas and jamas of, 130
— subterranean water-courses of, 150
— fungi of the caves of, 158
— iron of, 358
Carlsbad, hot springs of, 43
Carmel, Mount, grotto of the prophet Elijah on, 188
— church on, 188
Carniola, dollinas and jamas of, 130
— subterranean water-courses of, 150
Carnon, near Falmouth, tin-stream of, 338
Carrara marble, origin of, 4
— quarries of, 465
— situation of the quarries, 465
— the town of, 465
Carron iron-works established, 350
Carson river, silver mines near the, 314
Cass, General. his report on the copper mines of Lake Superior, 328
Cassiterides, or tin islands, Herodotus' mention of, 333
Cassotis, at Delphi, antiquity of the, 44
Castro, John di, his manufacture of alum at Tolfa, 458
Catacombs of Rome, 205
— gallery with tombs, 206
— sepulchral inscriptions, 209
— Bosio's discovery of the catacombs, 209
— Cavaliere de Rossi's researches, 210
— those of Naples and Syracuse, 210
— those of Paris, 210
Catania threatened by the lava-stream from Etna, 72
— partly destroyed by the lava, 73

CAT

Catorce, Alamos de, silver mine of, 303

Caucasus, mud volcanoes of the, 93, 95

— earthquakes of, 100

Cavern-roofs, falling in of, causing land-slips, 129

Caves, in general, 133

— their various forms, 133

— natural tunnels, 133, 134

— dimensions of, 135

— discovery of, 135

— the various rocks in which they occur, 136

— marine, 142

— volcanic, 146

— rivers, 169

— vegetation, 156

— subterranean animals, 159

— as places of refuge, 169

— hermit caves and rock-temples, 178

— subterranean places of worship, 181

— ice-caves and wind-holes, 192

— rock-tombs and catacombs, 202

— with bones of extinct animals, 213

— subterranean relics of prehistoric man, 221

— troglodytes, or cave-dwellers, 231

— cave of St. Peter's Mount, near Maestricht, 470

Celsius, his observations of the rise of the land in Sweden, 35

Cemeteries, rock-hewn, of Egypt, 204, 205

Cenis, Mont, railway tunnel through, 238–240

— machines for boring the, 238, 239

— mode of proceeding, 238–240

Cervus megaceros, the, of Ireland, 28

Ceylon, rock-temples of, 184

Chalcedony, 497

Chaldæa, silver mines of, 298

Chalk group, star fish of the, 18

Charlemagne, imperial mantle of, 478

Cheshire, salt mines of, 431

Chili, number of active volcanoes, 61

— great earthquakes of, in 1835, 79

— earthquakes of, generally, 100

— effects of the earthquake sea-wave after the shock, 108

— silver mines of, 313, 314

— copper mines of, 326

— lead mines of, 367

China-clay, or kaolin, how formed, 460

— mode of treating it, 460

— export of, from Cornwall and Devonshire, 461

'Chinaman's Hole,' gold diggings at, 292

Chinese, their use of springs of carburetted hydrogen, 90, 91

— at the Australian gold diggings, 291

— their discovery of gold near Mount Ararat, 291

COC

Choke-damp, or black-damp, 278

— destruction caused by, 281

Choquier, bones of extinct animals found in the cavern of, 214

Christians, tombs of the early, near Rome, 207, 208

Chrome, uses of, 385

— discovery of, 386

— whence obtained, 386

Chrysoberyl, or oriental chrysolite, 491

Chuquibamba, height of the volcano of, 54

Cinnabar, uses of, in early ages, 370

Cirknitz Lake, the Proteus first discovered in the, 164, 165

Clara, Boveda de Santa, at Almaden, 372

Cleveland district, iron manufacture of the, 354

Clausthal, length of the argentiferous veins of, 247

— great adit levels of the mines of, 270

Clodius, Roman prætor, defeated by Spartacus at Vesuvius, 82

Coal and coal mines, 245, 246

— age of, 390

— plant of the Carboniferous age, 391

— extent of the coal seams, 395

— vast time required for the formation of the coal-fields, 395

— the probable mode of formation, 396

— derangements and dislocation of coal beds, 397, 398

— separation of a coal-field into small areas by dykes or faults, 399

— bituminous and non-bituminous coals, 401

— chief coal-producing countries of the world, 402

— the coal-fields of Great Britain, 402–422

— the hewers and their work, 415, 418

— other workmen, below and above the pit, 416, 417

— early knowlege of coal, 419

— its use prohibited by Edward I. in London, 419

— the trade in coal in the middle of the seventeenth century, 420

— increase in the demand and supply, 420

— the question of the duration of our coal fields, 420

— coal-fields of foreign countries, 422–425

Coal-hewers of the North of England, 414

— at work, 415

— how they are paid, 416

Coalbrookdale, iron manufacture in, 349

Coal-cutting machines, 415

Cobalt, name of, 384

— uses of, and whence obtained, 384

Coca, stimulating properties of, 311

Cochin China, rock temples of, 184

COI

Colos, the oldest known gold, 287
Collieries, casualties in, 245
— drainage of the water in, 272
Colossochelys Atlas, gigantic proportions of the, 24
Columbia, mud volcanoes of, 93
Columbia, British, gold-fields of, 293
— coal-fields of, 424
Consolidated mines in Cornwall, amount of sinking in the, 251
Conto, Monte, landslip of the, 127
Copal-tree, resin at the foot of the, 451
Copiapo, in Chili, discovery of silver at, 248
— silver mines of, 313
Copper, name and antiquity of, 315
— how found, 315
— its uses and compounds, 315
— mines of Cornwall, 316, 317
— ores and process of smelting, 320, 321
— mines of Sweden, Germany, and Russia, 322–326
— those of America, 326–329
— and of Australia, 329
— history of some of our copper mines, 329
— lodes of Cornwall, 337
Copperopolis, copper mines of, 328
Coquimbo, copper mines of, 326
Corals, primeval, 16
Corneale, Cave of, colossal stalagmites of the, 140
Cornwall, mines of, 316
— tin mines of, 336
— persons employed in them, 343
— zinc produced in, 382
— China-clay, 415
Corsica, marine caves of, 145
Cort, Mr., his improvements in iron manufacture, 350
Corundum, 489
Cosiguina, phenomena of an eruption of, 65, 67
— destruction caused by the eruption of 1835, 67
Cosmo III., Grand Duchy of Tuscany, burns a diamond, 479
Cotopaxi, shape of, 53
— enormous stones hurled by an eruption of, 66
— phenomena of the eruption of 1803, 69
— noises heard 109 miles off during an eruption of, 104
Cretaceous period, fossils of the, 19, 22, 23
— causes of landslips in the, 129
Crete, labyrinth of, 174, 175
— consecrated caves and grottoes to Zeus in, 187
Crimea, mud volcanoes of the, 93
Crinnis Copper Mine, Old, abandoned but reworked, 329, 330
Crinoids, or sea-lilies, fossil, 17

DEN

Crœsus, his enormous wealth, 286
Crookes, Mr., his discovery of thallium, 388
Crowe, Mr., of Hammerfest, forms a copper-mining company in Norway, 324
Crustaceans of the Silurian seas, 11, 12
— cavern, 163, 167
Cuba, copper mines of, 329
— iron ores of, 363
Cumana, destruction of the town of, by an earthquake, 102
— sounds accompanying the shocks, 103
Curtis, Thomas, his difficult work in the Huel Wherry tin mine, 339, 340
Cuthbert, St., his cave on the coast of Northumberland, 180
— account of him, 180
— 'beads of St. Cuthbert,' 180
Cyclops, troglodytic caverns of the, at the base of Mount Etna, 232
Cyprus, ancient silver mines of, 298
Cyrus, enormous treasures accumulated by, 286
Cyzicus, the oldest known specimen of a gold coin of, 287

DAHRA, French atrocities at the caves of the, 176
Dalecarlia, iron ores of, 360
Dalmatia, dollinas and jamas of, 150
— subterranean water-courses of, 150
Dalton-le-Dale, drainage of the coal-mine of, 272
Dambool, rock-temple of, 184
Dammara australis, masses of resin at the base of the trunk of the, 451
Dana, Professor, his views respecting volcanoes, 79
Dannemora, iron-works of, 360
Dantzig, amber found near, 449
Darien, platinum discovered at, 382
Darius Hystaspis, his enormous wealth, 286
Davy, Sir Humphry, his safety-lamp, 280
— his discovery of aluminium and magnesium, 387
— and of sodium, 388
Delgada, Punta, in the Island of San Miguel, 147
Delphi, subterranean hollow under the tripod of the priestess of, 187
Demidoff, Prince, his copper mines, 326
— his iron mountain in the Oural, 357, 358
Denbighshire, lead mines of, 366
Derbyshire, lead mines of, 366
Denmark, enormous antiquity of the peat mosses of, 221
— remnants of a former vegetation and articles of human workmanship found in the mosses of, 222.

DEN

Denmark—continued.
— the 'shell-mounds' of, 222
Depressions, subterranean, 34, 36
— submarine forests in various places, 36
— evidence of depression, 36, 37
— probable causes of, 38
Derbyshire spar, 469
Devon Great Consols Mines, success of the copper mines of, 330
Devonian period, fishes of the, 13
Devonshire, tin mines of, 336
— miners and wages of, 343
— china-clay of, 460, 461
Diablerets, falls of the, 121
— escape of a peasant from his living tomb in the, 122
— causes of the phenomenon, 123
Diamond, the, 477
— diamond-cutting, 478
— rose diamonds and brilliants, 479
— destroyed by heat, 479
— stones of India and Brazil, 480
— the Russian diamond, and the Pitt or Regent diamond, 485
— that of the Grand Duke of Tuscany, 485
— the Koh-i-Noor, 486
— diamonds and diamond-dust used for industrial purposes, 489
Dinornis, size of the, 28
— Professor Owen's resuscitation of the, 217
Dinotherium, size and characteristics of the, 23
Diodorus Siculus, his account of the tin trade of Britain, 333
Diving rod, the, 248
— how used, 249
Dolcoath tin mine, 337
Dolores, mine of, 303
Domingos, San, in Portugal, Roman mines of, 448
— now worked by the Baron of Pommerão, 448
— account of the works, 448
Donati, Vitaliano, his account of the fall of a mountain, near Sallenches, 122
Doncaster, gigantic fungus in a tunnel near, 158
Donegal, bursting of bogs in, 131
Droitwich, salt-works of, 432
Drontheim, or Tronyem, city of, 324
Dudley, Lord, establishes iron-works near Stourbridge, 349
Dufan, in Arabia, sulphur of the, 446
Dukinfield colliery, depth of, 247
Dunfermline, monastery of, obtains a licence to dig coals, 419
Durham, coal-fields of, 403, 407
Dyeing, use of tin in the processes of, 335

EARTHQUAKES, preceding volcanic eruptions, 65

ENC

Earthquakes—continued
— volcanoes considered as the safety-valves of, 78, 79
— but sometimes accompany volcanic eruption, 79
— extent of misery caused by, 97–99
— the horrors of, increased by man, 99
— the progress of civilisation retarded by, 99, 100
— regions to which they are confined, 100
— duration of the shocks, 101
— indications of coming, 102
— sounds accompanying, 103
— sounds unaccompanied by movement of the earth, 104
— vertical or undulatory motion of shocks, 104
— extent and force of the seismic wave motion, 105, 106
— movements of the sea in, 106, 107, 117
— extent of the wave motion, 109
— changes caused by in the configuration of the soil, 109, 110
— causes of, 111
— probable depth of the focus, 111, 112
— opinions of Sir C. Lyell and Mr. Poulett Scrope, 112
— effects of an, on man and animals, 112, 113
— account of the great, at Lisbon, 114
Egg, Isle of, atrocities of the Macleods in the cave of the, 171
Egypt, rock-temples of, 184
— tombs of the kings in Thebes, 202–204
— compared to an iron furnace, 347
— quarries of, 474, 475
Ehrenberg, his discovery of the animated dust of the Harmattan, 156
Eifel, volcanic district of the, 58
— carbonic acid gas springs of the, 88
— crateriform hollows, or maars, in the, 131
Eileithyia, rock-hewn cemeteries of, 205
Eimeo, hole in the island of, 133
— tradition respecting this hole, 133
Elba, iron industry of, 362
Elephanta, rock-temples of, 184
Elevations of the land produced by earthquakes, 111
Elfdal, porphyry of, 467
Elura, rock-temples of, 183
Emerald, or beryl, 491, 492
Emery, whence obtained, 463
— total production of, 463
Emmanuel, St., church of, in Abyssinia, 187
Ems, hot springs of, 43
Enamel, materials used for, 335
Encrinites lily, called 'St. Cuthbert's beads,' 181
Encrinus liliiformis, fossil, 17, 18

ENG

Engihoul, human remains in the cavern of, 226
— Dr. Schmerling's explorations, 226
— Sir C. Lyell's, 227
Engis, human bones discovered in the cavern of, 226
Engines, stationary, used in mines, 263
England, subsidence of the land on the west and east coasts of, 36, 37
— effects of a violent earthquake in, 100
— shocks felt in, at various times, 100, 101
— effects of the great earthquake of 1755 in, 118
— the extinct hyena of, found in caves, 214
— ossiferous caves in, 227
— flint instruments found in, 231
— main causes of the prosperity of, 245
— copper mines of, 316
— manufacture of iron in, 349
— lead mines of, 365, 366
— zinc produced in, 381, 382
— vast deposit of coal of, 402, 403
— and their convenient distribution, 403
— extent of the Great Central Coal-field, 406
— quarries of, 464
Eozoon canadense, the only fossil found in the Laurentian rocks, 10
— its extreme antiquity, 10
Epomeo, volcano of, its long periods of rest, 58
Erasinos, in Greece, antiquity of the spring of, 44
Ernst August Stollen, in the Harz, 271
Erzberg, or iron mountain, in Styria, 358
— works at, and produce of, 359
Esquimaux, their iron implements, 347
Estrello do Sul, or Star of the South, diamond, 487
Etna, Mount, M. Houel's exploration of the crater of, 55
— streams of lava in the eruption of 1669, 70
— numbers of parasitic cones on the flanks of, 72
— rate of progress of the lava-stream of 1699, 72
— retention of heat in the lava-stream of 1832, 73
— the Fossa della Palomba on, 147
— ice-caves of, 198
— troglodytic caverns of the Cyclops at the base of, 232
Euripides, his triumph, 476
Europe, volcanoes of, 61
Eurypterids, of the Silurian seas, 12

FUN

FAHLUN, horses used in the copper mines of, 262
— narrow escape in the mine, 264
— copper-mine of, 322
— ore of the mine, 323
— the preserved body found in, 323
Ferdinand, Archduke, his visit to the Cave of Magdalene, 166
Fez, effects of the earthquake of 1755 at, 118
Fingal's Cave, Sir W. Scott's description of, 143
Fino, Don Andrea del, narrative of, in an earthquake, 99
Fire, its eternal strife with water, 1, 2
— the subterranean forces, 7
Fire-damp, or carburetted hydrogen, 278
— fatal explosions caused by, 281
Fish disgorged by volcanoes from caverns, 69
— blind cavern of the Mammoth Cave of Kentucky, 168
— of the Upper Silurian group, 13
— destruction of vast numbers of, by volcanic eruption, 15, 16
Flintshiro, lead mines of, 366
Flores, Padre, his silver mine of 'La Bolsa de Dios Padre,' 304
Florins, or fiorini, origin of, 287
Fontaine-sans-fond, the, near Sable, 149
Footprints of former ages, preservation of, 28, 29
Forests, submarine, in various places, 36
Fossils, chronological importance of, to the geologist, 5, 6, 8
— extinction of species, 9, 10, 14
— those of the oldest and later periods, 10-29
Fountains, artificial, principle on which they are constructed, 42
— of lava, 71, 72
— of marine caverns, 146
Foxdale, lead mine, in the Isle of Man, 366
Frais Puits, phenomena of the, 150
France, effects of the great earthquake of 1755 in, 118
— tin mines of, 336
— consumption of coal in, 423
Frauenmauer Mountain in Upper Styria, ice-cave of the, 196, 197
Fredonia, town of, lit by springs of carburetted hydrogen, 90
Freiberg, drainage of the mines of, 270
French, their atrocities in the Cave of Longura, 170
— their cruelty in Algeria, 176
Frio, Serro do, diamonds of the, 480
Fuegians, 'shell mounds' of the, 222
Fumaroles or steam-jets of volcanoes, 63
— those of Jorullo of 1759 seen in 1803, 74
Fungi, subterranean, 157
— Scopoli's description of, 157

FUN

Fungi—continued.
— gigantic one at Doncaster, 158
— the artificial mushroom-beds near Paris, 158
Furnaces, reverberatory, 321

GALLICIA, salt mines of, 436
— Ganoid fishes of the Upper Silurian group, 13
Garnet, the, 494
Garnock river bursts into a colliery, 276
Gas-springs, 88
Gellivara, in Swedish Lapland, mounds of magnetic iron-ore at, 360
Gems, superstitious power of, 477
Geological revolutions, influence of, on the earth-rind, 1
— tabular geological profile, 3
— periods of geological formations, 5
— the same mineral substances in the oldest and newest formations, 5
— guidance of the geologist in ascertaining the periods of the formations, 5
— a continuous development to more highly organised species, 6
Georges, St., ice-cave of, 192
— entrance to the glacière of, 201
Georg Stollen, great adit levels of the, in the Harz, 270
Germain, St., artificial mushroom-beds at, 158
Germany, effects of the great earthquake of 1755 in, 118
— copper mines of, 325
— lead mines of North Germany, 365
— coal-fields of, 422
— consumption of coal in, 423
— quarries of, 464
Gibraltar, Rock of, monkeys of the, 24
Girgenti, town and trade of, 442
— the sulphur mines of, 442
Glass, stained, colours of, how formed, 335
Glenmalure, lead mines of, 366
Glyptodon, size and characteristics of the, 26
Goaves, or old workings in coal mines, fire-damp in, 279
Goeppert, Professor, his observations on the extinct amber-tree, 451
Goethe, his remarks on the great Lisbon earthquake, 119
Goffin, Hubert, his heroism, 275
— his future career, 276
Gold, antiquity of man's knowledge of, 285
— the story of the Golden Fleece, 285
— statues of gold in ancient temples, 285
— quantities of gold possessed by ancient monarchs, 286
— earliest use of the metal, 287

GYP

Gold—continued.
— auriferous land of the Iberian peninsula, 283
— California and Australia, 288, 289
— British Columbia and other places, 293
— localities in which gold is deposited, 293, 294
Goldau, Vale of, devastated by a land-slip, 123
— destruction of the village of, 124
Golden Fleece, story of the, 285
Golubinas, or pigeon-holes, in Dalmatia and Carniola, 130
Goniatites of the Carboniferous period, 18
— extinction of the, 18
Good Hope, Cape of, upheaval of the land, at the, 34
Gorobladgodat, Kuschwa, platinum of, 382
Gortyna, in Crete, labyrinth of, 174, 175
Gosforth Colliery, 409
Gothard, Mount St., proposed tunnel through, 241
Gower, bone-cave of, 228
Grace Dieu, glacière of, 192
— stalagmites of ice in the, 193, 194
Graham's Island, volcanic formation of, 59
— its disappearance, 59
Granada, New, extent of the wave-motion of an earthquake, 105
Granada, in Spain, destruction of the Moors of, 173
Graphite. See Plumbago.
Grasshopper, wing of, of the Carboniferous period, 15
Greece, subterranean water-courses of, 150
— consecrated caves and grottoes of, 187
Greenhouses kept warm by water from Artesian wells, 50
Greenland, evidence of subsidence of the land at, 37
Grenelle, heat of the Artesian well of, at various depths, 32, 49
Grenier, Mount, landslip of, 127
Grosmont, iron manufacture of, 355
Guncharo, the Cueva del, 160, 161
Guacharo, a troglodytic bird, 160, 161
— wholesale slaughter of, by the Indians, 161, 162
— where found, 160, 162
Gualgayoc, the ventanillas of, 133
Gualgayoc, Cerro de San Fernando de, silver mines of, 309, 311
Guanaxuato, subterranean noises heard at, without earthquake, 104
Guanaxuato, rise of the town of, 302
Guatemala, volcanoes near the town of, 61
Guadiana, engulfment of the river, 150
Gunpowder, amount of, used in blasting in mines, 260
Gwennap, copper mines of, 317
Gypsum, origin of, 4

HAG

HAGGAR Silsilis, in Egypt, quarries of, 475

Haiti upheaval of the land at, 34

Hann, Professor, a coal-hewer in early life, 419

Hanover, iron manufacture of, 357

Harmattan, animated dust of the, 156

Hartlepool, export of coal from, 413

Hartley, Colliery accident in the, 253

Harz Mountains, subterranean flora of the, 158

— ice-caves of the, 197

— great adit levels of the mines in the, 270

Hausemann, Professor, his visit to the Norwegian copper mine of Röraas, 324

Hawaii, effect of the eruption of Manna Loa in 1840, 76

— effects of an earthquake sea-wave at, 109

Heat, subterranean, 31

— zone of invariable temperature, 31, 32

— increasing temperature at a greater depth, 32

— rate of increase, 32

— proof everywhere of a subterranean source of heat, 32

Hentou, accident at the colliery of, 273

Herculaneum, destruction of the town of, 81-85

— the mud-stream which caused the destruction, 85

— discovery of the buried town, 86

Hermits, caves of, 178

Hermits, numbers of, in rock-caves and huts in the East, 179

Herodotus, his mention of the Cassiterides, 333

Hetton Colliery, ventilation of the, 278

Hiera, volcanic island of, 60

Hilda, St., collieries and galleries of, 410

Himmelfürst, in Saxony, silver-fields of, 299

Hindostan, coal-fields of, 424

Hoffmann, G. F., his description of the subterranean flora of the Harz Mountains, 158

Holland, earthquakes felt in, 101

— effects of the great earthquake of 1755 in, 118

Homer, tin ornaments mentioned by, 332

Honduras, coal-fields of, 424

Horses used in mines, 262

Hot-springs in the frozen lands as well as in the tropics, 33

— as a vent of subterranean heat, 33

Houal, M., his dangerous exploration of the crater of Mount Etna, 55

Hewitt, William, his description of shipping coal on the Tyne, 412

Huancayo, the Franciscan monk of, 313

Huatuleo, fountains of marine caverns in, 146

IRO

Huel Wherry, rise and fall of the tin-mine of, 339

Humboldt, M, his visit to the volcano of, Rucu-Pinchincha, 55

— his treatise on subterranean fungi, 153

Huancavelica, quicksilver mine of, 378

Hungary, ice-caves of, 197

— salt mines of, 436

Hutton, Dr., a coal-hewer in early life, 419

Hydraulic mining in California, 295, 296

Hydrostatic laws regarding the flow of springs, 40, 41

Hyena, remains of, found in caves, 213, 214

IBARRA, supposed cause of a fever at, 70

Iberian peninsula, auriferous land of, 288

Ice, effect of the meeting of a lava-stream with, 74

Ice-caves and their phenomena, 193-201

Iceland, volcanic formation of, 4

Iceland, geysirs of, 45-48

— Bunsen's theory of the causes of the geysirs of, 47, 48

— volcanoes of, 61

— mud-volcanoes of, 93

— ice-caves of, 198

Ichthyosaurus communis, characteristics and size of the, 20, 21

— where found, 22

Idria, fungi of the mines of, 153

— quicksilver mines of, 373-376

Iguanodon, size and characteristics of the, 22

Iktis, island of, mentioned by Diodorus Siculus, 333

Ilezk, rock-salt deposit of, 438

Illinois, coal-fields of, 424

India, mud-volcanoes of, 93

— rock-temples of, 181

Indiana, coal-fields of, 424

Indies, West, earthquakes of, 100, 101

Insects enclosed in amber, 452-455

Ipsamboul, rock-temple of, 184

— Warburton's description of it, 184-186

Ireland, effects of the great earthquake of 1755 in, 118

— coal-fields of, 404

— why they are so little worked, 404

Iron, its value, 345

— its wide diffusion, 345

— meteoric iron, 347

— ancient knowledge of, 347

— extension of its uses in modern times, 348

-- British iron production, 348

— smelting, 349

— the hot blast, 353

— the Cleveland district and the trade of Middlesborough, 354 c

IRO

Iron—*continued.*

Iron, amount and value of the British iron trade, 355

Iron, other statistics of the trade, 356

— production of foreign countries, 357–363

rtysch, copper and coal near the, 326

Isalco, formation of the volcano of, 59

— in a constant state of eruption, 62

Iscalonga, in Basilicata, cave-dwellings of, 234

Iserlohn, in Westphalia, discovery of a cavern at, 135

Ispica, Val d', cave-dwellings in the, 232

Istria, subterranean water-courses of, 150

Italy, mud-volcanoes of, 93

— earthquakes of, 100

— effects of the earthquake of 1755 in, 118

— maare, or crateriform hollows of, 132

— cave-dwellings of Southern, 234

— iron industry of, 362

Iwogasima, or Sulphur Island, of Japan, 444

JAPAN, sulphur of, 444

Java, number of active volcanoes of, 81

— the 'Valley of Death,' or Poison Valley, of, 89

— mud-volcanoes of, 93

— maare, or crateriform hollows, of, 132

— sulphur of, 445, 446

Jesuits, their intrigues during the earthquake at Lisbon, 116

Jet, formation of, 429

— found at Whitby, 429

— manufacture of jet ornaments, 430

John the Evangelist, St., his cave in the Isle of Patmos, 188

the cave converted into a chapel, 188

Jorullo, formation of the volcano of, 58

— length of time the heat was retained in the lava-stream of 1759, 74

Judd, Dr., his dangerous visit to the crater of Kilauea, 66

Jura Mountains, cauldron-shaped depressions in the, 130

KAB, El, in Upper Egypt, rock-hewn cemeteries of, 205

Kamtschatka, energy of the volcanoes of, 81

— earthquakes in, in 1737, 79

Kan, rock-hewn cemeteries of, 205

Kanara, rock-temples of, 181–183

Kaolin, or china-clay, how formed, and where, 460–462

Karli, rock-temples of, 183

Kea, Mount, tranquillity of the eruption of, in 1843, 76

Kentucky, coal-fields of, 424

Kertsch, mud-volcanoes near, 93

LAU

Kilauea, the lava lakes of, 64

— length of the lava-stream in the eruption of 1840, 70

— amount of lava thrown out by the eruption of 1840, 70

Killingworth Colliery, 410

Kingston, in Jamaica, effects of an earthquake sea-wave at, 107

Kinsale, effect of an earthquake sea-wave in the harbour of, 118

Kirghise hordes, their salt-works, 438

Kirkdale, Dr. Buckland's account of the ossiferous cave of, 214, 215

Klaproth, his discovery of uranium, 385

— works at Joachimsthal, 385

— his discovery of rutile, or titanium, 386

Kljutschewskaja Skopa, eruption of the volcano of, 79

Koh-i-Noor, or Mountain of Light, diamond and its history, 486

Kohl, uses of, 383

Konsberg, in Norway, silver mines of, 299

— nuggets found at, 299

Kopperberg, iron manufacture of, 360

Kötlingia, effects of the eruption of, in 1758, 69

Kremnitz, discovery of the gold mines of, 248

Krisuvick, in Iceland, solfatara of, 444

Kupferschiefer, or copper-slate, of Thuringa, fossils of the, 16

LABUAN, coal-fields of, 424

Lacustrine dwellings of Switzerland, 315

— discovery of the, 223

— ancient iron weapons found in the, 317

Laibach, Upper, river traversing the caves of, 150

Lalibala, rock-churches of, 186

— town of, and country round, 187

Landslips, effects of earthquakes in producing, 110

— that of Putley, in Hertfordshire, 110

— igneous and aqueous causes of landslips, 121

— cases of landslips, 121–128

— caused by the falling in of cavern-roofs, 129

Lanuto volcano, lake formed in the extinct crater of the, 57

Lapis lazuli, 494

Lapland, auriferous veins in, 293

Laureacum, on the Danube, Roman iron manufactures of, 358

Laurentian rocks, 2, 3

— their thickness, 2

— the only fossil found in the, 10

Laurium, ancient silver mines of, 298

LAU

Laurium—*continued*.
— amount of lead in the scoriæ of the ancient silver mines of, 367
Lava, formation of fiery streams of, during volcanic eruptions, 70
— phenomena attending the flow of a lava-stream, 72
— effect of the meeting of a lava-stream with the sea, 73
— and with ice, 74
— vast dimensions of lava-streams, 74-76
— waste of desolation of lava-fields, 77
— progress of lava-streams, 77
Laxey lead mine, in the Isle of Man, 366
Lead, mine of, in Cardiganshire, section of a, 252
— its property and uses, 364
— its antiquity, 361
— the mines of, in Europe, 365
— production of, in foreign countries, 366-368
— preparation of the ores, 368
— Pattinson's process, 368
Leadhills, in Lanarkshire, lead mines of, 366
Lebadeia, in Bœtia, cave of Trophonius near, 187
Lepidodendron elegans, 302
Leptodirus Hochenwartii, of the Cave of Adelsberg, 193
Levant, Cornish copper mine of, 319
Levels, in mining, 251
— extent of the works in some cases, 251
— drainage by adit levels, 269, 270
Lias, fossils of the, 19
Liège, depth of the coal mines of, 217
— accident in a colliery at, 263
Life, organic, progress of, on earth, 28, 29
— everywhere present on the earth, 156
Lignite, or wood-coal, 401
Lima, frequency of earthquakes at, 104
— displacement of stones of obelisks by earthquake shocks, 105
— effects of the earthquake sea-wave of 1746, 108
— indifference of the inhabitants of, to earthquakes, 113
Limestone, magnesian, or Permian group, animal remains of the, 15
Limestone caves, 136
— causes of their excavation, 138
— stalactites and stalagmites, 139, 140
— origin and slow formation of limestone, 141
Lisbon, great earthquake of, 114
— effect of the shock, 114
— fire and thieves in the city, 115, 116
— total loss of life from all causes, 116
— effects of this earthquake in various parts of the world, 117-119
Little Bounds, copper mine of, 319
Livres, St., ice-caves of, 192

MAM

Livres, St.—*continued*.
— lower glacière of, 193
— upper glacière of, 195
— ice-streams of the upper glacière, 195, 196
Lizards, oldest known fossils of, 14, 15
— the enormous species of the Mesozoic ocean, 20, 21
Llandegai, slate quarries of, 469, 470
Locke, his remark respecting iron, 345
Lomond, Loch, sea-shells found on the banks of, 34
— effects of the great earthquake of 1755 in, 118
London shaken by the Lisbon earthquake of 1755, 118
— subterranean wonders of, 237, 238
Long, Major, his report on the copper mines of Lake Superior, 328
Longarn, Cave of, massacre by the French in the, 170
Lorca, sulphur mine of, 444
Lowerz, destruction of the village of, by a landslip, 121
Lugnuure, lead mines of, 366
Lumm, Pope, waterspout of, 146
Lycopolis, rock-hewn cemeteries of, 205

M

MAARE, or crateriform hollows, of the Eifel, 131
— in other places, 132
Maculuba, mud volcano of, 94
— known to the ancients, 94
Madaua, in Santa Cruz, height of the volcano of, 54
Madeira, volcanic formation of, 4
Madfunch, rock-hewn cemeteries of, 205
Magdalena Grotto, or 'Black Grotto,' Festoi of the, 165, 166
— visit of the Archduke Ferdinand to the, 166
Magnesium, discovery and uses of, 387
Magnetic mountain in Russia, 367
Maina, marble of, 467
Malacca, limestone of, 335
Malaga, effects of the earthquake of 1755 at, 118
Malmesbury, section of the coal field south of, 398
Malta, troglodytes of, 234
Malwah, rock-temples of, 184
Mammalia, geological period of its prominence in life, 23
Mammoth, or primitive elephant size and characteristics of the, 26
— Professor Owen's skeleton of the, 217
— Gray's Inn Lane an ancient hunting-ground for, 231
Mammoth Cave, in Kentucky, vast dimensions of the, 135, 136
— Professor Silliman on the, 139

MAM

Mammoth Cave—*continued.*
— clusters of bats in the, 159
— animals of the, 167
Man, Isle of, lead mines of, 366
— zinc produced in, 382
Man, prehistoric, subterranean relics of, in Denmark, 221
— in Switzerland, 223
— age of human relics in caves, 225
Manchester Coal-field, 403
Manganese, ores of, 386
Mansfeldt, in Prussia, silver and copper mines of, 325
Marble of Derbyshire and Devonshire, 465
— that of Carrara, 4, 465
— that of Pentelikon and Parian, 466
— Rosso antico and Verde antico, 467
Marennes, upheaval of the chalk cliffs at, 36
Marpena, Mount, in Paros, marble of, 466
Marquette, American town of, its iron industry, 362
Marshall, James, his discovery of gold in California, 288
— his subsequent life, 289
Marsupites ornatus, fossil, 18
Martinique, Island of, destructive earthquake in the, 101
Maryland, copper mines of, 328
Masaya, volcanoes of, constant eruption of the, 63
Massachusetts, copper mines of, 328
Master-borers in the North of England, 250
— their charges per fathom, 250
Mastodon, where the fossils of, are mostly found, 27
— size and characteristics of the, 27
Matlock, thermal springs of, 43
Mauna Loa, in Hawaii, shape of the volcano of, 53, 54
— Dr. Judd's visit to, 56
— growth of ferns on, 63
— the lava-lakes of, 64
— length of the lava-stream of an eruption of, 70
— parasitic cones of, 71
— volume of the lava-stream of 1840, 75
Mauuch Chunk (or Bear Mountain), in Pennsylvania, enormous coal-field of, 423
Mauritius, fountains of marine caverns in, 140
Mediterranean Sea, upheaval of the land on the shores of the, 34
— marine caverns of the coasts of the, 144
Medellin, the proprietor of the mine of Dolores, 303
Moorfeld, crateriform hollow and lake of, 132
Megatherium, size and characteristics of the, 24, 25
Molidoni, cave of, Turkish massacre in the, 175

MIN

Mequinez, effects of the earthquake of 1755 at, 118
Mercado, Cerro del, of Mexico, 363
Mercury, its properties and uses, 370, 371
— known to the Greeks and Romans, 370
— mines of Almaden, 371
— those of Idria, 373
— diseases to which the miners are liable, 373
— mines of America, 378
Metallic veins, how generally found in mines, 246
— how ores collected or precipitated in, 247
Metamorphic rocks, origin of, 4, 5
Meteoric iron, 347
— the mass found at Otumpa, 347 *note*
Mettler, Bläsi, story of the escape of him and his wife, 124
Mettler, Sebastian Meinhardt, his escape from destruction, 125
Mouse, ossiferous caverns of the valley of the, 226
Mexico, silver mines of, 300–308
— iron ores of, 363
Michael's Mount, St., in Cornwall, 333
Middlesborough, its rapid rise, 354
— its iron manufacture, 355
Miguel, Island of, the Punta Delgada of the, 147
Milagros, his silver mine in San Luis de Potosi, 303
Miller, Hugh, his account of a coal forest, 393
Milo, Island of, mud volcanoes of, 93
— sulphur caves of the island of, 416
Mina Grande, lead mine of, 367
Minardo, Monte, near Bronte, volcanic formation of, 67
— height of, 71
Mines, in general, 244
— labours and perils of the miner, 244, 245
— casualties in mines, 245
— life in a mine, 245, 246
— length and depth of mines, 247
— discoveries of lodes, 248
— the divining-rod, 248
— boring, 249
— divisions in coal mines, 255
— long-wall working, 257
— general view of mining operations, 257
— tools employed in Cornwall, 258
— mode of blasting, 258
— heroism of miners, 259, 274
— mode of loosening hard stones, 260, 261
— tramways underground, and the conveyance of minerals, 261, 262
— methods of descending, 263–266
— man-engines for ascending or descending, 267
— timbering and draining, 268–272
— inundation, or drowning of mines, 273
— evolution of foul gases, 276, 277

MIN

Mines—*continued.*
— ventilation, 277
— choke-damp, fire-damp, and blowers, 278, 279
— the safety-lamp, 280
— burning mines, 283
— habits of the Mexican miners, 302
Minnesota mine, copper of, 327, 328
— enormous nugget of copper found near, 328
Miocene period, animals of the, 23
Mirrors of silver among the Romans, 298
— substance used for making, 335
Mississippi, ancient mounds in the valley of the, 224
Missouri, 'iron mountains' of, 362
Moa, the great extinct bird of New Zealand, 216, 217
— the cave of the Moa, 219
Moeris, Lake, hermits near the, 179
Molinos of the silver mines of Mexico, 306
Molybdenum, discovery and uses of, 387
Monarchs, vast treasures of in ancient times, 286
Monkeys, fossil, of South America, characteristics of the, 24
— small species of, on the Rock of Gibraltar, 24
Monk Wearmouth Colliery, 408
Montaño, Francisco, his descent into the crater of Popocatopetl, 446
Monte Real del, silver mines of, 304
— present yield of, 305
Monte Video, upheaval of the land at, 34
Montgomeryshire, lead mines of, 366
Montmartre, gypsum and alabaster of, 468
Montrouge, artificial mushroom-beds at, 158
Moors of Granada, destruction of, by Philip II. of Spain, 173
Moren, silver mines of, 304
Morocco, earthquakes of, 100
— effects of the earthquake of 1755 at, 118
Morran, in Algeria, Artesian well in the desert of, 51
Mososaurus, size and characteristics of the, 23
— skull of the, found, 473, 474
Moulin de la Roche, artificial mushroom beds at, 158
Mountain Ash, in South Wales, coal workings of the New Navigation Pit at, 406
Mud-streams caused by volcanic eruptions, 69
— destruction of Herculaneum and Pompeii by, 85
— those of Oba, 95
Mud-volcanoes, 93–96
— in various places, 93-95
— origin of, 95, 96

NOR

Murchison, Sir Roderick, his surmises respecting gold in Australia, 289
Mürtschenstock, tunnel in the, 134
Mushrooms, subterranean, 157
— the artificial mushroom-beds near Paris, 158
Musk-ox, food of the, 26
Mylodon, size and characteristics of the, 24, 25
— Professor Owen's skeleton of the, 217

NAPLES, earthquake in, in 1857, 98
— catacombs of, 210
Nassau, iron manufacture of, 357
Nativity, grotto of the, at Bethlehem, 188
— church of the, 188
Nauheim, carbonic acid gas spring of, 88
Naxos, consecrated caves to Dionysos in, 187
Naxos, emery of the island of, 463
Neilson, Mr., his discovery of the hot blast for iron furnaces, 353
Nomi, Lake of, the crateriform hollow forming the, 132
Nent Force Level, great drain of, 270
Nertschinsk, in Transbikalia, copper mines of, 326
— lead mines of, 367
Nettuno, Antro di, in Sardinia, 144
Nettuno, Grotta di, in Sicily, 145
Neusalzwerk, temperature of the well of, at various depths, 32
Nevada, state of, silver mines of the, 314
Newcastle, coal-fields of, 407
— their extent, 408
— the various seams of coal, 408
— human activity of the coal-fields, 411, 412
— appearance of the town, 413
— first license to dig coals given to the town, 419
Newfoundland, gradual upheaval of the land of, 36
— fountains of marine caverns in, 146
Niagara, carburetted hydrogen evolved near the falls of, 93
Nicaragua, Lake of, volcanoes near the, 61
— mud-volcanoes of, 93
— earthquakes of, 100
Nicholas, St., rock-chapel of, in Crete, 189
— legend of, 190
Nickel, name of, 384
— uses of, and whence obtained, 384
Nicolas d'Aliermont, St., aquiferous layers or beds of stone at, 40
Noises, subterranean, accompanying earthquakes, 103
Normandy, traces of depression of the land on the coast of, 37
Nore Lake, emptied by a landslip, 130
Northumberland, coal-fields of, 403

NOR

Northwich, salt mines of, 431
Norway, copper mines of, 324
Noss, islet of, its marine caves, 142
Notornis, Professor Owen's reconstruction of the, 217
Nuovo, Monte, in the Bay of Baiæ, volcanic formation of the, 67

OBERSTEIN, rock-chapel of, 190
— legend of the chapel, 190, 191
Obregon works the silver mine of Guanaxuato, 301, 302
— his title and urbanity of character, 302
Obu, eruption of the, 95
— mud-streams of, 95
Oche, Dent d', landslip of the, 127
Oosterby, iron-works of, 360
Ohio, ancient mounds in the valley of the, 224
'Oil harvest' of Caripe, 161
Olm, or Proteus, discovery of the, 164, 165
— various places in which it has been found, 166
— description of the animal, 164–167
Oloune, Island of, upheaval of the land round the, 36
Onyx, the, 497
Ontanagon district, in America, ancient copper mines, 327
Oolite rocks, their thickness, 2
Oolitic period, fossils of the, 19, 22
Opal, precious, 405
— mines of, in Hungary, 435
Ophir, seat of, 287
Ores, how generally found in mines, 246
— how they have been collected or precipitated, 247
'Orkneyman's Harbour, The,' the marine cavern so called, 143
Oroomiah, Lake, in Persia, salt of the hills and plains of, 437
Orthoceratites of the primitive seas, 18
— extinction of the, 18
Oscillatory movements of the earth, 34–37
— probable causes of, 38
Otero, a shopkeeper, joins in working the mine of Guanaxuato, 302
Owen, Professor, his memoir and skeleton of the great Moa of New Zealand, 217
Owls, cave-haunting, 160

PACHUCA, silver mines of, 304
Pachyderms, remains of large extinct, 26, 27
Palæotheriums, size and characteristics of, 23
Palæopteryx, Professor Owen's reconstruction of the, 217
Palladium, discovery and uses of, 388
Palestine, Southern, hermits in, 179

PHA

Phiomba, Fossa della, on Etna, 147
Papalardo, Baron, his efforts to divert the lava-stream from Catania, 72, 73
Parian marble, or Lychnites, 466
Paris, artificial mushroom-beds near, 158
— catacombs of, 210, 211
— old cemeteries of, 211
— plaster of, 468
Parsees, their worship of fire, 91
— their legend of the devil, 91
— their occupation and abandonment of Baku, 92
Pasco, Cerro di, silver mines of, 309
Pasco, mining town of, 310
— mines of, 310, 311
Patmos, cave and church of St. John the Evangelist, 188
Paul, St., of Thebes, the first hermit, his cave, 178
Pausilippo, Grotto of, 241
— origin of the, 242
Paviland, ossiferous caves of, 215
Peak, in the Island of Timor, blown up and replaced by a cavity, 68
Pecopteris adiantoides, 391
Peniscola, fountains of marine caverns in, 146
Pennsylvania, copper mines of, 328
— coal-fields of, 425
— petroleum springs of, 427
Pentacrinus briareus, fossil, 17, 18
Pentelikon, or Mount Penteles, marble of, 466
Prpandajan, in Java, the volcanic cone of, blown to pieces, 67
Percy, Dr. John, his discovery of aluminium-bronze, 387
Perdix, Bartholomew, his manufacture of alum, 458
Permian period, fishes of the, 13
— fossils of the, 15
Peroxide of tin, or tin-stone, 335
— richest deposits of, 335
Porticara di Talamella in Italy, sulphur mines of, 444
Peru, active volcanoes of, 61
— earthquakes of, 100
— indifference of man in, to earthquakes, 113
— silver mines of, 300
— iron furnaces of, 353
Peter's Mount, St., near Maestricht, quarries of, 470
— visit of Faujas de Saint-Fond, 472
Petroleum, formation of, 426
— old springs of, in Europe, 426, 427
— production of the springs of America, 427
Potrospongidæ, or stone sponges, 17
Pfeiffer, Ida, her visit to the diamond mines of Borneo, 480
Phuraohs, rock-tombs of the, in Thebes, 202–204

PHI

Philip II. of Spain, his destruction of the Moors of Granada, 173
Philip, Port, town of, 290
Philothous, St., his cave on Mount Ponteles, 466
Phœnicians, their tin-trade, 333
— their traffic in and uses of lead, 304
Piotra Mala, burning springs of, 90
Pigeons, cave-haunting, 160
Pilot Knob 'iron mountain,' 302
Pines, black, of Trinidad, 420
Pitt, or Regent, diamond, 485
Pittasphalte, formation of, 426
Piuka Jama, cave of, 154, 156
— the river Poik flowing below the, 154
Piz Mountain, destructive effects of a landslip of the, 127
Planina, river traversing the Cave of, 150
— explored by Adolph Schmidl, 151
— abundance of Protei in the, 166
Platinum, discovery of, 312
— where obtained, 382
— its qualities, 383
Playfair, his observations as to the rise of the land in Sweden, 35
Plesiosaurus, size and characteristics of the, 21
— where found, 22
Pleurotomaria carinata, fossils of, 15
Pliny the Elder, death of, as described by his nephew, 82–84
Pliocene period, animals of the, 24
Plumbago, graphite, or black lead, former trade in, in Cumberland, 462
— the mine exhausted, 462
— places where found at present, 462
Pluns, town of, buried by a landslip, 127
Polistena, effects of an earthquake at, 98
Poik River, engulfment and re-appearance of the, 150
— a subterranean canoe voyage on the, 151–154
— the river flowing beneath the Piuka Jama, 154
Pombal, Marquis of, his conduct in the great earthquake of Lisbon, 116, 117
Pompeii, destruction of the town of, 81–85
— the mud-stream which effected the destruction, 85
— present state of the Roman town of, 87
Pontus, hermits in, 179
Popocatepetl, depth of the crater of, 54
— Montaño's visit to the crater of, 446
Porphyry of Elfdal, 467
— of the Altai, 468
Portugal earthquakes of, 100
Potosi, San Luis de, silver mines of, 303, 309
Precious stones, 477
Proteus anguinus, discovery of the, 104, 165
— description of the animal, 165–167

RIO

Proteus anguinus—continued.
— its abundance in the Cave of Planina, 166
— different caverns in which it has been found, 166
Prussia, iron manufacture of, 357
— production of zinc in, 381
— salt works of, 438
Pterichthys Milleri of the Old Red Sandstone of Scotland, 13, 14
Pterichctyli, size and characteristics of the, 21
Pterygotus acuminatus, 12
Pulveromaar of Gillenfeld, lake or maare of, 132
Puzzuoli, solfatara of, 414

QUARRIES, celebrated, 464
— those of France, 464
— those of England and Germany, 464
— of Carrara and the Pentelikon, 465, 466
— porphyry, 467, 468
— alabaster and plaster of Paris, 468
— slate, 469
— of St. Peter's Mount, near Maestricht, 470
— of Egypt, 474
Quicksilver. See Mercury.
Quito, active volcanoes of, 61
— tradition respecting them, 67
— earthquakes of, 100

RADOBOY, sulphur mines of, 444
Rain-prints of former ages, preservation of, 29
Rammelsberg, in the Hartz silver mines of the, 299
— discovery of the lode of the, 248
— burning hard mineral stone in, 260, 261
— copper found in, 325
Rat, blind cavern, of the Mammoth Cave of Kentucky, 167
Rathlin, island of, massacre by the English under Sir John Norris in the, 172
Rarinazzo Mountain, landslip of the, 127
Red lead, how made, 365
Redruth, copper mines of, 317
Regla, Conde de la. See Terreros.
Reptiles, oldest known fossils of, 14, 15
— enormous marine fossil reptiles of the Mesozoic ocean, 20
— footprints of reptiles of the Cambrian formation, 29
Rhodium, discovery and uses of, 388
Rhondale, iron manufacture of, 365
Riobamba, destruction of the town of, 78, 79
— destroyed by an earthquake, 104
— remarkable displacement of objects during the shocks, 104

RIO

Riobamba—*continued.*
— silence during the shocks, 104
Ripple-marks of former ages, preservation of, 28, 29
Rivers, cave, 149–151
— explorations of Adolph Schmidl in the Cave of Planina, 151
Rochello, La, upheaval of the land at, 36
Rock-tombs and catacombs, 202
Rock-crystal, 498
— the Grotto of Galenstock, 499
Roebuck, Dr. John, his improvements in iron manufacture, 350
Romanus, the monk, feeds St. Benedict in his cave, 180
Rome, wealth of, after the third Punic war and in the time of the Cæsars, 286, 298
— gold coins of, 287
Ronciglione, Lake of, formed in the extinct crater of a volcano, 57
Roquefort cheese, 198
Röraas Mountains, copper mines of, in Norway, 324
Rosa, Sierra de Santa, 301
— silver mines of the, 301
Rosalia, St., rock-church of, in Sicily, 188, 189
— story of, and of the discovery of her bones, 188, 189
Rossberg, or Rufi, landslip of the, 123
— causes of the catastrophe, 124
Rossi, Monte, height and area of, 71
Rossi, Cavaliere de, his researches in the catacombs of Rome, 210
Rosso antico, 467
Roth, natural ice-cave of, 198
Röthen, villages of Upper and Lower, destroyed by a landslip, 124
Royale, Isle, ancient copper mines of, 327
Ruby, the Oriental, 489
— in the crown of England, 490
Rucu-Pinchincha, Humboldt's view down the volcano of, 55
Russia, copper mines of, 326
— iron manufacture of, 357, 358
— salt-works of, 437, 438
— amber ornaments of, 457
— the Imperial diamond of, 485
Rutile, or Titanium, discovery and uses of, 386, 387

SAARBRÜCK, oldest known reptiles found in the coal-field of, 14
— other wonders of the coal-field of, 15
— vast time required for the formation of the coal-fields of the, 397
Sable, in Anjou, the Fontaine-Sans-fond near, 149
Sabrina, island of, volcanic formation of the, 59

SCO

Sabrina, island of—*continued.*
— its disappearance, 59
Sacrée Madame, near Charleroi, depth of the colliery of, 247
— mode of ventilation in the mine of, 277 *note*
Sacro, Monte, marble mountain of, 465
Safety-cages, used in descending mines, 264, 265
Safety-lamp, Davy's, 280
— improvements in the, 281
Sahara, wells of the inhabitants of the, 48, 50
— future importance of Artesian wells to, 51
Salamis, fleet which gained the battle of, 298
Salcedo, silver mine of, 311
— tragical end of its proprietor, Don José Salcedo, 312
Sallee, effect of an earthquake sea-wave at, 118
Sallenches, fall of a mountain near, 122
Salt, geological position of, 431
— the mines of Northwich, 431
— those of Droitwich and Stoke, 432
— that of Wieliczka, 433–436
— in other places, 426–439
— method of preparing it, 439
— origin of rock-salt, 440
Salza, manufacture of salt at, 436
Salzburg, salt mines of, 436
San Francisco, its rapid rise, 289
Santorin, submarine volcano of, 60, 61
Sapphire, red, 489
— oriental, 490
Sardinia, cave-dwellers of, 234
— the dwellings of the Sardo shepherds of the present day, 234
— lead mines of, 366
Saviour's, Our, tomb, at Jerusalem, church built over the, 198
Saxony, tin mines of, 336
— coal-fields of, 404
Shafloch, ice-cave of, 194, 195
Schmerling, Dr., his investigations respecting the antiquity of man, 226
Schemnitz, fungi of the mines of, 168
— discovery of the rich mines of, 248
— produce of silver in the mines of, 300
Schmidl, Adolph, his explorations of the subterranean river Poik and Cave of Planina, 151
Schneeberg, large block of silver found at, 299
— use made of the burning vapours, 283
— bismuth of, 383
Scilano, village of, buried by a landslip, 127
Scilla, Prince of, his death, 107, 108
Scopoli, his description of subterranean fungi, 157

SCO

Scoriæ, length of time the liquid fire is retained in the interior of a lava-stream, 73

Scotland, lead-mines of, 366
— coal-fields of, 403

Scott, Sir Walter, his visit to 'The Orkneyman's Harbour,' 143
— and to Fingal's cave, 143

Scrope, Mr. Poulett, his description of the Volcano of Stromboli, 62

Sea-shells found on the Andes and Alps, 34, 37
— and in other places at present removed from the sea, 34

Sea, movements of the, in earthquakes, 106, 107
— extent of the wave-motion, 105, 106
— cases of the destructive effects of the earthquake waves, 107-109, 117-119

Segeberg, deposit of salt at, 439

Senegal, deposits of the Area senilis on the banks of the river inland, 34

Sequoia, gigantic trees of the, 28

Seven Pagodas, rock-temples near Madras so called, 184

Seville, effects of the earthquakes of 1755 at, 118

Shelas, cave of, Colonel St. Arnaud's massacre in the, 176

'Shell-mounds' of Denmark, 222
— those of the Fuegians, 222

Shetland, marine caves of, 142

Shields inlaid with silver in Homer's time, 298

Siberia, auriferous land of, 288
— lead mines of, 367
— emeralds of, 491

Sicanians, cave-dwellings of the ancient, 232

Sicily, earthquakes of, 100
— marine grottoes of, 145
— sulphur mines of, 441

Sickingen, Count, his experiments with platinum, 382

Sidi Rascheed, in Algeria, Artesian well of, 51

Sigillaria oculata, 392

Silurian period, crustacea of the, 10-12
— brachiopods of the, 12, 13
— fishes of the, 13

Silver, discoveries of lodes of, 248
— antiquity of the discovery of, 297
— most ancient silver mines, 298
— European silver-fields, 289, 300
— mines of Mexico and Peru, 300-314
— mode of crushing and decomposing the ores, 306-307
— law of Peru respecting the silver mines, 311
— mines of Chili and Nevada, 313, 314

Singapore, antimony of, 383

Siphnos, ancient silver mines of, 298

STA

Siphon, principle of a, 44 note

Siphonia costata, fossil, 16

Sioa, constant state of eruption of the volcano of, 63

Sivatherium giganteum, size and characteristics of, 27, 28

Skaptar Jökul, in Iceland, lava-stream of the eruption of, in 1783, 70
— that of 1787, 75

Skerries, water-spouts or fountains of the, 146,

Slate quarries of North Wales, 409

Smeaton, John, his improvements in iron manufacture, 350

Sodium, discovery and uses of, 388

Solway Moss, appearance and area of the, 130
— bursting of the, 130, 131

Sommatino, conflagration of the solfatara of, 443

Somme, flint implements of the Valley of the, 230, 231

Spain, gold coins of the Visigoths of, 287
— auriferous land of, 288
— ancient silver mines of, 298
— tin mines of, 336
— iron industry of, 361
— lead mines of, 365
— coal-fields of, 423
— cupriferous pyrites of, 447

Spartacus, revolt of, 82
— his defeat of Clodius at Vesuvius, 82

Sperenberg, deposit of salt at, 439

Sphenopteris affinis, 391

Spider, eyeless, of the Cave of Adelsberg, 163

Spinel, the, 490

Spirifer princeps, 12

Spiriferidæ, 13

Sponges, fossils of the primitive seas, 16

Springs, always warmer than the air in the locality where they gush forth, 32
— hydrostatic laws regarding the flow of, 40-42
— temperature of the water of, 43
— geological phenomena favouring the production of thermal springs, 43
— mineral particles in springs, 43, 44
— intermittent springs, 44
— geysirs of Iceland, 45-48
— Artesian wells, 48-52
— carbonic acid gas springs, 88-90
— of carburetted hydrogen, 90-93

Staffordshire, burning mines of, 283
— the Burning Hill of, 283, 284

Stag, Professor Owen's skeleton of the primeval, 217

Stalactites and stalagmites, formation of, 139, 140
— their varieties of form and slow formation, 140

STA

Stalactites and stalagmites—*continued*.
— Dr. Schmidl's 'Stalactital Paradise,' 152
— in the cave of Guacharo, 162
-- of the Cave of Mulidoni, 176
— in the Norwegian copper mine of Röraas, 325
Stalita tænaria, the eyeless cavern spider, 163, 164
Stamping-mill in the silver mines of Mexico, 306
Star fish, of the Chalk group, 18
Stassfurt, mines of, 438
Steam, important part played by, in volcanic phenomena, 41
Steam-jets, or fumaroles, of volcanoes, 63
— those of the eruption of Jorullo in 1759, seen in 1803, 74
Stephenson, George, a coal-trapper in early life, 419
Stikeen, gold-fields of, 293
Stockton-on-Tees, export of coal from, 413
Stoke, in Worcestershire, salt mines of, 432
Stone implements of Denmark, 222
— of the Brixham caverns, 227
— of the Valley of the Somme, 230
Stromboli, diameter of the crater of, 54
— constant activity of the volcano of, 62
— Mr. Poulett Scrope's description of it, 62
Stromeyer, his discovery of cadmium, 386
Strontian, in Argyleshire, lead mines of, 366
Styria, iron of, 358
Subiaco, St. Benedict's Cave near, 180
Suffioni of the Florentine lagoons, 460
Sulphur of the mines of Sicily, 441
— exports of, 443
— conflagration of a sulphur mine, 443
— mines of Ternel and Lorca, 444
— combinations of sulphur with metals, 447
Sumatra, deposits of tinstone in, 335
Sunderland, export of coal from, 413
Superior, Lake, copper scattered near the shore of, 325, 327
-- ancient copper mines near, 327
Surtshellir, in Iceland, formation of the, 148
Sutherland, gold-fields of, 293
Swallows, cave-haunting, 160
Swansea, copper-works of, 320-321
Sweden, effect of the great earthquake of 1755 in, 118
— mode of descending mines in, 264
— copper mines of 322
Swifts, cave-haunting, 160
Switzerland, subterranean relics of pre-historic man in, 223
— ancient iron implements found in, 347
Swoszwice, sulphur mines of, 444

TIN

Syene, rock-hewn cemeteries of, 205
Syracuse, catacombs of, 210
— city of, 475
— the Latomiæ of, 475
Syria, earthquakes of, in the reign of Tiberius, 97, 100
Syout, in Upper Egypt, rock-hewn cemeteries of, 205

TAGILSK, Nishne, platinum of 382
Taman, mud-volcanoes of the peninsula of, 93, 95
Tamelhat, in Algeria, Artesian well at, 51
Tangiers, effects of an earthquake sea-wave in, 118
Tap cinders of the iron puddling furnaces, 355
Tasmania, coal-fields of, 424
Tauretunum, Roman town of, destroyed by a landslip, 127
Tees, importance of the river, 407
Teïr, Djebel, height of the volcano of, 54
Temboro, cone of the volcano of, blown to pieces, 67
Temenitz, engulfment and reappearance of the river, 150
Temples, rock, of India, 181
Teneriffe, Peak of, shape of the, 53
— ice-caves of, 198
— solfatara of the, 445
Tenger, Gunong, in Java, diameter of the crater of the volcano of, 54
Terebratulæ of the Silurian seas, 13
— hastata, fossils of, 15
Ternel, sulphur mine of, 444
Terni, Æolian caverns of, 198
Terranuova, effects of an earthquake at, 98
Terreros, Don Pedro, his silver mine of La Regla, 304
Tertiary period, mammalia of the, 23
Thallium, discovery and uses of, 388
Thaur, Mount, Mahomet's refuge in a cave of, 169
— Moslem miracle of, 169
Thebes, hermits in the desert of, 179
— the royal tombs of, 202-204
Themud, rock city of the, 236
— legendary tale respecting the, 236
Thomson, Dr., his cave explorations in New Zealand, 218-220
Tin, antiquity of the knowledge of, 332
— mentioned in the Bible, 332
— Phœnician trade in, 333
— uses and importance of, 334
— the two ores of tin, 335
— lodes of Cornwall, 337
— number of mines in Devon and Cornwall at work, 338
— smelting of tin, 342
— number and wages of the miners, 343
— nature of the miner's work, 343

TIN

Tin-foil, 334
Tino, island of, Rosso antico, 467
Titanium, or rutile, discovery and uses of, 386, 387
Titus, the Emperor, his benevolence to the survivors of Herculaneum and Pompeii, 85
Tjerimai, Guonong, in Java, extinct volcano of, 55
— its height and depth, 55
Tlalpujahua, silver mine of, 305
Toeplitz, hot springs of, 43
Tofua, constant activity of the volcano of, 63
— lake formed in the crater of the extinct volcano of, 57
Tolfa, manufacture of alum at, 458
Tombs, rock-hewn, 202
Topaz, the, 492
Töplitz, effect of the Lisbon earthquake of 1754 on the mineral waters of, 118
Torgatten, in Norway, natural tunnel of the grotto of, 134
Trajan, the Emperor, in an earthquake at Antioch, 97
Transbaikalia, iron of, 357
Transgariep country, cannibal caverns of the, 234
Transylvania, iron of, 358
— salt mines of, 436
Travers, Val de, asphalte mine of the, 428
Trebich Cave, near Trieste, 134
Tresavean Copper Mine, wealth of, 329
— tin mine, 337
Treviso, three villages of, buried by a landslip of the Piz Mountain, 127
Triassic rocks, their thickness, 2
— fossils of the, 22
Trilobites, 10, 11
— eye of, magnified, 11
— gradually vanish in the Carboniferous period, 14
Trinidad, mud volcanoes of, 93, 94
— the great Pitch Lake of, 428
— the black pines of, 429
Tripolite, its composition and uses of, 463
Troglodytes, or dwellers in caves, 232
Trophonius, Cave of, 187
— visitors to the cave for information, 188
Trou-aux-Moutons, vast ice-cave of the Rothhorn, 194, 195
Tschudi, silver ornaments of the, 298
Tuileries, principle on which the grand fountain of the, is supplied, 42
Tungstate of soda, uses of, 385
Tungsten, discovery of, 384
— uses of, 385
Tunnels, natural, 113, 134
Turks, their atrocities in modern warfare, 176
— their use of amber in pipes, 456

VES

Turquoise, the, 493
Turrilites, tuberculatus, 19
Tuscany, suffioni of the lagoons of, 460
Tyne, importance of the river, 407
— shipping coal on the banks of the, 411, 412
Tynemouth Priory, view from the, 414

UBES, St., nearly destroyed by an earthquake sea-wave, 117
United States of America, copper mines of, 328, 329
Upheavals, subterranean, 34
— taking place at the present day, 34
— slow elevation of the land in Sweden, 35
— and in other places, 35, 36
— marks chiselled on the Swedish rocks, 37
— probable causes of, 38
Upsala, iron manufacture of, 360
— the old city of, and the burial-places of Odin, Thor, and Freya, 360
Ural, or Oural, Mountains, the iron of the, 357
— copper mines of the, 326
Uranium, discovery and uses of, 385

VALENCIANA, Conde de, his silver mine and fortune, 302, 304
Valenciennes, depth of the coal mines of, 247
Valentinus, Basilius, his mention of antimony, 383
Valdivia, extent of wave motion of an earthquake at, 105, 106
Valparaiso, upheaval of the land at, 34
— copper mines of, 326
Vaucluse, celebrated fountain of, 149
Vegetation, subterranean, 157
— mushrooms or fungi, 157
Velleja, Roman town of, buried by a landslip, 127
Venetian gold coins, 287
Ventriculites, fossil, 16
Verde antico, 467
Vesuvius, its long period of rest, and resumption of its activity, 58
— the lava-stream of the eruption of 1822, 70
— lava-fountains of the eruption of 1794, 72
— advance of a lava-stream into the sea near Torre del Greco, 73
— vast dimensions of the lava-stream of, 75
— state of the volcano previous to the eruption of 79 A.C., 81
— first indication of the catastrophe, 82
— account of Pliny the Younger, 85

VET

Veta madre, silver mine of, 247, 301
Victoria, colony of, 290
-- gold-fields of, 294
Villaroel, Don José, his silver mine of the Corro di Potosi, 309
Vincent, Island of, volcanic eruptions on the, 66
-- disappearance of a mountain at, 68
Virgil, tomb of, 242
— belief in his incantations, 242
Visigoths of Spain, their gold coins, 287
Virarmis, carbonic acid gas springs of the. 88
Vivian's copper-works at Swansea, 321
Volcanoes, heat required for the production of the lava of the, 33
-- extent of the action of, 33
-- important part played by steam in volcanic phenomena, 41
— extinct and active volcanoes, 53
— their shapes and heights, 53, 54
— their craters, 54
— desolation near them, 54
— dimensions and heights of various craters, 54
— dangers of crater explorations, 55, 56
— lakes in the craters of extinct volcanoes, 57
— line of demarcation between active and extinct volcanoes, 58
— volcanoes still constantly forming, 58
— submarine volcanoes, 59
— formation of volcanic islands, 59
— number of known volcanoes, 60
— unequal distribution of, 61
— in a constant state of activity, 61, 62
— steam-jets, or fumaroles, 63. 74
— phenomena of volcanic eruption, 65
— stones and ashes thrown out, 65
— explosion of cones, 67
— disastrous effects of showers of sand, pumice, and lapilli, 68
— mud-streams formed, 69
— torrents formed by melted snow, 69
— formation of fiery streams of liquid lava, 70
— parasitic cones of eruption, 70
— wooded volcanic craters, 71
— phenomena attending the flow of a lava-stream, 72, 73
— effect of the meeting of lava and the sea, 73
— and of lava and of ice, 74
— vast dimensions of lava-streams, 75, 76
— waste of desolation in lava-fields, 77
— considered as safety-valves, 78, 79
— probable causes of. 79. 80
-- destruction of Herculaneum and Pompeii, 81–87

WAT

Volcanoes—continued.
— mud-volcanoes, 93–96
— formation of volcanic caves, 146-148
Volterra, alabaster of, 463
Vultur, Mount, beauty of the forest scenery round the extinct craters of, 57

WALES, auriferous veins found in, 293
— lead mines of, 366
— time required for the formation of the coal-fields of South, 397
— their superficial extent, 405
— the coal-fields of North, 403
— total number of pits in the South, 406
— state of quarries of North, 469
— New South, copper mines of, 329
— coal-fields of, 424
Walker colliery, on the Tyne, disaster in, prevented, 282
Wallsend colliery, drowned, 273
— attempt made to work a part of it, 278
Wanlockhead, lead mines at, 366
Warburton, his description of the rock-temple of Ipsamboul, 184-186
— his visit to the tombs of the Pharaohs at Thebes, 202-204
Wash, evidences of subsidence of the land on the shores of the. 37
Washoe silver mine 314
Water, its eternal strife with fire, 1, 2
— the waters of the Cambrian or Silurian ocean, 11
— filtered and made pure by the earth, 40
— temperature of the water of springs, 43
— subterranean distribution of the waters, 39
— hydrostatic laws regarding the flow of springs, 40, 41
— Bunsen's theory of the Geysirs, 47
— geological phenomena favouring the production of thermal springs, 43
— geysirs of Iceland, 45–48
— Artesian wells, 48–52
— effect of the meeting of a lava-stream and the sea, 73
— movements of the sea in earthquakes, 107
— action of water in limestone caves, 138, 139
— and in forming marine caves, 142
Water-spouts of caverns in the Skerries, 146
— in mines, 269
— modes of draining, 269
Waterfall, a subterranean, 153

WAT

Watt, James, his invention and its importance in iron manufacture, 351
Wear, importance of the river, 407
Wermlund, iron manufacture of, 360
Westphalia, coal-fields of, 405
Wheal Cock, copper mine of, 319
Wheal, Edward, Cornish copper mine of, 319
Wheal Vor, rise and fall of the tin mine of, 339
Whitehaven coal-basin, extent of the, 407
— excavations under the sea at the, 410
White-lead manufacture of the Brohl, 89
White-lead, how made, 365
Wicklow, lead mines of, 366
— iron pyrites of, 447
Wielitzka, salt mines of, 262, 433–435
— method of descending the, 253
— accident in the, 435
Wiesbaden, hot springs of, 43
Wind-grottoes, 198–200
— fables respecting, 199
Wirksworth, ossiferous caves of, 215, 216
Wissokaja Gora, the magnetic mountain of, 357
Wolfram, discovery of, 384
Wollaston, his discovery of palladium, 388
— and of rhodium, 388
Wood, Colonel, his discovery of a bone-cave at Gower, 228
Workington Colliery, drowned, 274
Worship, subterranean places of, 181–183
Worsley, in Lancashire, subterranean canals in, 263

ZWI

YEERMALIK, massacre by Genghis Khan in the cave of, 272, 273
— visit to the cave, 173
— as a natural ice-cave, 197
York, New, copper mines of, 328
Yorkshire, lead mines of, 366

ZACATECAS, silver mine of, 302
Zealand, New, effects of the earthquake of 1855 in upraising land, 111
— maars, or crateriform hollows of, 132
— the Apteryx australis of, 216
— the gigantic Moa of, 216, 217
— Professor Owen's memoir and skeleton of the bird, 217
— ossiferous caves of the country, 218
— gold-fields of, 293
— vegetation of, similar to that of the coal-fields, 395
— coal-fields of, 424
Zellerfeld, great adit levels of the mines of, 270
Zepeda, Don Barnobé de, his discovery of the silver vein of Catorce, 303
Zeus, Olympian, Phidias' ivory and gold statue of, 286
Zinc, not known to the ancients, 380
— production of, 180, 181
— the chief zinc producing countries, 381, 382
Zircon, the, 492
Zoroaster, religion of, restored by the Sassanides of Persia, 92
Zwickau, in Saxony, burning mine at, 283

PRINTED BY
SPOTTISWOODE AND CO., NEW-STREET SQUARE
LONDON